Results and Problems in Cell Differentiation

A Series of Topical Volumes in Developmental Biology

7

W0112403

Cell Cycle
and Cell Differentiation

Edited by J. Reinert and H. Holtzer

With Contributions by

T. W. Borun, Philadelphia   A. C. Braun, New York   S. R. Dienstman, Philadelphia
J. B. Gurdon, Cambridge   H. Holtzer, Philadelphia   R. K. Hunt, Philadelphia
R. C. King, Evanston   P. Lawrence, Cambridge   F. Meins, Jr., Princeton
R. E. Nelson, Los Angeles   S. E. Pfeiffer, Farmington   C. H. Phelps, Farmington
C. P. Selitrennikoff, Los Angeles   L. Shapiro, Bronx   R. W. Siegel, Los Angeles
R. Tsanev, Sofia   H. Weintraub, Princeton   N. B. Wood, Bronx

With 92 Figures

Springer-Verlag Berlin Heidelberg GmbH 1975

ISBN 978-3-662-21693-4   ISBN 978-3-540-37390-2 (eBook)
DOI 10.1007/978-3-540-37390-2

© by Springer-Verlag Berlin·Heidelberg 1975.
Originally published by Springer-Verlag Berlin Heidelberg New York in 1975.
Softcover reprint of the hardcover 1st edition 1975

Library of Congress Cataloging in Publication Data. Main entry under title: Cell cycle and cell differentiation. (Results and problems in cell differentiation; 7). Includes bibliographies and index. 1. Cell cycle. 2. Cell differentiation. I. Borun, T. W. II. Reinert, J. III. Holtzer, H., 1923-IV. Series. QH607.R4 vol. 7 [QH605]. 574.8′61. 75-8623.

Preface

It is instructive to compare the response of biologists to the two themes that comprise the title of this volume. The concept of the cell cycle—in contradistinction to cell division—is a relatively recent one. Nevertheless biologists of all persuasions appreciate and readily agree on the central problems in this area. Issues ranging from mechanisms that initiate and integrate the synthesis of chromosomal proteins and DNA during S-phase of mitosis to the manner in which assembly of microtubules and their interactions lead to the segregation of metaphase chromosomes are readily followed by botanists and zoologists, as well as by cell and molecular biologists. These problems are crisp and well-defined.

The current state of "cell differentiation" stands in sharp contrast. This, one of the oldest problems in experimental biology, almost defies definition today. The difficulties arise not only from a lack of pertinent information on the regulatory mechanisms, but also from conflicting basic concepts in this field. One of the ways in which this situation might be improved would be to find a broader experimental basis, including a better understanding of the relationship between the cell cycle and cell differentiation.

The contributors to this volume were not selected because they shared a common viewpoint regarding the interrelationship of the cell cycle and cell differentiation. For some of the authors differentiation occurs quite independently of the events of the cell cycle, for others some events associated with differentiation may be initiated only during a particular phase of the cell cycle, whereas still others predict that basic epigenetic changes can occur only during a quantal cell cycle. The internal coherence of the articles in this volume stems from the fact that each contributor in the past (1) has given serious thought to the relationship between the two subjects and (2) has provisionally accepted the definition of differentiation as "those mechanisms that make information not readily available in a particular mother cell readily available in its daughter cells."

February 1975

W. BEERMANN
H. HOLTZER
J. REINERT
H. URSPRUNG

Contents

The Cell Cycle, Cell Lineage, and Neuronal Specificity
by R. K. Hunt

Neurogenesis and the Cell Cycle
by C. H. Phelps and S. E. Pfeiffer

The Cell Cycle and Cell Differentiation in the Drosophila Ovary
by R. C. King

Contributors

BORUN, THADDEUS W., Fels Research Institute, Temple University School of Medicine, Philadelphia, Pennsylvania (USA)

BRAUN, ARMIN C., The Rockefeller University, New York, New York (USA)

DIENSTMAN, STEVEN R., Departments of Biology and Anatomy, University of Pennsylvania, Philadelphia, Pennsylvania (USA)

GURDON, J. B., Medical Research Council, Laboratory of Molecular Biology, Cambridge (England)

HOLTZER, HOWARD, Department of Anatomy, University of Pennsylvania, Philadelphia, Pennsylvania (USA)

HUNT, R. K., Institute of Neurological Sciences, University of Pennsylvania. School of Medicine, Philadelphia, Pennsylvania (USA)

KING, ROBERT C., Department of Biological Sciences, Northwestern University, Evanston, Illinois (USA)

LAWRENCE, PETER, Medical Research Council, Laboratory of Molecular Biology, Cambridge (England)

MEINS, FREDERICK, Jr., Department of Biology, Princeton University, Princeton, New Jersey (USA)

NELSON, R. E., Department of Biology, University of California, Los Angeles, California (USA)

PFEIFFER, S. E., Departments of Anatomy and Microbiology, University of Connecticut Health Center, Farmington, Connecticut (USA)

PHELPS, CREIGHTON H., Department of Anatomy and Microbiology, University of Connecticut Health Center, Farmington, Connecticut (USA)

SELITRENNIKOFF, C. P., Department of Biology, University of California, Los Angeles, California (USA)

SHAPIRO, LUCILLE, Albert Einstein College of Medicine, Bronx, New York (USA)

SIEGEL, R. W., Department of Biology, University of California, Los Angeles, California (USA)

TSANEV, R., Institute of Biochemistry, Bulgarian Academy of Sciences, Sofia (Bulgaria)

WEINTRAUB, HAROLD, Department of Biochemical Sciences, Princeton University, Princeton, New Jersey (USA)

WOOD, NANCY B., Albert Einstein College of Medicine, Bronx, New York (USA)

Myogenesis: A Cell Lineage Interpretation

S. R. DIENSTMAN and H. HOLTZER

Departments of Biology and Anatomy, University of Pennsylvania
Philadelphia, PA, USA

I. Introduction

This review is based on four interrelated assumptions: (1) The central problem in differentiation is understanding how coupled bits of information unavailable in a mother cell become available in daughter cells (HOLTZER et al., 1973); (2) The genetic controls regulating rates of synthesis of constitutive molecules such as glucose-6-phosphatase or RNA polymerases, or organelles such as mitochondria or ribosomes are, for the most part, independent of the regulatory controls for differentiation; (3) All embryonic cells—zygote, blastula, gastrula, stem-cells, etc.—exist as differentiated cell types, each committed to a unique and limited program of synthesis (HOLTZER, 1968, 1970); and (4) The number of options for transition in a developmental program available to any one cell, whether from an embryo or mature organism, is never greater than two (ABBOTT et al., 1974).

From a molecular point of view, not only are there no "undifferentiated" cells, but there is no such entity as a "multipotential" or "totipotential" cell. No zygote or blastula cell or teratocarcinoma cell (PIERCE, 1967; STEVENS, 1967, 1970) *itself* has the option to activate the coordinated sets of transcriptional, translational, and post-translational controls that characterize, for example, terminal myogenic or erythrogenic cells. No inducing molecule will transform a given zygote, or blastula, or teratocarcinoma cell, into a myoblast synthesizing myoglobin, and organizing myosin, actin and tropomyosin into striated myofibrils; or, alternatively, induce a zygote, or blastula, or teratocarcinoma cell to transform into an erythroblast synthesizing hemoglobin, carbonic anhydrase, and the glycoproteins characteristic of the surface of normal erythroblasts. The *only* option open to the zygote is to produce cleavage-stage cells; they in turn only synthesize and segregate those unique, but as yet unidentified molecules (e.g. pole plasm and its derivitives, MAHOWALD, 1971) that lead *only* to the early blastula cell types. Irrespective of information stored in their DNA, the only option open to early blastula cells is to produce late blastula cells with their specific developmental programs. That some of the *descendents* of the zygote, or blastula cell or teratocarcinoma cell may generate a myogenic or erythrogenic lineage does not mean that they, as replicating cells, have a greater number of choices at any one time than any other replicating cell with respect to differentiation.

This view stresses the importance of cell lineages in differentiation. It predicts that: (1) All embryonic cells at all times are members of some functional lineage; (2) The cells in one "compartment" of a lineage have an ongoing program of synthesis distinct from programs of synthesis in cells in antecedent or subsequent compartments; (3) The sequence of compartments in a lineage is obligatory, that is, a terminal cell type appears only as the result of the appearance of an invariable line of precursors; and (4) No single cell type has the option of producing more than 2 new cell types as *immediate progeny*. According to this view induction of competent cells is a permissive not an instructive event for the transition from one compartment to the next in a lineage (HOLTZER, 1968). Only if there are 2 or more generations can a collection of "bipotential" cells exhibit the properties of a "multipotential" system. For a "totipotential" zygote to elaborate a system of say 32 terminal cell types would require then a minimum of 5 generations.

Ideally, an embryonic cell in one compartment of a lineage may be doubly characterized: (1) by its unique ongoing program of synthesis, and (2) by its options, or potentiality, of yielding one, or maximally, two particular types of divergent progeny. It has been proposed that when transitions occur in a lineage that lead to daughter cells with programs of synthesis different from that of the mother cells they are cell-cycle-dependent events (HOLTZER, 1963, 1970; HOLTZER et al., 1972; ABBOTT et al., 1974).

II. Definition of the Ultimate and Penultimate Compartments of the Myogenic Lineage

Analysis of lineages on a cellular level is severely hampered by the absence of reliable cytological markers which immediately identify cells in the different compartments of a given lineage. In the myogenic lineage only the zygote and the terminally differentiated myoblast can be unambiguously recognized. No known biochemical or cytological marker currently permits distinguishing even between primitive mesenchyme cells (Ms cells) and their descendents, the presumptive myoblasts (PMb cells), presumptive chondroblasts (PCb cells) and presumptive fibroblasts (PFb cells). Only those cells in the myogenic lineage that have exercised their option to synthesize and organize myosin, actin, and tropomyosin into myofibrils (HOLTZER, 1961; HOLTZER and SANGER, 1971) or that display acetylcholine receptors (FAMBROUGH and RASH, 1972) can be termed "myoblasts" (Mb cells). Skeletal Mb cells have two other properties—they withdraw from the cell cycle, never again entering S and, they can fuse with each other (HOLTZER, et al., 1957; HOLTZER, 1958; STOCKDALE and HOLTZER, 1961). Myoblasts, either as mononucleated cells, or as part of a multinucleated myotube, constitute the terminal compartment of the myogenic lineage. Myoblasts, in turn, are the descendents of replicating cells that have been defined operationally as "presumptive myoblasts" (PMb). Though indirect, there is compelling evidence from grafting experiments (HOLTZER and DETWILER, 1953; HOLTZER, 1958) and tissue culture experiments that replicating PMb do not have the option of yielding cartilage cells, nerve cells, etc. As cells in the penultimate compartment of the myogenic lineage, PMb possess a unique, very limited set of options. In terms of developmental programs, PMb populations have the option to replicate to produce *either* more PMb or Mb. The transition from PMb to Mb populations has been studied extensively and while many agents appear to enhance or depress the frequency of this transition, there is no evidence that movement from one compartment to the next requires mysterious exogenous inducing agents or even known hormones (HOLTZER, 1961, 1970; HOLTZER and SANGER, 1972; FISCHMAN, 1970). There is no critical evidence that molecules such as collagen or unknown diffusible factors play any more unique a role in myogenesis than, for instance, concentration of ambient ions, pH, etc.

One of the more critical problems in myogenesis is to learn more about the unique molecules synthesized by PMb cells and to determine whether the capacity to synthesize such molecules is or is not incorporated into the ongoing program of daughter cells in the Mb compartment. For example, there is evidence that actin, heavy and light myosin chains, tropomyosin, hyaluoronic acid and collagen chains are synthesized by PMb and Mb. What is not yet clear is whether any of these molecules synthesized by PMb are identical to, rather than being alternative (allelic) forms of, those synthesized by cells in the Mb compartment (ISHIKAWA, et al., 1968, 1969; HOLTZER et al., 1973, 1974; RUBINSTEIN et al., 1974). Equally critical, though even more difficult to approach experimentally will be the task of delineating those unique molecules synthesized by cells in the various compartments of the myogenic lineage antecedent to the PMb compartment.

III. Is Movement from the PMb to Mb Compartment Dependent on DNA Replication and/or the Cell Cycle?

Brachial somites of early 3-day chick embryos consist of approximately 2×10^4 mesenchyme-like cells. Their cell cycle is probably less than 8 hrs, and the vast majority cycle for the next several days (BISCHOFF and HOLTZER, 1968; HOLTZER and MATHESON, 1970). At least half the cells are in the myogenic lineage and of these a modest, though unknown, number are in the PMb compartment. Some somite cells are in the chondrogenic and/or fibrogenic lineages. These 2×10^4 cells are the only progenitors to the many millions of definitive muscle, cartilage, and connective tissue cells that emerge from this region. By using fluorescein-labeled antibodies against myosin or tropomyosin on whole mount squashes, it is possible to count accurately the number of antibody-positive myoblasts/somite. There are between 10^2 and 10^3 post-mitotic, mononucleated, striated myoblasts in a single 3 day brachial somite (HOLTZER et al., 1957; HOLTZER and SANGER, 1972). Accordingly, with this technique it is possible to monitor the increase in numbers over a period of time of myoblasts/somite and, in so doing under a variety of conditions, determine what might block the movement of cells from the PMb to the Mb compartment.

The following experiments with the DNA synthesis inhibitor, 5-fluorodeoxyuridine (FUdR), and the mitotic inhibitor, colcemide, suggest that DNA synthesis and/or some event associated with cell division are obligatory for moving PMb

Number of Striated Myoblasts/Posterior 3-Day Myotome

	Control	Fudr-treated
	960	520
	1220	830
	810	410
(a)	1350	750

	Fudr-treated, not reversed	Fudr-treated, then reversed
	380	840
	650	1050
(b)	430	1010

Fig. 1. (a) Three-day trunks were transected into right and left halves and reared as organ cultures for 10 hrs in normal medium alone or with 10^{-6}M Fudr and 2×10^{-6}M Udr. After treatment with fluorescein-labeled antimyosin, the myotomes were squashed and the individual striated, mononucleated myoblasts counted. (b) Trunks were transected into right and left halves and organ cultured in Fudr plus Udr for 10 hrs. Either the right or left half was removed, washed, and then grown in normal medium with excess Tdr for an additional 15 hrs; the other half remained in Fudr. It is to be stressed that the number of myoblasts in the controls were essentially the number of myoblasts present at the time the somites were explanted. In experiments of this kind only one pair of somites can be used from a single embryo. The variation in numbers of myoblasts between control somites reflects the modest differences in stages of morphogenesis in the seven different embryos used in these experiments. The only significant comparison in these experiments is the difference in numbers between a control somite and its contralateral treated somite

cells into the Mb compartment: A right or left somite was placed in normal growth medium for 10 hrs, or for slightly more than one cell cycle. The contralateral somite was placed in growth medium plus FUdR. Squashes were prepared and the number of myoblasts counted under the fluorescence microscope. The data are shown in Fig. 1a (taken from HOLTZER et al., 1973). More new cells were recruited into the myoblast compartment in the controls than in the series in which DNA synthesis was blocked. FUdR for protracted periods can be cytotoxic. The observation (Fig. 1b) that removal of FUdR was followed by renewed recruitment into the Mb compartment suggests that the results observed were not due to the cytotoxic effects of the drug on cycling PMb cells. Similar results have been obtained using cytosine-arabinoside to block DNA synthesis. In other experiments it has been shown that these inhibitors of DNA synthesis do not directly block the synthesis of contractile proteins (HOLTZER et al., 1974; RUBINSTEIN et al., 1974).

Replicating PMb cells in colcemide can be reversibly blocked in metaphase for at least 10 hrs (BISCHOFF and HOLTZER, 1968). PMb cells in somites that were incubated in colcemide for 10 hrs did not move into the Mb compartment.

These experiments with FUdR, cytosine-arabinoside, and colcemide lead to two conclusions: (1) Recruitment of new cells into the Mb compartment requires DNA synthesis and at least passage through M, and (2) PMb cells arrrested at the G1-S interface, or in M, do not have the option to begin to synthesize and to organize myosin, actin, and tropomyosin into striated myofibrils. This second conclusion is worth emphasizing, for it suggests that options of synthesis open in G1 of the PMb are distinctly different from those open to its progeny in G1. Elsewhere we have termed a cell cycle that produces daughters with options of synthesis not open to the mother cell, a quantal cell cycle (HOLTZER, 1970; HOLTZER et al., (1973). It is this type of cell cycle that we postulate moves cells from one compartment to the next in a lineage. The more frequently studied cell cycle is the proliferative cell cycle, which increases the numbers of cells within a given compartment by producing daughters with the same developmental options that are open to the mother cell. In terms of control in the developmental program, the PMb-Mb transition requires a quantal cell cycle such that the withdrawal from the cell cycle in the Mb compartment cannot be imitated by any inhibition of DNA replication in the PMb compartment.

IV. Is Cytokinesis Obligatory to Move Cells from the PMb to the Mb Compartment?

Though the pleiotropic effects of Cytochalasin-B are not well understood (CARTER, 1967; WESSELS et al., 1971; HOLTZER et al., 1973; SPOONER, 1971, MIRANDA et al., 1974), all investigators agree that in addition to blocking amoeboid movement and the uptake of sugars, Cytochalasin-B at appropriate doses blocks cytokinesis. SANGER and HOLTZER (1972) have shown that in many kinds of cells the drug does not block the initiation of cleavage, but it does block its completion. PMb cells grown in the drug undergo normal nuclear division and cleavage begins, but since the cleavage furrow relaxes before the daughters separate, binucleated cells are produced. As described in detail elsewhere (SANGER et al., 1971; HOLTZER and SANGER, 1972; HOLTZER et al., 1973) replicating PMb cells in the drug yield bi-nucleated "myotubes" dense with thick and thin filaments. These bi-nucleated myotubes contract spontaneously as do normal myotubes *in vitro*. As long as such bi-nucleated cells remain in the drug they do not fuse. Failure to fuse may indicate that amoeboid movement is essential to the fusion processes, or it may reflect some readily reversible alteration at the cell surface. While the alignment of the myofibrils and the formation of Z bands are perturbed in these bi-nucleated myotubes, the hexagonal stacking of thick myosin and thin actin filaments is quite normal. Failure of normal alignment of the myofibrils could be due to the absence of normal insertion points at the cell surface for myofibrils, or to the failure of Z bands to form normally. Either of these events could involve sugar moieties at the cell surface that are known to be effected by the drug. The characteristic control of DNA replication operates in these bi-nucleated myotubes, for their nuclei do not re-enter the mitotic cycle.

These experiments with cytochalasin-B demonstrate that cytokinesis *per se* is not an obligatory event for moving cells from the PMb to the Mb compartment. It is possible that some event normally occurring *after* mitosis in the G1 of nascent daughter cells is the prerequisite condition for initiating the program of Mb cells.

V. How Much Time is Required for a Myogenic Cell to Demonstrate a Newly Available Option for Synthesis?

OKAZAKI and HOLTZER (1966) incubated somite cells in 3H-TdR for 2, 4, 6, 8, 10, etc. hours. The cells were glycerinated, treated with fluorescein-labeled antibody against myosin and against tropomyosin. The question posed was, when would double-labeled myoblasts first be detected, i.e. myoblasts with labeled thymidine in their nuclei, and bound antibody in their cytoplasm? Such double-labeled cells would provide information as to when in the cell cycle a given cell becomes antibody positive. Double-labeled cells first appeared in those series kept in 3H-TdR for 10 or more hours. With average S and G2 periods of 4.5 and 2 hrs, respectively, these results indicate that myogenic cells in S, G2, or M, or even in the early hours of G1, are antibody negative. If we assume the last half hour of S in a PMb cell to early G1 in the daughter Mb cell requires 3 to 4 hrs, then roughly a minimum of 6 to 7 hrs is required to go from a negative PMb in mitosis, to a myosin or tropomyosin antibody positive daughter myoblast. The rapidity with which newborn Mb cells express their unique and newly acquired program is not peculiar to myogenic cells. For example, first generation erythroblasts only several hours old, in contrast to their metaphase, mother hematocytoblasts, have synthesized sufficient quantities of hemoglobin to be benzidine-positive (WEINTRAUB et al., 1971).

VI. When in the Cell Cycle Do Myogenic Cells Fuse and Are They PMb or Mb Cells?

That the multinucleated myotube is the product of fusing myogenic cells and that the nuclei in myotubes are post-mitotic, was demonstrated some years ago (HOLTZER et al., 1957; HOLTZER, 1958; LASH et al., 1958; STOCKDALE and HOLTZER, 1961). The experiments of BISCHOFF and HOLTZER (1968) demonstrated that myogenic cells in S or G2 or M do not have the option to fuse; myogenic cells must be in G1 to fuse. This leads to the following question: If nuclei in myotubes are post-mitotic, and if fusion is confined to G1 myogenic cells, do cells withdraw from the cell cycle before expressing the option to fuse, or do they lose the capacity to enter S as a consequence of fusion? This issue is not trivial. If cells lose the ability to enter S as a consequence of fusing, then fusion would be a property of replicating myogenic cells (i.e. PMb cells), whereas if withdrawal from the cell cycle is an obligatory precondition to fusion, then fusion is a unique property of myoblasts (i.e. Mb cells).

Experiments correlating diminished DNA polymerase activity with fusion led O'NEILL and STROHMAN (1969) to conclude that fusion rendered myogenic nuclei post-mitotic. In contrast, HOLTZER (1961; 1970) argued that the decision to withdraw from the cell cycle might be a precondition to fusion. This argument was based on the observation that mononucleated, cross-striated, myoblasts do not enter the cell cycle, showing that withdrawal is not dependent on fusion. The

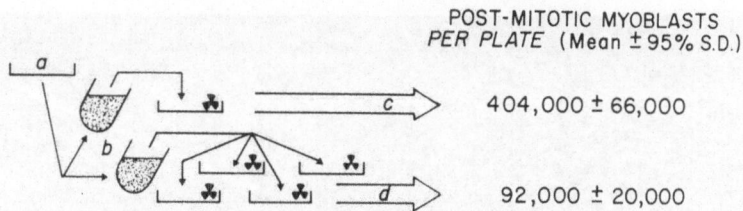

Fig. 2. O'NEILL and STOCKDALE subcultured 6-hr-old primary muscle cultures *a* to obtain two original inocula containing post-mitotic myoblasts and mitotic presumptive myoblasts *b* which were identical except for subsequent treatment in which one was plated at $^{1}/_{4}$ the density of the other *(c ~ 4d)*. Did the lesser density in d prompt the post-mitotic myoblasts to forego fusion, enter S, become labeled, and divide before fusing? No, for regardless of subsequent treatment, an estimate of the frequency of Mb in *b* from *c* or *d* is about the same

$$(b \sim c \sim 4d, b \sim 404000 \pm 66000 \sim 368000 \pm 80000).$$

Did higher density in *c* inhibit otherwise cycling PMb from dividing and yielding Mb which could fuse? Yes, the authors report *less* muscle formed as a *percentage* of all cells per plate in *c* than in *d*. In another trial, the statistical variation inherent in the experimental manipulations resulted in $4\bar{d} > \bar{c}$, the opposite of those authors' contention. (Source: Experiment No. 1, Table 1, STOCKDALE and O'NEILL, 1972)

recent paper by PATERSON and STROHMAN (1972) would also suggest that myoblasts pull out of the cell cycle without fusing.

Early mesenchyme cells in the somite or limb bud do not have the capacity to fuse. Since these cells replicate and pass through intermediate compartments of the myogenic lineage we proposed that eventually they undergo a particular cell cycle, yielding daughter cells that for the first time acquire the option to fuse (HOLTZER et al., 1972). Question: if a myogenic cell that has just acquired the option to fuse in a particular G1 should fail to fuse, can it enter S again? If this happens, then the option to fuse following a particular quantal cell cycle is a property that subsequently "waxes and wanes" (OKAZAKI and HOLTZER, 1966) as these cells go through additional *proliferative* cell cycles.

O'NEILL and STOCKDALE (1972) have recently described a statistical analysis of experiments involving subculture of myogenic cells at low or high density which they claim supports the notion that myogenic cells capable of fusion enter S, if they fail to contact one another. We would question their interpretation of the data presented on two counts: (1) There is a significant statistical variation between *any* primary and secondary culture, and between cultures set up at different times; thus significant differences observed in comparisons of myogenic primary and secondary cultures are not readily interpretable; (2) Their interpretation of the data rested on the assumption that "low density" would provoke otherwise post-mitotic cells (e. g. Mb) to divide, while ignoring the alternative assumption that "high density" would inhibit otherwise mitotically competent cells (e. g. PMb) from entering S and dividing. As shown in their Table 1 (STOCKDALE and O'NEILL, 1972) these two factors are critical in their cultures.

Because of the theoretical importance of these interesting experiments of O'NEILL and STOCKDALE (1972), one such experiment is summarized in Fig. 2. Their data, in fact, suggest that Mb do not re-enter the cell cycle as a result of low density.

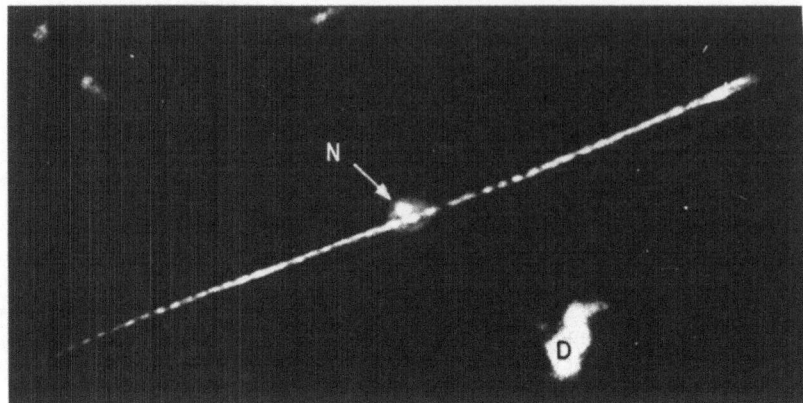

Fig. 3. A post-mitotic, cross-striated, mononucleated myoblast in a chick breast muscle culture grown for 5 days in EGTA. Glycerinated preparations were treated with fluorescein-labeled antimyosin and then nuclei stained with fluorescent label. This permits the simultaneous identification of myofibrils and the single nucleus (N) in one photograph; however, non-specific staining of some debris (D) also results. 40 × obj. mag., UV fluorescence microscopy

Recently DIENSTMAN and HOLTZER (in press) have approached this issue by blocking fusion with the calcium chelator, EGTA (STROHMAN and PATERSON, 1971). EGTA was added to myogenic cultures established from 10-day chick embryo breast muscle. After 2, 3, 4, or 5 days the cultures were treated with fluorescein-labeled antimyosin. Small numbers of elongated, mononucleated, striated myoblasts were first unequivocally observed in 3 day cultures and by day 5 approximately 30% of the elongated cells were antibody positve (Fig. 3). These findings indicate that in breast muscle, as in somite muscle, fusion is not a requirement for movement from the PMb to the Mb compartment. It also confirms the view that there is no obligatory coupling between fusion and the expression of other properties of the terminal program of Mb cells: myoblasts may organize myofibrils before or after fusion (HOLTZER et al., 1957; HOLTZER, 1970; HOLTZER and SANGER, 1972). These findings are not in accord with data presented by YAFFE and DYM (1973), PATERSON and STROHMAN (1972), EASTON and REICH (1972), or BUCKLEY and KONIGSBERG (1974).

Further experiments by DIENSTMAN and HOLTZER (in press) demonstrated that with the appropriate inoculum by the end of the 4th day in culture containing adequate calcium, most of the fusible cells will have been incorporated into myotubes, and further culturing only results in an increase in the number of mononucleated cells. Cultures reared in EGTA for 4 days contain only mononucleated cells. Many are fusion-blocked muscle cells collected in the mononucleated state. Extending growth in normal medium containing calcium breaks the block, and an intense period of fusion ensues. By introducing 3H-TdR at the time of the break it was found that the nuclei in the myotubes that subsequently formed were *unlabeled*. Thus the fusible cells are post-mitotic cells (i.e. Mb) and the cells that replicated after day 4 were either fibroblasts or myogenic cells that did not enter the Mb compartment.

It might be argued that myoblasts did not divide because of the circumstances of relatively high density of cells at the end of the 4th day of culture. To investigate this possibility, following 4 days in primary culture in EGTA, the cells were subcultured and plated at the same density at which primary cultures were established, in medium with 3H-TdR. There was extensive DNA synthesis and cell division among the mononucleated cells, yet the myotubes that formed were almost entirely from non-labeled, and therefore post-mitotic, cells. One remaining possibility might be that EGTA is *selectively* toxic to PMb, and thus only Mb survive to fuse. That this was not the case, was shown by removing EGTA after only 2 days in primary culture, and subculturing in the presence of 3H-TdR. In this case, the majority of nuclei in the myotubes that subsequently formed were labeled. Therefore, during the first days in EGTA, myogenic cells divided, allowing the muscle lineage to progress, so that by the time of "reversal", the PMb compartment (somewhat within 2 days, mostly by 4 days) was exhausted, having generated the Mb, postmitotic compartment. The data for these experiments are shown in Fig. 4.

That mononucleated myoblasts are programmed to withdraw from the cell cycle prior to fusion and that as mononucleated cells they will organize thick and thin filaments into myofibrils, also can be demonstrated by experiments using Cytochalasin-B (see also Section IV). This drug causes Mb cells to become sessile and spherical in shape. Fusion is blocked. As a consequence, rounded-up, mononucleated Mb cells which can contract appear in myogenic cultures treated with cytochalasin-B. As shown in Fig. 5, treated post-mitotic mononucleated Mb cells organize thick and thin filaments into unaligned, stacked hexagonal arrays.

Others have argued that skeletal myogenesis may be a "2-phase" process, simply involving a population of replicating, mononucleated, myogenic cells which are induced by proper exogenous conditions of medium or density to fuse, then withdraw from DNA replication, and form myofibrils. The experiments of O'NEILL and STOCKDALE (1972) are ambiguous, and in fact may indicate that the population of Mb cells which fuses is insensitive to changes in density. KONIGSBERG (1971) has claimed that "conditioned medium" can affect the time of appearance *in vitro* of the first myotubes. DOERING and FISCHMAN (1974) have attempted to quantitate this phenomenon by measuring "% fusion" in their cultures. Using this measurement, they also have concluded that most of the replicating myogenic cells could be prompted to fuse instead of dividing, either by growth in "conditioned medium", or by artificially inhibiting replication with cytosine arabinoside and allowing the arrested cells to "condition" their own environment. However, the index "% fusion" was not a measure of the actual percentage of cells which fused, because it excluded counting in the ratio many unfused, mononucleated cells. The authors rationalized this exclusion on 2 untested assumptions: (1) the excluded mononucleated cells were all non-myogenic, (2) the included mononucleated cells were all myogenic and could eventually fuse. In addition, these authors ignored two contradictions to their hypothesis and conclusions in their results: (1) in every case, one third to one half of the mononucleated cells assessed as myogenic never did fuse; (2) the failure of these supposedly myogenic cells to fuse when arrested in the presence of "medium factors" which should "cue" fusion easily might mean that only their Mb progeny can fuse, but not they themselves.

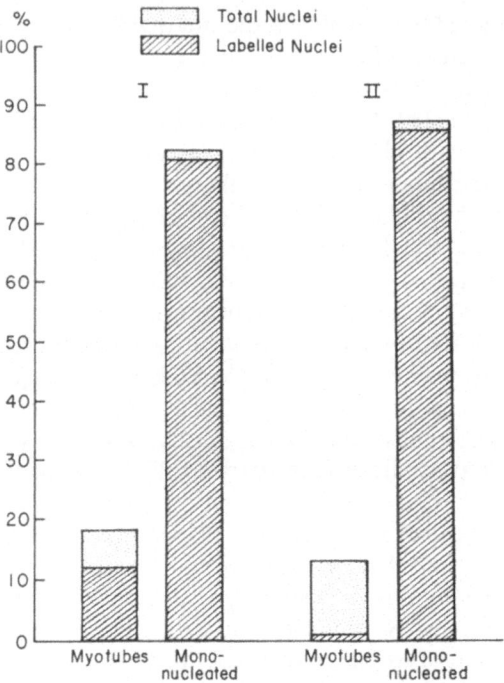

Fig. 4. Ten-day chick breast muscle cells were grown in the fusion inhibitor EGTA as primary cultures for 2 (I) or 4 (II) days. During this time, myogenic and fibrogenic cells replicated, expanding their population sizes, although the myogenic cells underwent the PMb-Mb transition producing many unfused, post-mitotic cells by day 4. Cultures I and II were subcultured at the original density, released from the EGTA-fusion block and grown in fresh medium containing labeled thymidine for 4 more days. A smaller percentage of PMb had given rise to Mb after 2 than after 4 days; lower density at the outset of secondary culture did not prompt Mb to forego fusion and enter S phase; as a percentage of all cells, the post-mitotic muscle population fell greatly behind in size compared to the constantly dividing fibroblasts

We would suggest that such cells which show no overt sign of terminal differentiation (e.g. fusion, myofibril formation) in fact cannot be classified, and that the index "% fusion" of DOERING and FISCHMAN (1974) is unreliable and highly misleading. To accept the latter authors' conclusions, it would have to be demonstrated that: (1) the history of proliferation behind the population of cells which did fuse was in fact altered by the addition of conditioned medium, (2) media or mitotic inhibitors did not themselves alter the morphologies of "myogenic" or "fibrogenic" cells, and (3) a straightforward accounting of all cells was not inconsistent with those conclusions.

BUCKLEY and KONIGSBERG (1974) have observed that continuous growth of myogenic cultures in ^{3}H-TdR results in 100% labeling of the mononucleated cells after 24 hrs. These authors interpreted this to mean that post-mitotic mononucleated cells—myoblasts (Mb cells)—do not exist and that the withdrawal from the cell cycle is a condition restricted to nuclei in multinucleated myotubes. We, on the other hand, would suggest this observation simply means that all the Mb

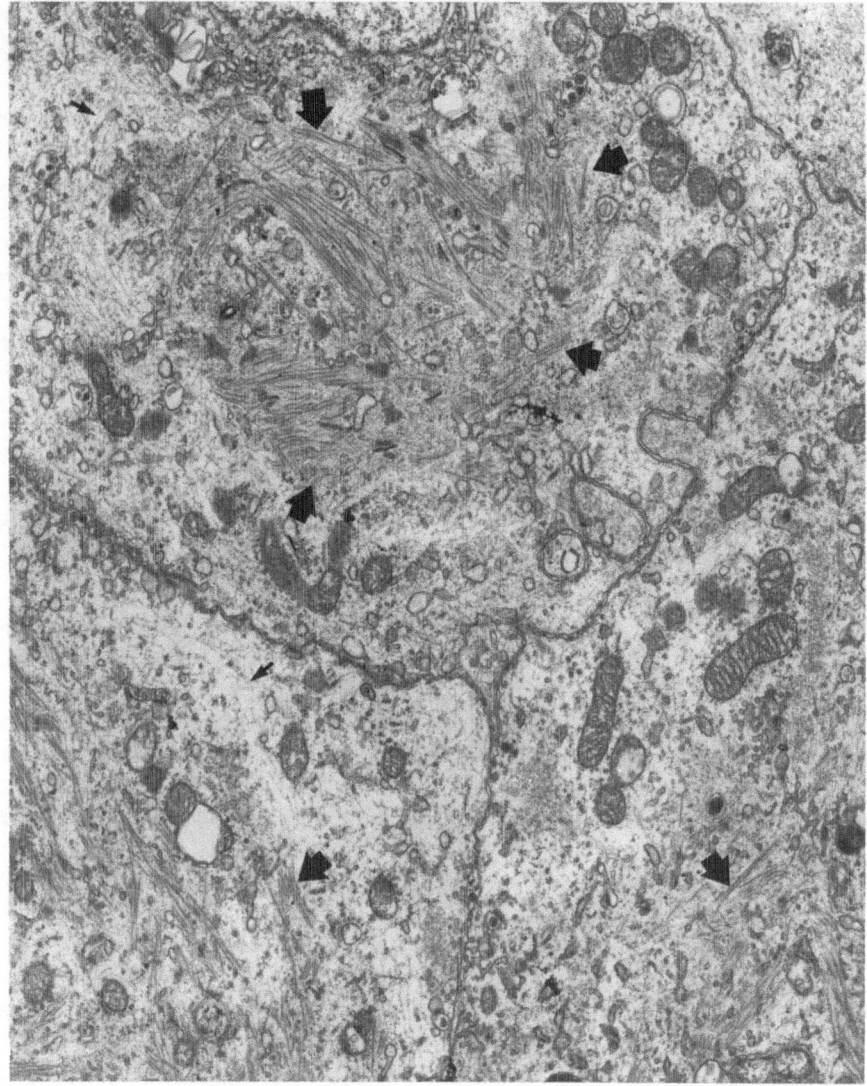

Fig. 5. Section through a cluster of 3 juxtaposed, mononucleated myoblasts that have been grown in Cytochalasin-B (10 µg/ml) for 2 days. Note the numbers of randomly oriented but hexagonally stacked thick myosin and thin actin filaments (→) in the 3 cells. Discrete Z-bands are not evident but dense bodies are frequent (HOLTZER et al., 1973). The cells are rich in intermediate 100 A° filaments (→) and small vessicles perhaps associated with the formation of the transverse tubules (ISHIKAWA, 1969; ISHIKAWA et al., 1968)

cells which were generated prior to the incorporation of label had exercised the option to fuse within that 24 hrs period, and by fusing, allowed the labeled cells to reach 100% of the mononucleated population. Furthermore, nothing in their data excludes the likelihood that many of the labeled cells that appeared during this 24 hrs period were in fact post-mitotic myoblasts.

Our experiments, taken together, show that the Mb compartment is real, identifiable, and separable from the PMb compartment. These experiments are: (1) the accumulation in EGTA of fusible, post-mitotic myoblasts that do not re-enter the cell cycle even when subcultured at low density in fresh medium; (2) the production of myosin in Mb cells in medium containing EGTA or CB; (3) the failure to produce such myoblasts in medium containing inhibitors of DNA synthesis.

In summary: fusion and the organization of striated myofibrils are independent options in the program of cells in the Mb compartment and one can be expressed independently of the other. PMb do not have the option to do either. We postulate that within 4 hrs after a particular PMb completes a quantal cell cycle, one or both daughters will have withdrawn from the cell cycle and have the option to fuse, and/or organize striated myofibrils.

VII. Is There Evidence for a Developmental Program Controlling the Different Compartments within the Myogenic Lineage?

At moderate concentrations the thymidine analogue, 5-bromo-2-deoxyuridine (BUdR), applied to cultures blocks fusion and the emergence of myoblasts (STOCKDALE et al., 1964; OKAZAKI and HOLTZER, 1965; COLEMAN and COLEMAN, 1968; BISCHOFF and HOLTZER, 1970). LOUGH and BISCHOFF (1973) have found that incorporation of the analogue at the beginning of S is more critical than incorporation at the end of S. At lower concentrations fusion is blocked, but not the formation of striated myofibrils by mononucleated Mb (BISCHOFF and HOLTZER, 1970). In these experiments it is still unclear whether all the myoblasts with myofibrils in fact had the analogue incorporated in the previous cell cycle; if they did it would mean that at a low percent BUdR-substitution the program for myofibril formation was expressed but the program for fusion was blocked.

BUdR-suppressed myogenic cells have been maintained as mononucleated, replicating, non-fusing cells for many subcultures. If allowed to replicate in normal medium, the progeny of the suppressed cells will enter the terminal Mb compartment. BUdR did not cancel or detectably alter the genetic commitment of the suppressed cells, and even though they incorporated the BUdR into their DNA, the cells were operationally indistinguishable from cells in the PMb compartment. On the biological level, BUdR permitted PMb to undergo proliferative cell cycles, keeping them in the penultimate compartment, rather than permitting them to enter the quantal cell cycle that would move them into the Mb compartment. Furthermore, whatever the actual mechanism of the BUdR effect, it serves to distinguish the operation of the developmental program from all the genetic or physiological regulators for synthesis essential to viability in eukaryotic cells.

BUdR blocks myogenic cells in still earlier compartments of the myogenic lineage. If early somites are grown in BUdR for 48 hrs and then removed and grown in normal medium, their descendents are healthy replicating mononucleated cells. These cells, however, never enter the terminal compartments of the myogenic or chondrogenic lineages (ABBOTT et al., 1972). The apparent "irreversibility" of BUdR on cells in early compartments is difficult to interpret. Primitive mesenchyme cells ordinarily are unable to move into the PMb and then Mb compartments in our cultures (DIENSTMAN et al, 1974). Therefore, the BUdR-suppressed early myogenic cells, even if they "recovered", may not pass through the next 2 quantal cell cycles owing to the absence *in vitro* of a relatively "trivial" environmental factor (e. g. Ca^{2+} concentration, cyclic AMP, hormone, etc.).

VIII. What Is the Minimal Number of Compartments within the Myogenic Lineage?

If the myogenic lineage includes all cell types leading to the terminal compartment, then the zygote, some blastula and some gastrula cells must be included as cells in such a lineage. Still later there is probably a primitive mesenchyme compartment (Ms). According to the embryological literature (BELLAIRS, 1971) the primitive mesenchyme compartment consists of cells that are ancestral to the myogenic, chondrogenic, and fibrogenic lineages proper. Question: If there is a discrete compartment of Ms cells, how does this single cell type give rise to myoblasts, chondroblasts, and fibroblasts? In other words, what is the minimum number of compartments and what is their relationship in the course of the differentiation of the progeny of Ms cells into definitive muscle, cartilage and connective tissue cells? Is there, for example, one or more compartments in the myogenic lineage between the Ms and the PMb compartments?

Recently we described experiments suggesting the existence of at least one compartment between the Ms and PMb compartments (ABBOTT et al., 1974). These experiments were based on the observation that a mononucleated cell isolated from developing limbs can yield a "myogenic clone". Such clones, however, contain large numbers of cells that are indistinguishable from muscle fibroblasts. The single cell that serves as progenitor for myogenic and fibrogenic cells will not yield chondrogenic cells. In short, a cell has been isolated that has the option to yield myogenic and fibrogenic, but not chondrogenic, progeny. The experiments were as follows: Muscle cultures were prepared from 8-day leg muscles. After 4 days, these cultures, rich in myotubes and mononucleated cells, were dissociated; with a micropipette, under the microscope, single cells were removed. The single cells were deposited in the wells of a microtest slide. Three to four weeks later over 90% of the large surviving clones were myogenic. These myogenic clones consisting of myotubes and mononucleated cells, were dissociated and all the cells of a single clone plated into a 100 mm petri dish. Without exception these subcultures produced large numbers of mononucleated cells that were cytologically indistinguishable from muscle fibroblasts. The biochemical similarity between these mononucleated cells and authentic fibroblasts was indicated by the fact that their output of hyaluronic acid and Type I collagen chains was indistinguishable from normal fibroblasts from leg muscles. In other experiments the mononucleated cells that could yield myogenic and fibrogenic cells were challenged to yield chondrogenic progeny. When grown under different *in vitro* and *in vivo* conditions known to permit movement from the penultimate into the ulti-

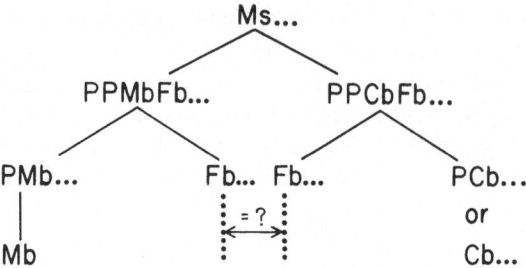

Fig. 6. A proposed lineage based on binary transitions beginning with cells in the "primitive mesenchyme" (Ms) compartment that leads to myoblasts (Mb), to fibroblasts (Fb), and to chondroblasts (Cb). Ellipses indicate that cells within the compartment are replicating and capable of either increasing the population within that compartment, or of yielding progeny for the next compartment. Thus the presumptive myoblast (PMb) compartment consists of cells yielding either post-mitotic Mb cells or more replicating PMb cells. It is not clear whether a given PPMbFb cell actually divides asymmetrically yielding one PMb and one PFb cell or whether such cells only divide symmetrically yielding 2 PMb or 2 PFb cells. As terminally differentiated Fb cells may emerge very early in development and readily replicate, there may be no self-renewing PFb compartment. Currently there is no way of identifying presumptive chondroblasts (PCb). While some small myogenic clones described in the literature (e.g. Fig. 4 in Dienstman et al., 1974) may have been derived from cells within the PMb compartment, the larger myogenic clones are believed to be from the PPMbFb compartment. The arrow and question mark indicate uncertainty about the possible relationships of fibroblasts that may emerge via different lineages (see also Schiltz et al., 1973; Mayne et al., 1974). Fibroblasts, however, do not form myoblasts (Holtzer, (1970b)

mate compartment of the chondrogenic lineage these cells did not deposit metachromatic chondroitin sulfate nor synthesize Type II collagen chains.

Though alternative explanations are possible, the simplest would be that the cell giving rise to a myogenic clone was in fact ancestral to both myogenic and fibrogenic lineages, and that this single cell did not have the option to produce cells in the chondrogenic lineage. This postulated compartment of cells has been designated as the PPMb Fb compartment. Figure 6 is a tentative lineage detailing the proposed relationship between Ms, PPMbFb, Fb, PMb, and Mb, as well as accounting for the emergence of chondroblasts and their associated fibroblasts.

If lineages are to be useful, their nodal points must be translated into specific generations of cells and the restricted synthetic options open to cells in any one compartment explained. We can ask, for example, where in Fig. 6 is cell division facultative or obligatory in moving cells from one compartment to the next? PMb cells have been shown not to have the option of assembling cross-striated myofibrils, although their daughter cells do. On the other hand, is the Ms cell obligated to divide asymmetrically always yielding one PPMbFb and one PPCbFb cell, or is the event a probabilistic one, each division being symmetrical, but cell divisions, in response to exogenous fluctuations, lead either to 2PPMbFb or 2PPCbFb cells?

The concept of lineage is, to large degree, incompatible with the concept of "multipotentiality". The population of cells in Fig. 6 has the property of a "multipotential" *system*, but any one cell is at most "bipotential". A given Ms cell does not have the option itself of directly transforming into, or within one generation of giving rise to myoblasts, or chondroblasts, or fibroblasts. The only option in terms of phenotypic diversification open to a given Ms cell is to yield PPMbFb or PPCbFb cells. Our prediction is that no eukaryotic cell is multipotential in the sense that a single cell has many options itself of diversifying into many phenotypes, and that stepwise diversification is the sum of many binary decisions or quantal cell cycles (HOLTZER et al., 1972; ABBOTT et al., 1974). The similarities between this scheme for generating diversity in mesenchymal tissues and those proposed for hematopoietic cells and for cells mediating antigen-antibody reactions requires no elaboration here (EDELMAN and GALL, 1969; TILL et al., 1967).

It is to be stressed that in the context of this discussion, "bipotential" refers to cells, *combinations* of nuclei *and* cytoplasm. Elsewhere it has been suggested that the initiating events for programming transitions in a lineage probably depend upon nuclear-cytoplasmic interactions (HOLTZER and ABBOTT, 1968; HOLTZER, 1970; HOLTZER et al., 1972). The binary transitions postulated in Fig.6 would require (1) cycling precursor cells with an initial cytoplasmic state and limited nuclear program for synthesis (2) exchange of information, such that neither nucleus nor cytoplasm retains the previous synthetic state, (3) progression of the cell cycle so that the "new" cytoplasmic state permits the induction of a further option, and (4) expression of the new nuclear program for synthesis in the next generation. Experiments with nuclear or cytoplasmic transplanation would be useful to test bipotentiality and the degrees of restriction in cytoplasmic cues.

IX. When Do PMb Cells First Appear during Embryogenesis?

According to the scheme in Fig.6, "myogenic clones" can be initiated by cells in the Ms, PPMbFb, or PMb compartments. The presence of cartilage cells in addition to Mb and Fb cells should distinguish clones derived from the Ms compartment from clones derived from cells in later compartments. If all the cells in a clone entered myotubes or develop cross-striations, then such a clone would have been derived from a PMb cell. In the absence of this condition, which is occasionally observed (Fig.5 in KONIGSBERG, 1963; Fig.4 in DIENSTMAN et al., 1974), it would be impossible to distinguish clones derived from PPMbFb from PMb, as it is currently impossible to discriminate cytologically between replicating fibrogenic and replicating myogenic cells. In spite of these ambiguities, challenging cells from earlier and earlier chick limb buds or chick somites to form myogenic clones should provide information about the time at which the first PPMbFb or PMb cells emerge (see ABBOTT et al., 1972, for this technique as applied to somitic chondroblasts). HAUSHKA and WHITE (1972) and DIENSTMAN et al. (1974) have found that small numbers of myogenic clones can be recovered *in vitro* from dissociated stage 21–22 limb buds. Using culture conditions which permitted the PMb-Mb and the presumptive chondroblast (PCb)-chondroblast (Cb) transitions, DIENSTMAN et al. (1974) estimated that approximately 5% of limb mesoderm cells were in either the penultimate or ultimate myogenic or chondrogenic compartments by stage 21–22. In short, about 95% of all cells in stage 21–22 limb buds can only be classified as precursors in compartments antecedent to either PMb or

PCb compartments. Another observation from these experiments was that mixed "chondrogenic and myogenic clones" did not emerge. These observations suggest that in stage 21–22 limb buds either (1) the Ms compartment has been depleted, or (2) more likely, the culture conditions which permit terminal differentiation from penultimate cell types do not permit Ms cells to yield progeny that move into and through subsequent compartments of the myogenic or chondrogenic lineages.

Dienstman et al. (1974) found that though dissociated stage 17–18 limb buds did not yield myogenic or chondrogenic clones, explants of such limb buds after 3 days in culture did form modest numbers of myoblasts, and after 4 days formed readily recognizable clusters of chondroblasts. A rough estimate would be that of all cells in these very young limb buds about 2% entered the terminal myogenic or chondrogenic compartments. Thus, though stage -17 or -18 limb buds at the beginning of the experiment did not contain any identifiable Mb or Cb, after 4 days in culture in the total absence of "inducing" agents some cells moved into the terminal compartments of one lineage or the other.

Cultures of explants or monolayers of succeedingly older limb bud stages (21–22, 24–25) permitted the terminal differentiation of a steadily increasing percentage of limb cells. Because the culture conditions used are only known to permit terminal differentiation from the penultimate cell types (PMb, PCb), the increased frequency of terminal muscle and cartilage cells with culture of older stages tends to reflect the increased frequency of PMb and PCb cells placed in culture. On the other hand, the declining majority of cells which do not differentiate terminally reflects the mesoderm population distributed in compartments antecedent to PMb and PCb. However, it remains to be determined whether these precursors to the PMb or PCb are Ms cells, or have differentiated into intermediate compartments for muscle *or* cartilage *or* connective tissues.

X. Is the PMb Compartment Ever Depleted?

Post-mitotic myoblasts that will never again synthesize DNA begin to accumulate in 48-hr chick embryos. The number of myotube nuclei steadily increases over the next 2 to 3 weeks, and then the rate of their accumulation markedly drops around hatching. Muscle regeneration, however, occurs in adult chicks and mammals. Question: If myoblasts cannot be induced to re-enter the cell cycle, what is the source of new cells responsible for muscle regeneration? Claims that nuclei in mature muscle or in myotubes may form new cell membranes and then duplicate their DNA (Holtzer, 1958; Hay, 1970) are based on static histological procedures that cannot prove this points (see Holtzer, 1970b). Changes in myotube nuclei following viral infection or injury, including uptake of 3H-TdR (Yaffe and Gershon, 1967; Fogel and DeFendi, 1967) or the appearance of chromosomes, are likely to be due to DNA repair enzymes (Hahn et al., 1971; Stockdale and O'Neill, 1972b) and chromosomal condensation (Rao and Johnson, 1971) rather than to entering and *successfully* completing an S period (Holtzer and Sanger, 1972). It is highly probable that the source of cells for muscle regeneration are the "satellite" cells, which operationally behave as mitotically quiescent PMb. When mature muscle is injured these satellite cells are induced to undergo a number of proliferative cell cycles and many of their progeny move into the terminal Mb compartment. Muscle regeneration in adults simply recapitualtes myogenesis in the embryo (HOLTZER, 1958).

The challenging theoretical importance of regeneration which also relates to myogenesis in the embryo is the kind of control that permits *staggered* transition from the PMb into the Mb compartment over great periods of time. As depicted in Fig. 6, it is likely that a given PMb divides in a 72-hr limb bud. One daughter may be programmed to withdraw irreversibly from the cell cycle and within 4 hrs assemble a detectable striated myofibril with typically mature sarcomeres. The other daughter, in effect, may serve as progenitor for literally tens of generations of PMb cells. Most of these descendent PMb will, in the next 2 to 3 weeks, undergo that quantal cell cycle that moves them into the Mb compartment. The remainder of the descendents of the PMb cell that divided in the 72-hour limb bud may persist as quiescent PMb or satellite cells until activated by injury or until death of the animal.

Regulating the frequency of PMb-Mb transitions is one of the few places the microenvironment probably plays a role in myogenesis. This exogenous influence could involve (1) a generalized mitogenetic component, as well as (2) one that specifically promotes one particular quantal cell cycle, or conversely, selectively promotes proliferation within one particular compartment in the lineage. Erythropoietin, for example, probably serves such a role in the erythrogenic lineage and there is some evidence for growth hormone functioning in this manner in the development of the skeletal system.

XI. Does the Ongoing Program of Mb Cells Differ Qualitatively from That of PMb Cells?

The concept of lineages developed in this review postulates that the ongoing program within a given compartment must differ from the programs of cells in earlier or later compartments. Muscle cells, we suggest, are not "determined" early in myogenesis by a one-step event, nor do they retain this determined state, transmitted unchanged through many generations of proliferative cell cycles.

Rather we propose the occurrence of a multi-compartmented lineage in which each compartment is characterized by having different bits of coordinated information transcribed. Different bits of coordinated information would be made available by successive quantal cell cycles. Accordingly, information should be available in PMb cells that is not available in Mb cells, and information should be available in Mb cells that was not available in PMb or still earlier precursor cells. Differentiation of embryonic cells is not simply due to intracellular changes within one cell associated with the passage of time; differentiation involves quantal cell cycles that move proliferating populations from one compartment to the next.

One bit of coordinated information not available in Mb cells but present in all precursor cells obviously relates to DNA synthesis. Whether this loss is referable to information no longer available for (1) the cytoplasmic factors required to initiate DNA synthesis (Rao and Johnson, 1971; Harris, 1970; Gurdon, 1969) or (2) the synthesis of DNA polymerases, etc., is not known.

Question: What specific molecules are synthesized only in Mb cells, but not in PMb, or other kinds of cells? Work with fluorescein-labeled antibodies against skeletal myosin and skeletal tropomyosin has stressed the absence of these specific antigens in PMb and non-muscle cells (Holtzer et al., 1957; Holtzer et al., 1973; Masaki, 1972). On the other hand, work in the past 5 years makes it abundantly clear that information for the transcription and translation of actomyosin-like molecules is not confined to muscle cells, but probably some forms of contractile proteins are constituitive molecules to be found in all eukaryotic cells. Ishikawa et al., (1969) reported that not only PMb, but Fb, Cb, and even nerve and intestinal cells treated with H-meromyosin form arrow-head complexes in association with the cell surface; thus demonstrating the availability of an actin gene in many kinds of cells (see also Schroeder, 1973; Perry et al., 1971). Similarly Adelstein et al. (1973) working with blood platelets and Bray (1973) working with nerve cells have demonstrated myosin-like molecules in non-muscle cells, thus demonstrating the availability of genes for different allelic forms of myosin in many kinds of cells.

Antibody against skeletal myosin or skeletal tropomyosin does not cross-react in agar gels with the myosins or tropomyosins from fibroblasts, chondroblasts, or even PMb cells (Holtzer et al., 1973, 1974; Holtzer, 1970). This suggests that there are polymorphic forms of myosin and tropomyosin, and that in different compartments of the myogenic lineage, different structural genes for myosin and tropomyosin are transcribed. Recently, Rubinstein et al. (1974) and Holtzer et al. (1974) have shown that PMb, Cb, Fb, and BUdR-suppressed PMb cells synthesize actomyosin-like molecules. Clearly, central to the problem of developmental programming is the mechanism by which choices are made among kinds of myosin heavy and light chains during myogenesis, and how some molecules are "repressed" and another set is "derepressed" in one compartment.

It is likely that all cells synthesize actin and that the conservative nature of the actin molecule renders it likely that the structural gene(s) for actin will prove to be similar in base sequences and perhaps equally available for transcription and translation in many cell types (Elzinga and Collins, 1973). However, the finding that BUdR-suppressed myogenic cells do not assemble thick or thin filaments for myofibrils (Bischoff and Holtzer, 1970) though they assemble *cortical filaments* that can be decorated with H-meromyosin suggests the possibility of at least two programs of synthesis involving actin: one that is BUdR-sensitive, and one that is BudR-insensitive. This raises the possibility of there being two identical structural genes for actin, one for skeletal myofibrils and one for cortical filaments. Alternatively, one structural gene may subserve two distinct programs, one for the cortex in all cell types (including muscle cells), and one for thin filaments that function only in cells of the Mb compartment. If there is but one structural gene for actin it might be "leaky" in the presence of BUdR and its product preferentially polymerized in assocation with the cell surface. It is also possible that actin for myofibrils will only be translated if myosin monomers are at the same time being polymerized into thick filaments (HOLTZER et al., 1973).

There is now considerable evidence for allelic forms of tropomyosin and myosin. The tropomyosins found in nerve and fibroblasts are probably transcribed from genes that differ in base sequence from those active in muscle (COHEN and COHEN, 1972; FINE et al., 1973). SRETER et al. (1972) demonstrated that the kinds of myosin light chains synthesized *in vitro* by myotubes which formed from Mb cells obtained from an embryonic fast muscle, display molecular weight and type-specificity characteristic of adult fast, rather than slow muscle. HOLTZER et al. (1974) have shown that the actomyosins from replicating non-muscle cells and PMb cells have very different half-lives. In skeletal muscle cultures from embryonic fast muscles, containing both myotubes and their attendant fibroblasts, 3 different myosin light chains are found. Yet cultures of PMb, PCb, or Fb, or BUdR-suppressed myogenic cells yield only 2 light chains. The different light chains are distinguishable on a molecular weight basis.

PMb and Mb cells synthesize collagen as well as such glycosaminoglycans as hyaluronic acid and a type of chondroitin sulfate (MAYNE et al., in press). The fact that the chondroitin sulfate (CS) synthesized by myotubes is not deposited as metachromatic matrix suggests the possibility that it differs—either in sugars, sulfation, or protein core—from that synthesized by cartilage cells. While preliminary work demonstrates that myotubes do in fact secrete Type I collagen chains in a 2:1 ratio, there is some suggestion that in addition they synthesize another type as well. The cell-type specificity of mucopolysaccharides or collagens probably reflects both transcriptional and post-transcriptional differences in synthesis, i.e. differences in both mRNA's for protein cores *and* the enzymes for glycan synthesis or glycosylation. For example, the extracellular sulfated glycosaminoglycans secreted by somite precursor or cardiac precursor cell types differ qualitatively from those synthesized by terminal cartilage or heart cells (O'HARE, 1973; PALMOSKI and GOETINCK, 1972; MANASEK et al., 1973). These characteristic differences, in part, probably reflect isozymic differences.

XII. What Is the Mechanism of the Quantal Cell Cycle?

Elsewhere we have stressed that currently available data on a variety of experimental systems precluded one simple mechanism for a "quantal cell cycle". Many obvious schemes could be formulated *a priori* to result in cell cycles leading to daughter cells with synthetic options different from those of the mother cell. Briefly, these schemes fall into two, not necessarily mutually exclusive, groups: (1) Causal, at the level of DNA synthesis and/or chromosomal structure; an (2) Permissive, at the level of a cytoplasmic or cell membrane cue that normally operates only in a particular phase of the mitotic cycle. Similar to the cellular reprogramming that occurs during viral infection, the first kind of quantal cycle might begin with a cytoplasmic factor acting on a unique nuclear locus only as that locus replicates during S. This event could simply involve blocking the binding of histone. This altered locus would not be a structural gene but would be a higher order genetic unit (a regulator gene) associated with the transition to the next program of synthesis in the next compartment of a lineage, and it could not be transcribed in a G2 or M cytoplasm. Expression of this higher genetic unit could only occur in the G1 cytoplasm of daughter cells. This kind of regulatory scheme postulates that reprogramming is not simple stimulation or inhibition of an ongoing activity—e.g. depress rRNA synthesis or augment cytochrome oxidase synthesis—but requires the act of DNA replication to initiate a train of events that leads only to the next "predetermined" program in the lineage.

The second kind of scheme for quantal cell cycles still permits a central role for the cell cycle, but emphasizes the individual reprogramming event in a cell. It relegates the concept of higher order genetic control to, at best, a circumstantial phenomenon, and the role of S phase to secondary importance in simply providing for growth and expanded expression. In this type of schema (HOLTZER, 1963; HOLTZER and ABBOTT, 1968; HOLTZER, 1970) the nucleus only responds in terms of reprogramming to cytoplasmic inducers during very restricted periods of the cell cycle. For example, it has been proposed that the actual inductive interaction between spinal cord and somites only occurs in G1, and that otherwise responsive somite cells can not respond during S, G2 or M (HOLTZER et al., 1973; HOLTZER and MAYNE, 1973). Similarly the decision to withdraw from the cell cycle and initiate the program for terminal myogenesis—or to move from any compartment in the lineage to the next—may be triggered by a "trivial" or a "specific" cytoplasmic molecule, that becomes active after the condensed chromosomes relax in the daughters of a PMb cell. What is of importance is that either the "trivial" or "specific" factor would lead to terminal myogenesis only in a cell in the proper compartment of the myogenic lineage. The "trivial" cytoplasmic factor could exert one unique effect in a cell genetically prepared to become an Mb cell, but a very different effect or none at all in a cell genetically prepared to become an erythroblast. By such a definition ecdysone or progesterone could be such "trivial" molecules. Alternatively, it may be that a "specific" molecule is uniquely synthesized in PMb cells, and therefore is present to activate the program for "terminal myogenesis" only in cells subsequent to PMb cells, i.e. Mb cells.

Skeletal myogenesis has been used in this article to indicate three broad aspects of cell differentiation. First, the terminal cell types appear through the differentiation of an entire lineage of precursor cell types. Second, the cells of each compartment of such a lineage share a common program for the synthesis of a characteristic set of macromolecules, some of which may be allelic forms of constituitive gene products. The sequence of lineage compartments reflects, on a molecular level, a sequence of programs of synthesis. Third, the options available to any cell, at any time, in any lineage, are strictly circumscribed. Transitions in a sequence (e.g. PMb-Mb) are linked, causally or permissively, to cell cycle events. Inactivation of an ongoing program and activation of a subsequent one result from nuclear-cytoplasmic interaction. These points essentially encompass questions of *control* in differentiation, and together define what is termed the "developmental program". Figure 7 gives an overview of these concepts applied to skeletal myogenesis.

SKELETAL MYOGENESIS

DEVELOPMENTAL PROGRAM

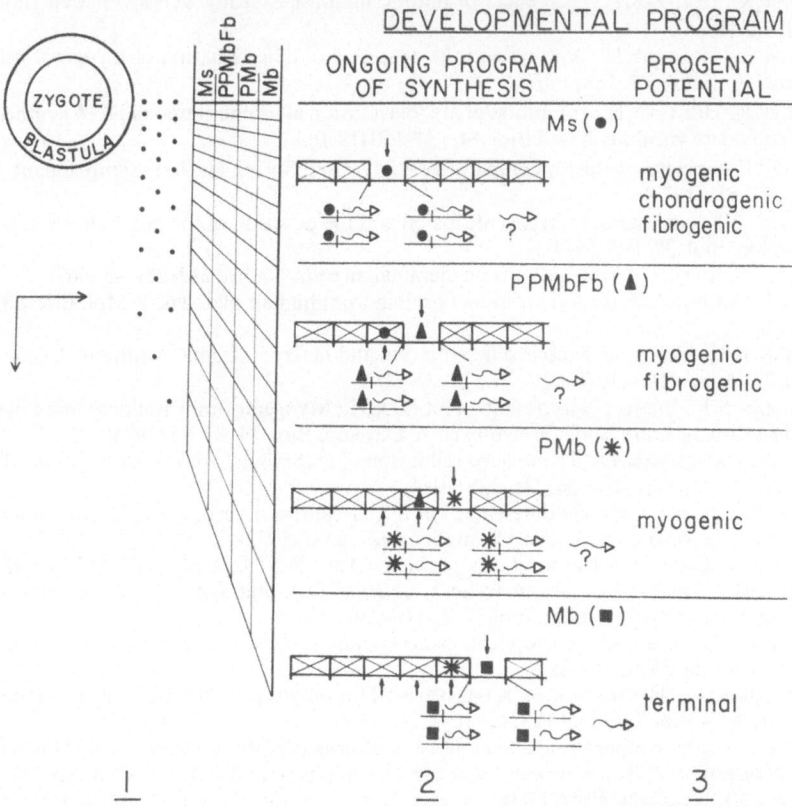

Fig. 7. A schematized version of three aspects of the developmental program for skeletal myogenesis: (1) Rapid and extensive growth of muscle occurs through a fixed sequence of cell types in which the transitions in the lineage represent a constant number of quantal cell cycles which are interspersed with a variable number of proliferative cell cycles. The horizontal arrow is the direction of progression of lineage compartments; the vertical arrow represents the direction of growth with time. Note the first Mb cell may appear in a 48 hrs embryo, whereas other Mb cells appear after hatching; but all Mb cells have resulted from the same number of quantal cell cycles. (2) Each compartment cell type has a unique program for synthesis (e.g. Ms, ⌐•⌐) available which specifies a coordinated set of genes; some gene products, such as the myosin heavy and light chains, are probably allelic forms. (Gene transcript,〜〜▷; blocked, inactive program of synthesis in the sequence, ▷◁▷◁; active, ongoing program, ⌐↓⌐; cytoplasmic cueing, ↑↓). (3) Each precursor cell type is specified for its progeny potential, as well as its ongoing program

References

Abbott, J., Mayne, R., Holtzer, H.: Inhibition of cartilage development in organ cultures by the thymidine analogue, 5-bromo-2 deoxyuridine. Develop. Biol. **28**, 430–442 (1972).

Abbott, J., Schlitz, J., Dienstman, S. R., Holtzer, H.: Proc. Natl. Acad. Sci. (in press) (1974).

Adelstein, R. S., Conti, M.: The characterization of contractile proteins from platelets and fibroblasts. Cold Spring Harbor Symp. Quant. Biol. **37**, 599–605 (1973).

BELLAIRS,R.: Developmental processes in higher vertebrates. London: Logos Press 1971.
BISCHOFF,R., HOLTZER,H.: The effect of mitotic inhibitors on myogenesis. J. Cell Biol. **36**, 111–128 (1968).
BISCHOFF,R., HOLTZER,H.: Mitosis and the processes of differentiation of myogenic cells *in vitro*. J. Cell Biol. **41**, 188 (1969).
BISCHOFF,R., HOLTZER,H.: Inhibition of myoblast fusion after one round of DNA synthesis in 5-bromodeoxyuridine. J. Cell Biol. **44**, 134–150 (1970).
BRAY,D.: Cytoplasmic actin: A comparative study. Cold Spring Harbor. Symp. Quant. Biol. **37**, 567–571 (1973).
BUCKLEY,P., KONIGSBERG,I.: Myogenic fusion and the duration of the post-mitotic gap (Gl). Develop. Biol. **37**, 193 (1974).
CARTER,S.B.: Effects of cytochalasins on mammalian cells. Nature **213**, 261–4 (1967).
COHEN,I., COHEN,C.: A tropomyosin-like protein from human platelets. J. Mol. Biol. **68**, 383 (1972).
COLEMAN,J., COLEMAN,A.: Muscle differentiation and macromolecular synthesis. J. Cell Physiol. **72**, (Supp. 1) 19–34 (1968).
DIENSTMAN,S.R., BIEHL,J., HOLTZER,S., HOLTZER,H.: Myogenic and Chondrogenic Lineages in Developing Limb Buds Grown *in vitro*. Develop. Biol. **39**, 83–95 (1974).
DOERING,J., FISCHMAN,D.: The *in vitro* cell fusion of embryonic chick muscle without DNA synthesis. Develop. Biol. **36**, 225–235 (1974).
EASTON,T., REICH,E.: Muscle differentiation in cell culture. Effects of nucleoside inhibitors and rous sarcoma virus. J. Biol. Chem. **247**, 6427–6431 (1972).
EDELMAN,G., GALL,W.: The antibody problem. Ann. Rev. Biochem. **38**, 415–466 (1969).
ELZINGA,M., COLLINS,J.: The amino acid sequence of rabbit skeletal muscle actin. Cold Spring Harbor Symp. Quant. Biol. **37**, 1–7 (1973).
FAMBROUGH,D., RASH,J.: Development of acetylcholine sensitivity during myogenesis. Develop. Biol. **26**, 55–68 (1972).
FINE,R., BLITZ,A., HITCHCOCK,S., KAMINER,B.: Tropomyosin in brain and growing nerves. Nature New Biol. **245**, 182–185 (1973).
FISCHMAN,D.: The synthesis and assembly of myofibrils in embryonic muscle. In: MOSCONA, A., MONROY,A. (Eds.): Current Topics in Developmental Biology, Vol. 5, pp. 133–179. New York: Academic Press 1970.
FOGEL,M., DEFENDI,V.: Infection of muscle cultures from various species with oncogenic DNA virus (SV40 and Polyoma). Proc. Natl. Acad. Sci. **58**, 967–973 (1967).
GURDON,J.: Intracellular communication in early animal development. In: 28th Symposium of the Society for Developmental Biology (ed. A.LANG), 59–82. New York: Academic Press 1969.
HAHN,G., KING,D., YANG,S.: Quantitative changes in unscheduled DNA synthesis in rat muscle cells after differentiation. Nature New Biol. **230**, 242 (1971).
HARRIS,H.: Cell fusion. Cambridge, Mass: Harvard University Press 1970.
HAUSCHKA,S., WHITE,N.: Studies of myogenesis *in vitro*. In: B.BANKER (Ed.): Research in Muscle Development and the Muscle Spindle, pp. 53–71. Amsterdam: Excerpta Medica 1972.
HAY,E.: Regeneration of muscle in the amputated amphibian limb. In: MAURO,A., SHAFIQ,S., MILHORAT,A. (Eds.): Regeneration of Striated Muscle and Myogenesis, pp. 3–24. Amsterdam: Excerpta Medica 1970.
HOLTZER,H.: The development of mesodermal structures in regeneration and embryogenesis. In: THORNTON,C. (Ed.): Regeneration in Vertebrates. Chicago: University of Chicago Press 1958.
HOLTZER,H.: Aspects of chondrogenesis and myogenesis. In: RUDNICK,D. (Ed.): The 19th Growth Symposium, pp. 35–88. New York: Ronald Press 1961.
HOLTZER,H.: Comments on induction during cell differentiation. In: Induktion und Morphogenese. 13th Mosbacher Colloquium. Berlin-Göttingen-Heidelberg: Springer 1963.
HOLTZER,H.: Induction of chondrogenesis: A concept in quest of mechanisms. In: BILLINGHAM,R. (Ed.): Epithelial-Mesenchymal Interactions, pp. 152–164. Baltimore: Williams and Wilkins 1968.

HOLTZER,H.: Myogenesis. In: SCHJEIDE,O., DEVILLIS,J. (Eds.): Cell Differentiation, pp.476–503. Princeton: Van Nostrand and Rheinhold 1970.

HOLTZER,H.: Discussion. In: MAURO,A., SHAFIQ,S., MILHORAT,A. (Eds.): Regeneration of Striated Muscle and Myogenesis, Vol.7. Amsterdam: Excerpta Medica 1970b.

HOLTZER,H., ABBOTT,J.: Oscillations of the chondrogenic phenotype *in vitro*. In: URSPRUNG,H. (Ed.): Results and Problems in Cell Differentiation, Vol.1, The Stability of the Differentiated State, pp.1–16. Berlin-Heidelberg-New York: Springer 1968.

HOLTZER,H., ABBOTT,J., LASH,J.: On the formation of multinucleated myotubes. Anat. Rec. **131**, 567 (1957).

HOLTZER,H., DETWEILER,S.: III. Induction of skelotogenous cells. J. Exp. Zool. **123**, 335–370 (1953).

HOLTZER,H., MARSHALL,J., FINCK,H.: An analysis of myogenesis by use of fluorescent anti-myosin. J. Biophys. Biochem. Cytol. **3**, 705 (1957).

HOLTZER,H., MATHESON,D.: Induction of chondrogenesis in the embryo. In: BALAZS,E.. FINEGOLD,M. (Eds.): Chemistry and Molecular Biology of the Intercellular Matrix, pp.52–64. Baltimore: Williams and Wilkins 1970.

HOLTZER,H., MAYNE,R.: Experimental morphogenesis: The induction of somitic chondrogenesis by embryonic spinal cord and notochord. In: PERRIN,E., FINEGOLD,M. (Eds.): Pathobiology of Development, pp.52–64. Baltimore: Williams and Wilkins 1973.

HOLTZER,H., RUBINSTEIN,N., CHI,J., DIENSTMAN,S.R., BIEHL,J.: On the phenotypic heterogeneity of myogenic clones and what cells synthesize which molecules. In: MILHORAT,A. (Ed.): Exploratory Concepts in Muscular Dystrophy and Related Disorders. Amsterdam: Excerpta Medica (in press) 1974.

HOLTZER,H., SANGER,J.: Myogenesis: Old views rethought. In: BANKER,B. (Ed.): Research in Muscle Development and the Muscle Spindle, pp.122–133. Amsterdam: Excerpta Medica 1972.

HOLTZER,H., SANGER,J., ISHIKAWA,H., STRAHS,K.: Selected Topics in Skeletal Myogenesis. Cold Spring Harbor Symp. Quant. Biol. **37**, 549–566 (1973).

HOLTZER,H., WEINTRAUB,H., BIEHL,J.: Cell-cycle-dependent events in differentiation. FEBS Symposium on Cell Differentiation. New York: Academic Press 1972.

HOLTZER,H., WEINTRAUB,H., MAYNE,R., MOCHAN,B.: The cell cycle, cell lineage and cell differentiation. In: MOSCONA,A., MONROY,A. (Eds.): Current Topics in Developmental Biology, Vol.6, pp.229–256. New York: Academic Press 1972.

ISHIKAWA,H.: Formation of elaborate networks of T system tubules in cultured skeletal muscle with special reference to T system formation. J. Cell Biol. **38**, 51–66 (1968).

ISHIKAWA,H., BISCHOFF,R., HOLTZER,H.: Mitosis and intermediate-sized filaments in developing skeletal muscle. J. Cell Biol. **38**, 538 (1968).

ISHIKAWA,H., BISCHOFF,R., HOLTZER,H.: Formation of arrow-head complexes with heavy meromyosin in a variety of cell types. J. Cell Biol. **43**, 312 (1969).

JOHNSON,R., RAO,P.: Nucleo-cytoplasmic interactions in the achievement of nuclear synchrony in DNA synthesis and mitosis in multinucleated cells. Biol. Rev. **46**, 97 (1971).

KONIGSBERG,I.: Clonal analysis of myogenesis. Science **140**, 1273–1284 (1963).

KONIGSBERG,I.: Diffusion-mediated control of myoblast fusion. Develop. Biol. **26**, 133–152 (1971).

KONIGSBERG,I., MCELVAIN,N., TOOTLE,M., HERRMANN,H.: The dissociability of deoxyribonucleic acid synthesis from the development of multinuclearity of muscle cells in culture. J. Biophys. Biochem. Cytol. **8**, 333–343 (1960).

LASH,J., HOLTZER,H., SWIFT,H.: Regeneration of mature skeletal muscle. Anat. Rec. **128**, 679–693 (1957).

LOUGH,J., BISCHOFF,R.: Inhibition of myoblast fusion by BUdR during the first half of DNA synthesis. J. Cell Biol. **59**, 202a (1973).

MAHOWALD,A.: Origin and continuity of polar granules. In: REINERT,J., URSPRUNG,H. (Eds.): Results and Problems in Cell Differentiation, Vol.2, pp.159–169. Berlin-Heidelberg-New York: Springer 1971.

MANASEK,F., REID,M., VINSON,W., SEYER,J., JOHNSON,R.: Glycosaminoglycan synthesis by the early embryonic chick heart. Develop. Biol. **35**, 332–348 (1973).

MASAKI, T., YOSHIZAKI, C.: The onset of myofibrillar protein synthesis in chick embryo *in vivo*. J. Biochem. **71**, 755–757 (1972).

MAYNE, R., SCHILTZ, J., HOLTZER, H.: Some overt and covert properties of chondrogenic cells. In: PIKKARAINEN, J. (Ed.): The Biology and Biochemistry of the Fibroblast. New York: Academic Press (in press) 1973.

MIRANDA, A., GODMAN, G., DEITCH, A., TANENBAUM, S.: Action of cytochalsin D on cells of established lines. J. Cell Biol. **61**, 481–500 (1974).

O'HARE, M.: An histochemical study of sulphated glycosaminoglycans associated with the somites of the chick embryo. J. Embryol. Exp. Morph. **29**, 197–208 (1973).

OKAZAKI, K., HOLTZER, H.: An analysis of myogenesis using fluorescein labeled antimyosin. J. Histochem. Cytochem. **13**, 726 (1965).

OKAZAKI, K., HOLTZER, H.: Myogenesis: Fusion, myosin synthesis and the mitotic cycle. Proc. Natl. Acad. Sci. **56**, 1484 (1966).

O'NEILL, M., STOCKDALE, F.: Differentiation without cell division in cultured skeletal muscle. Develop. Biol. **29**, 410–418 (1972).

O'NEILL, M., STROHMAN, R.: Changes in DNA polymerase activity associated with cell fusion in cultures of embryonic muscle. J. Cell Physiol. **73**, 61 (1969).

PALMOSKI, M., GOETINCK, P.: Synthesis of proteochondroitin sulfate by normal, nanomelic, and 5-bromodeoxyuridine-treated chondrocytes in culture. Proc. Natl. Acad. Sci. **69**, 3385–3388 (1972).

PATERSON, B., STROHMAN, R.: Myosin synthesis in cultures of differentiating chicken embryo skeletal muscle. Develop. Biol. **29**, 113, 138 (1972).

PERRY, M., JOHN, H., THOMAS, N.: Actin-like filaments in the cleavage furrow of newt egg. Exp. Cell Res. **65**, 249–253 (1971).

PIERCE, B.: Teratocarcinoma. In: MOSCONA, A., MONROY, A. (Eds.): Current Topics in Developmental Biology, Vol. 2. New York: Academic Press 1967.

RUBINSTEIN, N., CHI, J., HOLTZER, H.: Actin and myosin in a variety of cell types. Biophys. Biochem. Res. Commun. (in press) (1974).

SANGER, J., HOLTZER, H.: Cytochalasin B: effects on cell morphology, cell adhesion, and mucopoylsaccharide synthesis. Proc. Natl. Acad. Sci. **69**, 253 (1972).

SANGER, J., HOLTZER, S., HOLTZER, H.: effects of cytochalasin B on muscle cells in tissue culture. Nature New Biol. **229**, 121–123 (1971).

SCHILTZ, J., MAYNE, R., HOLTZER, H.: The synthesis of collagen and glycosaminoglycans by dedifferentiating chondroblasts in culture. Differentiation **1**, 97–107 (1973).

SCHROEDER, T. F.: Actin in dividing cells: Contractile ring rilaments bind heavy meromyosin. Proc. Natl. Acad. Sci. **70**, 1688–1692 (1973).

SPOONER, B., WESSELLS, N. K.: Microfilaments and cell locomotion. J. Cell Biol. **49**, 595–613 (1971).

SRETER, F., HOLTZER, S., GERGELY, J., HOLTZER, H.: Some properties of embryonic myosin. J. Cell Biol. **55**, 344–356 (1972).

STEVENS, L. C.: The biology of teratomas. In: ABERCHROMBIE, M., BRACHET, J. (Eds.): Advances in Morphogenesis, Vol. 6, pp. 1–32. New York: Academic Press 1967.

STEVENS, L. C.: The development of transplantable teratocarcinomas from intratesticular grafts of pre- and post-implantation mouse embryos. Develop. Biol. **21**, 364–382 (1970).

STOCKDALE, F., HOLTZER, H.: DNA Synthesis and Myogenesis. Exp. Cell Res. **24**, 508–520 (1961).

STOCKDALE, F., OKAZAKI, K., NAMEROFF, M., HOLTZER, H.: 5-Bromodeoxyuridine: Effect on myogenesis *in vitro*. Science **146**, 533 (1964).

STOCKDALE, F., O'NEILL, M.: Deoxyribonucleic acid synthesis, mitosis and skeletal muscle differentiation. In Vitro **8**, 212–227 (1972).

STOCKDALE, F., O'NEILL, M.: Repair DNA synthesis in differentiated embryonic muscle cells. J. Cell. Biol. **52**, 589–597 (1972 b).

STROHMAN, R., PATERSON, B.: Calcium-dependent cell fusion and myosin synthesis in cultures of developing chick muscle. J. Gen. Physiol. **51**, 244 (1971).

TILL, J. E., MCCULLOCH, E., PHILLIPS, R., SIMINOVITCH, L.: Analysis of differentiating clones derived from marrow. Cold Spring Harbor Symp. Quant. Biol. **32**, 461–464 (1967).

WEINTRAUB, H., CAMPBELL, G., HOLTZER, H.: Primitive erythropoiesis in early chick embryogenesis. I. J. Cell Biol. **50**, 652–668 (1971).
WESSELLS, N. K., SPOONER, B., ASH, J.: Microfilaments in cellular and developmental processes. Science **171**, 135–143 (1971).
YAFFE, D., DYM, H.: Gene expression during differentiation of contractile muscle fibers. Cold Spring Harbor Symp. Quant. Biol. **37**, 543–547 (1973).
YAFFE, D., GERSHON, D.: Multinucleated muscle fibers: Induction of DNA synthesis and mitosis by polyoma virus infection. Nature **215**, 421–424 (1967).

The Organization of Red Cell Differentiation

Harold Weintraub

Department of Biochemical Sciences, Princeton University, Princeton, NJ, USA

I. Introduction

In this Paper I concentrate primarily on the molecular events involved in the transition from a non-hemoglobin(Hb)-producing precursor cell to a Hb-producing red blood cell. Available evidence will be presented which indicates that this transition requires the "exposure" of the globin genes so that they become available for binding by RNA polymerase. Additional experiments will show that the proper rate of DNA chain elongation is required for the acquisition of the capacity to synthesize Hb and several related observations concerning the properties of the chromosome at the growing fork for DNA replication will be discussed. Lastly, a detailed analysis of the effects of the thymidine analogue, BUdR, will be presented. This analysis may provide an insight into the way the erythropoietic program is organized at the genetic level. Since space is limited, many related topics must, unfortunately, be neglected [1].

II. Primitive Chick Erythropoiesis: Background to a Model System

At about 35 hrs of incubation, Hb first appears in the yolk sac of the developing chick blastoderm. There are about 2×10^6 cells in the whole embryo at this time and of these, roughly 10^5 are precursors to the primitive line of red blood cells and are isolated in the peripheral regions of the embryo. The progeny from these precursor cells mature as a relatively homogeneous cohort of developing erythroblasts which synthesize Hb and continue to divide about 6 times until day 5 of incubation (Weintraub et al., 1971). At this time, the cells cease synthesizing detectable levels of RNA and DNA but continue to make Hb for several more days. Once Hb synthesis begins at 35 hrs, a characteristic sequence of developmental events follows in a relatively synchronous fashion until the 6th or 7th day. This sequence has been described both biochemically and morphologically

---

[1] The interested reader is therefore referred to a volume on hematopoiesis edited by A. Gordon (1971) for peripheral material dealing with such topics as erythropoiesis, the multipotentiality of hematopoietic stem cells, and the exogenous factors which modulate the development of different types of blood cells.

(WILT, 1966; LUCAS and JAMROZ, 1966; HASHIMOTO and WILT, 1966; WEINTRAUB et al., 1971; CAMPBELL et al., 1971; HAGOPIAN et al., 1972; SMALL and DAVIES, 1973). And it has been proposed that the changes occurring during this maturation period are based upon a program that is coupled to the events dictating the transition from the precursor cell (HOLTZER et al., 1972). Consequently, the very interesting changes taking place after the commencement of Hb synthesis will be discussed only in so far as they relate to the mechanisms responsible for instituting the erythropoietic program.

III. What Is To Be Explained?

In the light of recent developments, the basic question of how a cell initiates the synthesis of Hb can *probably* be reduced to the question: How does a cell make the Hb genes accessible? A related, but separate question involves the reason why this occurs at a specific time during embryogenesis. Evidence that active genes are more accessible is presented in Fig. 1. Here nuclei from 4 days erythroblasts were isolated and treated with pancreatic DNase (MIRSKY,

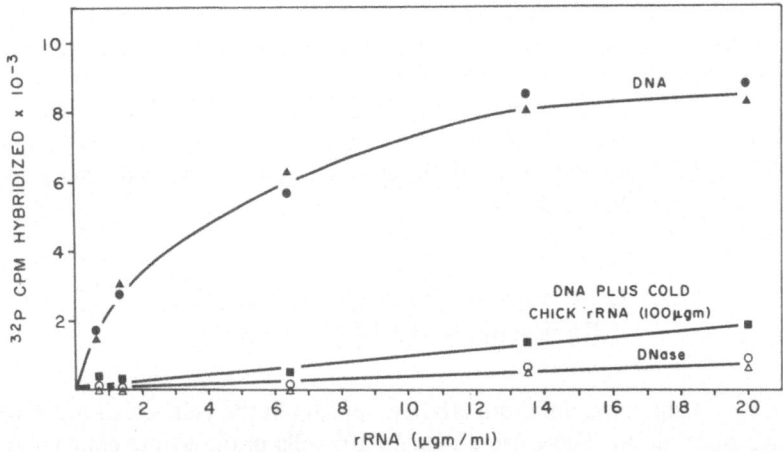

Fig. 1. Selective digestion of nuclear ribosomal genes by exogenous DNase. Nuclei were isolated and about 15% of the DNA digested by addition of pancreatic DNase as previously described (WEINTRAUB, 1973b). The remaining 85% of the DNA was isolated and hybridized to p^{32} labeled ribosomal DNA (RUBIN and SULSTON, 1973) obtained from Xenopus laevis tissue cultures. As a control, DNA was isolated from mature chick erythrocytes and used for the hybridization. To show that the interspecies hybridization was specific, rRNA from chick was competed against the labeled Xenopus rRNA and shown to completely inhibit its hybridization. The molecular weight of all DNA preparations was adjusted by limited DNase digestion to about 200000 daltons. Controls showed that about 70% of the DNA remained on the filters during the hybridization. rRNA was at a specific activity of 0.8×10^6 CPM/ugm. (●— ●) ▲—▲ hybridization to whole DNA; ○—○, △—△ hybridization to DNA obtained from DNase digested nuclei; ×—× hybridization to DNA in the presence of 100 μgm of chick rRNA. Additional experiments, using solution hybridization, have confirmed these findings, with no more than 30% of the rRNA remaining after limited DNase

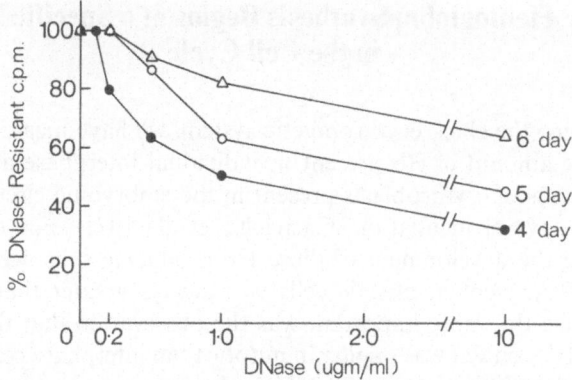

Fig. 2. Increased resistance of nuclear DNA to DNase with erythroblast maturation. 4-, 5-, or 6-day red blood cells were pre-labeled *in vivo* with H 3-TdR for several generations. Approximately 10^6 nuclei from each day were incubated in 1 ml of buffer containing 5 mM sodium phosphate, pH 6.7; 3 mM $MgCl_2$; 0.25 M sucrose and the appropriate concentration of pancreatic DNase. After incubation for 30 min, TCA precipitable counts were determined and expressed as a percentage of a control incubation in which no DNase was added

1971; ITZHAKI and COOPER, 1973; BILLINGS and BONNER, 1972; PEDERSON, 1972; WEINTRAUB, 1973a) to digest only the most sensitive nuclear DNA (about 15% of the total genome). The remaining DNA was then purified and hybridized to ribosomal RNA. In comparison to undigested nuclei, the nuclei partially digested with DNase had no detectable DNA sequences complementary to the rRNA. These results are paralleled by the findings of AXEL et al. (1973) who have shown that *E. coli* RNA polymerase transcribes the globin genes only from reticulocyte chromatin, but not from liver chromatin. Thus, it seems likely that a basic difference between a precursor red cell and a cell making Hb will prove to be the inaccessibility of the globin genes in the former. The finding that Hb mRNA is very low in populations of erythroid precursor cells lends further support to this idea (TERADA et al., 1972; LEDER et al., 1973; M. GRODINE, personal communication).

The use of DNase provides still another insight into erythropoiesis, Fig. 2 shows the dose response to DNase for 4-, 5-, and 6-day red blood cell nuclei. It is apparent that as maturation proceeds and as macromolecular synthesis declines, nuclear DNA becomes more and more resistant to DNase. The fast component of the digestion, which has tentatively been associated with active genes, gradually disappears. Associated with these changes in accessibility is a progressive increase in the chicken red-cell-specific histone, f2C (Moss et al., 1973; WEINTRAUB, unpubl.). Although the appropriate experiments are only now in progress, it is tempting to think that the f2C histone has been selected in order to make those essential genes which are available in all cells less accessible during the latter stages of erythroid development. This will be a difficult statement to prove since other changes, notably a decrease in non-histone chromosomal protein and an altered pattern of histone modification, also accompany the decrease in sensitivity to DNase during maturation.

IV. Hemoglobin Synthesis Begins at a Specific Time in the Cell Cycle

In describing the chick erythropoietic system, we have measured, cytophoto-metrically, the amount of Hb present in individual interphase and mitotic cells from populations of erythroblasts present in the embryonic circulation between 2 days and 6 days of incubation (CAMPBELL et al., 1971). For each generation characterizing the development of these Hb-producing cells, we found that the amount of Hb present in mitotic cells was always greater than that found in interphase cells. But more important was the observation that the coefficient of variation for Hb content was *smaller* in mitotic than interphase cells. The simplest explanation of these findings is that Hb synthesis begins at a specific time in the cell cycle of the first generation of Hb-producing erythroblasts. If Hb synthesis began randomly in the cell cycle, then we would have expected no differences in the coefficients of variation for mitotic and interphase cells of a given generation. Figure 3 shows the Hb levels per cell obtained cytophotometrically from the 6 successive mitoses of primitive chick erythropoiesis. The abscissa corresponds to the sequential division times, while the ordinate is the average amount of Hb present at any point in a given generation.

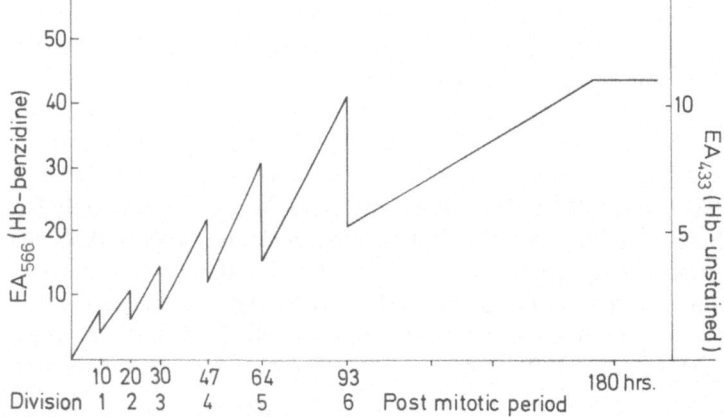

Fig. 3. Hb content per cell during successive stages of red blood cell maturation. The amount of Hb present in mitotic cells of primitive chick embryo erythroblasts was determined cyto-photometrically (CAMPBELL et al., 1971) between 35 hrs of incubation when the first cells appear and 9 days when Hb accumulation ceases. This was correlated with the established cell cycle times for the corresponding population. The accumulation of Hb from one mitosis to the next is represented by the line joining the Hb values and is therefore an average

V. DNA Synthesis Is Required for the Initiation of Hb Synthesis

Given the likelihood that gene activity is determined by local subtleties of chromosome structure, a problem arises regarding the stability of these structures during replication. One way of dealing with this problem is to monitor genetic activity during a perturbation of DNA replication. As mentioned before, Hb first appears in chick embryogenesis at 35 hrs of incubation. Previous experiments

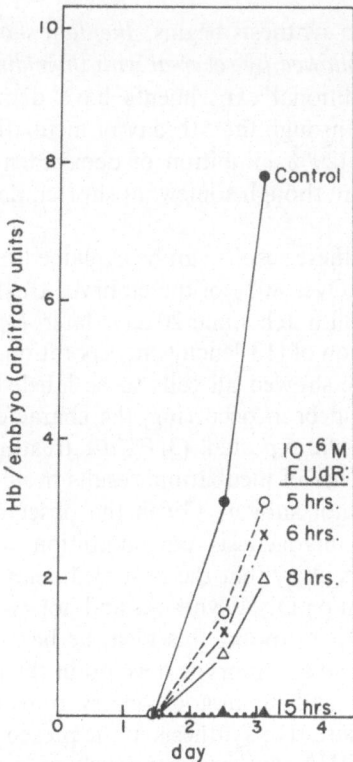

Fig. 4. Delayed appearance of Hb after transient exposure to sub-maximal doses of FUdR. 25 hr embryos were treated with 10^{-6} M FUdR (3 ml per embryo) for the stated periods of time. Thymidine (1.0 ml at 5 mg/ml) was then added to reverse the FUdR block to DNA synthesis. Hb was determined at various times thereafter. These assays were done on 25–30 embryos for each point according to methods previously described (WEINTRAUB et al., 1973). 10^{-6} M FUdR inhibits H 3-deoxyadenosine incorporation in OVO into DNA by about 60–70%. Controls were embryos treated simultaneously with FUdR and TdR. Uridine does not reverse the effects of FUdR

showed that if, at 25 hrs, DNA synthesis was inhibited by the addition of FUdR to these embryos, then no Hb appeared. A related observation has been made for erythropoietin-induced Hb synthesis in mouse liver cultures (PAUL and HUNTER, 1968). These types of experiments are, however, unsatisfactory for firmly establishing a requirement for DNA synthesis in the institution of the red cell program. The principal problem is the difficulty in distinguishing an effect that is due to a coupling of differentiation to DNA replication from one that is secondary to the obvious cell death that is occuring in the presence of FUdR.

By lowering the FUdR concentration to inhibit DNA replication by only 60–70% and by reversing the FUdR (by adding thymidine to the embryo) after only 5–8 hrs of inhibition, we think we have been able to define conditions that cause no detectable signs of cell death, but significantly *delay* the appearance of Hb. Figure 4 shows the time course of accumulation of Hb per embryo for the first 3 days of incubation, as well as the time course for embryos treated with low doses of FUdR (60–70% inhibition of H 3-deoxyadenosine incorporation) for increasing periods of time before Hb synthesis begins. *The data shows that a partial FUdR inhibition of replication, followed by reversal with thymidine, results in a lag in the accumulation of Hb.* Additional experiments have demonstrated that the decreased Hb levels persist through the 5th day of incubation and that even lower concentrations of FUdR (20% inhibition of deoxyadenosine incorporation for 5 hrs) produce a significant, though somewhat shorter, delay in the accumulation of Hb.

We do not think that these results can be explained in terms of cell death for the following reasons: (1) Over 80% of the embryos treated with 10^{-6} M FUdR or less for 6 hrs or less will hatch some 20 days later. (2) During treatment with FUdR, there is no inhibition of H3-leucine incorporation. Autoradiography using labeled leucine or uridine showed all cells to be labeled with either isotope. If "thymidine-less" death had been occurring, the characteristic presence of unlabeled cells would have been expected. (3) FUdR treatment several generations *earlier*, between 5 and 10 hrs of incubation, results in *no* detectable delay in the appearance of Hb (data not shown). Given the observation that inhibition of DNA synthesis prevents myogenesis but inhibition of cytokinesis does not (HOLTZER et al., 1972), it is likely that the observed delay in accumulation of Hb reflects an event dependent on DNA synthesis and not cytokinesis. Recent experiments by WHITTAKER (1973) support this idea; he has shown that a variety of enzymes appear when Ascidian embryos develop in the absence of cell division. PAUL and HUNTER (1968) made analogous observations in experiments showing that erythropoietin stimulates Hb synthesis in the presence of colchicine.

Is the dependence on DNA synthesis a property of Hb synthesis per se or is it only a requirement for its initiation? To answer this question we monitored the effect of FUdR on the synthesis of Hb in cells which had already started to make Hb. Four-day erythroblasts, which make Hb and actively synthesize DNA were incubated for 10 hrs *in vitro* and *in vivo* in the presence of doses of FUdR that block DNA synthesis by over 98%. Under these conditions, no inhibition of Hb synthesis was observed (Table 1). Lower doses of FUdR gave the same results. Longer exposures to high doses of FUdR led to the accumulation of unlabeled cells and obvious signs of cell death. Consequently, experiments were not contin-

Table 1. Lack of effect of FUdR on the synthesis and accumulation of Hb in cells that have already begun to make Hb. All assays were done on the fourth day of incubation. Embryos were pretreated with the stated concentration of FUdR for the indicated periods of time. For studies of synthesis, H3-leucine was added to 5 control embryos and 5 treated embryos, and incorporation per cell was measured into CMC-isolated Hb. The data above represent incorporation over the entire period of treatment with FUdR. Similar results are obtained if incorporation is measured only over the last hour of treatment. Hb per embryo was determined from 10 embryos in each series. 10 hrs before analysis, the Hb value was 2.8×10^{-8} moles; while 30 hrs before analysis, the HB value was 1.3×10^{-8} moles per embryo

Treatment	H3-leucine CPM in Hb (per cell $\times 10^{-6}$)	Hb per embryo (moles $\times 10^{-8}$)
Controls	13862	4.9
FUdR (10^{-5} M; 10 hrs)	14341	4.7
FUdR (10^{-6} M; 30 hrs)	—	4.8

ued beyond 10 hrs. In contrast, up to 30 hrs of exposure to low doses of FUdR—doses which cause a delay when given for only 5 hrs at 25 hrs of incubation—results in no signs of cell death and no decrease in Hb accumulation when administered at 4 days of incubation (Table 1). *These experiments show that a partial and transient inhibition of DNA replication inhibits only the initiation of Hb synthesis and not synthesis that has already been initiated.*

VI. The Cell Cycle Separating the Precursor Erythroblast from Its Progeny Is a Unique One

The FUdR experiments described above indicate that there is something special about the cell cycle that preceeds the inception of Hb synthesis—it is in some way coupled to the institution of the erythropoietic program. Once Hb synthesis begins, however, it is not affected by inhibitors of DNA synthesis. A parallel situation occurs with a different thymidine analogue, BUdR. Unlike FUdR which inhibits DNA synthesis by making thymidine rate-limiting BUdR is incorporated into the double-helix at the growing point for replication. When this occurs in a differentiating system, there is a selective, often reversible, inhibition of differentiation in the context of only minimal effects on viability (BISCHOFF and HOLTZER, 1970; STELLWAGEN and TOMKINS, 1971; MAYHE et al., 1971; TURKINGTON et al., 1971; COLEMAN et al., 1970; WEINTRAUB et al., 1972; ABBOTT et al., 1972). For chick erythropoiesis, BUdR prevents the inception of the erythropoietic program (MIURA and WILT, 1971; WEINTRAUB et al., 1972; HAGOPIAN et al., 1972), but like FUdR has *no* effect once Hb synthesis has begun. Levels of BUdR that partially inhibit Hb production per embryo (Fig. 5) have no effect on the levels of Hb present in the individual red cells that manage to differentiate under these conditions. Thus, at the cellular level Hb production in the presence of low levels of BUdR is "all-or-none" (WEINTRAUB et al., 1973). Consequently, it is possible to estimate the size of the BUdR-sensitive target for red cell differentiation using a target theory analysis of the experimental data (WEINTRAUB, 1973b). *Such an analysis has demonstrated that the absolute amount of BUdR-sensitive DNA coding for red cell specific differentiation is exceedingly small—about 4000 base pairs.*

The data also show that the substitution of one BUdR molecule is adequate to destroy the activity of an average of 300–500 base pairs of target; thus, *1 BUdR molecule can affect the biological integrity of an extremely large stretch of DNA.* It is not at all clear how this can occur. One explanation might be that the BUdR-sensitive target is maintained in some type of configuration which is altered upon substitution of BUdR. Another explanation is that an estimated target size of some 500 base pairs represents a compound target. Thus, the BUdR-sensitive target, *determined analytically*, might represent 50 subtargets, each 10 base pairs in length. 1 BUdR substitution in any one sub-target would not only inactivate that target, but would prevent the initiation of Hb synthesis if the integrity of *all* sub-targets were required. Such a scheme would correspond to the model of gene regulation proposed by GEORGIEV (1972), who has recently reviewed the evidence supporting his ideas. Besides explaining the quantitative data, the idea of a com-

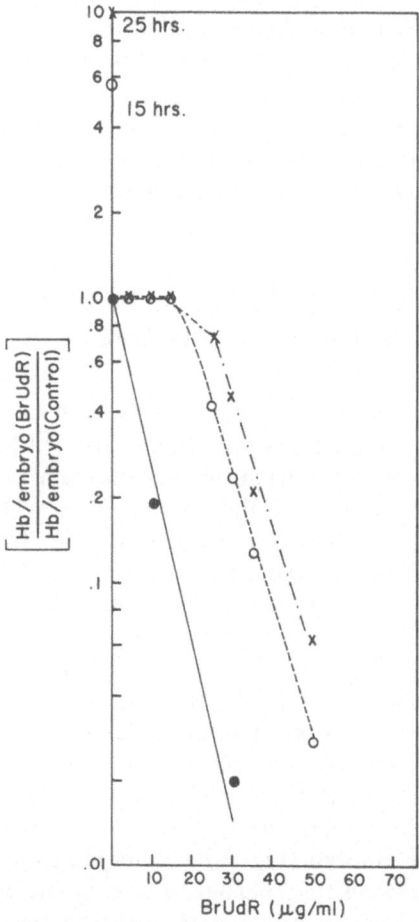

Fig. 5. Inhibition of the initiation of Hb synthesis by BUdR. 3 ml of BUdR at the indicated concentration was added to embryos at 5 (-●-), 15 (-○-), or 25 (-×-) hrs of incubation. Hb assays were done at 70 hrs of incubation. The data shows that a given dose of BUdR is more inhibitory if given earlier in development. Although not shown, a partial inhibition of Hb per embryo indicates a decreased number of red cells and not a decrease in the amount of Hb per red cell, since levels of Hb in cells obtained from embryos partially inhibited with BUdR were the same as those in cells from control embryos (WEINTRAUB et al., 1973). Concurrent treatment with BUdR and a five-fold excess of TdR abolishes the BUdR inhibition. A 15-hr treatment with maximal doses of BUdR is qualitatively reversible by the addition of TdR to the embryo

pound target may help in part to explain why BUdR is so specific for specialized functions: As proposed by LIN and RIGGS (1972), BUdR-subsituted DNA of higher cells may bind certain regulatory proteins differently from control DNA. An alternative proposal—that BU-DNA is a poor template for RNA polymerase—is not substantiated by recent experiments of A. TRAVERS (personal communication), who has found that BUdR-substituted T4 DNA has about the same dependence on temperature as control DNA when challenged with RNA polymerase. The temperature dependence has been interpreted in terms of a particular configuration of RNA polymerase binding sites and presumably this configuration is intact when BUdR is present in phage T4 DNA. In keeping with the LIN and RIGGS argument, if the binding site of all regulatory proteins were altered by BUdR, then the specificity that BUdR shows for specialized functions could, in part, be explained by the assumption that differentiated functions are controlled by *more* regulatory proteins (and hence, more binding sites) than genes coding for essential functions. It is unlikely that this type of explanation can account entirely for the preference that BUdR demonstrates for differentiated functions. Two additional components to this very complex effect are likely to be a differential longevity of certain mRNAs and proteins (STELLWAGEN and TOMKINS, 1971) and the possibility that a BUdR-induced defect in an essential enzyme may be by-passed if the enzyme product is present in the tissue culture medium.

VII. The Time of Replication of the BUdR-Sensitive Targets

Most evidence indicates that BUdR affects specific regions of the DNA. Further support for this notion comes from the experiments described in Fig. 6. 25-hour embryos were treated with BUdR for increasing periods of time before reversal with thymidine. The resulting inhibition of Hb accumulation was moni-

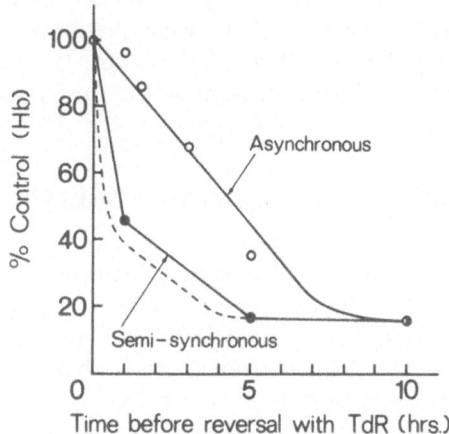

Fig. 6. Increasing inhibition of Hb accumulation with increasing time in BUdR. Maximal inhibitory doses of BUdR (100 µgm/ml; 3 ml per embryo) were added to embryos at 25 hrs. After the indicated times, TdR was added (1 mg/ml; 3 ml per embryo) to prevent any further developmental effect of BUdR. Hb per embryo was monitored on day 3. Each point represents the average results from 20 combined embryos. Controls were treated simultaneously with BUdR and TdR. The dotted line is the theoretical inhibition curve assuming every cell in S were sensitive to BUdR. Semi-synchronous populations were obtained by stopping DNA replication in 25-hr embryos with 3 ml 10^{-5} M FUdR. After 5 hrs 200 µgm/ml of BUdR and 60 µgm/ml TdR were added in a final volume of 3 ml. After the indicated period of time, TdR (2 mg/ml; 3 ml) was added and Hb assayed on day 3. Control embryos received 260 µgm/ml TdR (3 ml) at the indicated times. Each point represents the combined Hb from 20 embryos. All experimental control embryos were indistinguishable from untreated embryos

tored on day 3 and found to be proportional to the time under BUdR. Since our population is asynchronous, we assume that the inhibition curve reflects the passage of cells through a defined, BUdR-sensitive period in the cell cycle. If cells were sensitive at all times during S, a sharp initial inhibition of Hb accumulation would have been expected (dotted line in Fig. 6). This would represent the 40% of the cells in S at any one time. Since this was not observed, it is likely that the BUdR-sensitive period occurs at a unique time in the cell cycle for all precursor red blood cells. To determine this period, cells were pseudo-synchronized by treatment for 5 hrs with FUdR at concentrations which completely stop DNA replication. Autoradiography showed that after such treatment, over 95% of the cells were in the S phase. Of these, roughly 60% were probably at the G1/S boundary. Addition of thymidine after 5 hrs in FUdR reversed the block, and after a lag (Fig. 4) Hb began to accumulate. If, however, the FUdR was reversed for 1 hr with a 3:1 ratio of BUdR to TdR and this was followed by addition of TdR to a final ratio of 1:10 (BUdR:TdR), there was a rapid inhibition (closed circles in Fig. 6) of Hb accumulation relative to controls (which had been reversed initially with a 1:10 ratio of BUdR to TdR). These experiments indicate that the BUdR-sensitive DNA may replicate at the beginning of S. This is supported by control experiments in which the pseudo-synchronized cells are reversed with TdR for 2 hrs; BUdR is then added to a final ratio of 3:1 (BUdR:TdR); and after 5 hrs, TdR is added to a ratio of 1:10 (BUdR:TdR). When this protocol was used to incorporate high levels of BUdR during mid to late S for most of the population, no more than 10% inhibition of Hb accumulation was observed (data not shown). It is very likely therefore that the DNA targets for BUdR-sensitive, red cell differentiation replicate during the beginning of the S period.

VIII. The Genetic Organization of Erythropoiesis

What genetic elements are involved in the activation of the red cell program? From the available data on human thalassaemias, it appears very likely that at least one componet is a cis-acting regulatory locus closely linked to the alpha and beta structural genes. This follows from the fact that certain alpha and beta thalassaemia variants result in only a *partial* decrease in alpha or beta globin chain synthesis. *Moreover, this partial decrease occurs only in the cis position.* The globin chains produced from the thalassaemia genes are indistinguishable from normal alpha or beta globin. Although similar arguments could be put forth on the basis of data obtained from patients with hereditary persistence of fetal hemoglobin or from delta or delta-beta thalassaemia, these would be less conclusive, since the findings from these deficiencies could be explained in terms of a simple deletion of structural gene elements [2].

Epigenetic studies with BUdR may also prove useful in the genetic analysis of erythropoiesis. A prime question to be answered is: What is the target for BUdR action? This is almost certainly DNA (MAYNE et al., 1972; COLEMAN et al., 1971;

---

[2] An excellent detailed review of the thalassaemia syndromes is provided by WEATHER-ALL and CLEGG, 1972).

STELLWAGEN and TOMKINS, 1971; OSTERTAG et al., 1973). But in the case of Hb synthesis, is the target a locus distinct from the globin genes or is the target the globin genes themselves—or perhaps some cis-acting regulatory locus analogous to that affected in some thalassaemias? On the basis of two principal observations, it appears as if the target for the BUdR effect is a locus that is distinct from the globin genes and distinct from any cis-acting regulator of the globin genes:

1. The first observation supporting this statement is that a partial inhibition of Hb production by low concentrations of BUdR results in a decreased number of Hb-producing red cells and not a decreased level of Hb per cell. If BUdR were affecting the globin genes or a cis-acting regulator of them, we would then have expected to find cells with intermediate levels of Hb. These cells would represent that part of the population with only 1 or 2 active globin genes instead of the usual number of perhaps 8 (HASHIMOTO and WILT, 1966; D'AMELIO, 1966; SHALE-KAMPF et al., 1972). Since these types of cells were not observed over a wide range of BUdR doses, it follows that either active genes can compensate for inactive ones or that BUdR is altering a *separate* target that is required for globin production from *all* globin genes. Although compensatory mechanisms exist to some extent in human thalassaemias (SCHWARTZ, 1970; GILL and SCHWARTZ, 1973), these appear to operate over a limited range and seem more to be designed to keep the alpha to beta ratio at 1, than to keep a required amount of Hb in a given cell. Direct evidence against considering a compensation mechanism to be the explanation of the all-or-none effect with BUdR is that the globin chains produced under these conditions show the same pattern on 8 M-urea stacking acrylamide gels as do controls (WEINTRAUB, manuscript in preparation). If a compensation mechanism existed, at a level of BUdR at which each cell was functioning on perhaps 1 unsubstituted globin gene, it would be expected that for any particular cell, the active gene would be chosen at random; i.e. that stretch of DNA that happens not to be substituted with BUdR. Consequently, all globin chains obtained from a large population of cells should be in equimolar concentrations. This was clearly not observed, the relative ratios being approximately 3:2:1:0.3 or the same as the controls. It is therefore unlikely that the all-or-none effect can be explained in terms of a random inactivation of globin genes and a subsequent compensation by active genes. The most dramatic demonstration of this point, however, awaits a single-cell analysis of the globin chains from embryos treated with low doses of BUdR.

2. A second observation also supports the idea that BUdR is altering loci which act "trans" to the globin genes. We and others have shown that once Hb synthesis begins, BUdR has no detectable effect, even though it is substituted at some 80% replacement for thymidine. Data was presented that indicated that the resistance of Hb synthesis to BUdR was difficult to explain in terms of a stable Hb mRNA (WEINTRAUB et al., 1972). If these experiments can be extended to include the appropriate quantitative analysis of total Hb-mRNA, it would then be difficult to escape the conclusion that BUdR is altering some locus other than those coding for the globin genes and that a stable product of that locus confers resistance to BUdR during subsequent development.

Fig. 7 is my very preliminary conception of the genetic organization of erythropoiesis. In keeping with the observations of LIN and RIGGS, it is assumed

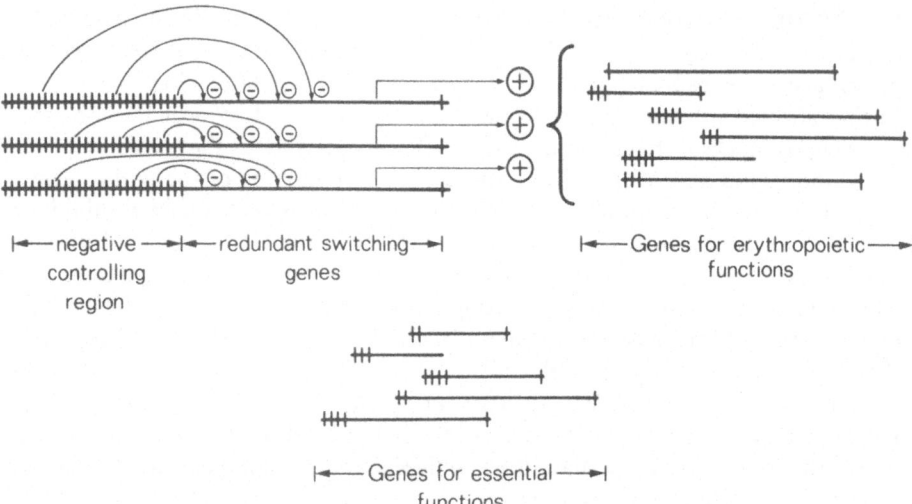

Fig. 7. A model for the circuitry involved in the control of erythropoiesis. The experimental data which led to the formation of this circuitry is explained in the text. (+) represents activation; (−) represents inhibition. Each small interval represents a regulatory binding locus on the DNA. The long intervals represent areas of DNA coding for structural genes

that the basic effect of BUdR is a generalized one which alters DNA-protein binding. Because more proteins control "switching-genes", these genes are thought to be most sensitive to the substitution of BUdR. The product of the switching gene acts "trans" to activate the erythropoietic genes. A BUdR-sensitive locus acts "cis" to negatively control the output from the "switching-gene". A cis-acting locus is also proposed for the globin genes primarily on the basis of the data from thalassaemia patients. A similar locus associated with other genes may explain the slight amount of inhibition of "essential" enzymes observed by STELL- WAGEN and TOMKINS (1971). Several *redundant* "switching genes" are indicated because the BUdR inhibition curve gives an extrapolation number greater than 1; however, the extrapolation number could just as well be explained if the BUdR target were much larger and required several sub-targets to be inactivated before the "switching-gene" was inactivated. A basic postulate of this analysis of the action of BUdR is that "switching-genes" for differentiation are controlled in a negative fashion by *multiple, cis-acting* binding sites for regulatory proteins. These were hypothesized to explain the large quantity of DNA inactivated by 1 BUdR molecule. If this interpretation of the data is correct, we must then confront the problem of why the system is designed in this way. In the terminology of BRITTEN and DAVIDSON (1969), the multiple binding sites, acting as multiple "sensors", may serve the function of determining when many variables are precisely tuned before a commitment to erythropoiesis is made. Alternatively, if the multiple binding sites are redundant, they may serve as an extremely efficient mechanism for keeping the erythropoietic genes under very rigid control. In this regard it is of some interest that even though the lac operator appears to be singular in nature (GILBERT and MAIZELS, 1973), the binding sites for the lambda repressor are 6-fold redundant (MANIATIS and PTASHNE, 1973).

IX. How Might DNA Replication Function in the Activation of the Erythropoietic Program?

Because of the demonstration in Fig. 4 that DNA synthesis was required for red cell differentiation, I have been trying to understand certain aspects of DNA replication that could contribute to an explanation. Two main lines of investigation are (1) the structure of the DNA at the replication fork, and (2) the structure of chromatin at the replication fork. The first possibility might lead to the type of proposal that would implicate Okazaki fragments, nicks, regions of single-strandedness, etc.—all apparently associated with the growing point for DNA replication—in the activation of the red cell program. The second possibility considers (a) the stability of protein-DNA complexes as the replication apparatus copies the DNA and (b) the way in which proteins are added to the daughter DNA double-helices after replication.

The particular features of the DNA at the growing fork have been studied in a number of laboratories (see Cold Spring Harbor Symposia for 1968 and 1973 for a general review) and it appears that, in general, these characteristics are very similar to those found in viral and bacterial systems.

In trying to understand how proteins become incorporated into the chromatin of the daughter strands, I have measured the sensitivity of newly replicated DNA to pancreatic DNase. Cells were pulsed with H 3-TdR for increasing periods of time; the nuclei were then isolated and digested with different concentrations of DNase. As these experiments have recently been reported (WEINTRAUB, 1973 a), I will only summarize them here: For all doses of DNase and all periods of exposure to label, the sensitivity of DNA in chromatin was much less than that of free DNA. Pulses less than a minute were even less sensitive than longer pulses. This was attributed to the increased protection of the DNA at the growing fork provided by the replication apparatus. Since no period of exposure to label could detect DNA that was relatively sensitive to DNase, it was concluded that protein became associated with the daughter strands within a minimum of one minute of replication. If replication were to generate regions of uncovered DNA—regions that might be available for transcription—the DNase assay certainly does not detect them. An additional point can also be made: When DNA is replicated in the absence of protein synthesis, the daughter strands are found to be *more sensitive* to DNase. This demonstrates that newly made histone probably enters the new chromosome at the growing fork for DNA replication. In addition, evidence was presented that in the absence of protein synthesis, one daughter double-helix was normally covered by protein, presumably recycled from the original parental chromosome, while the other double-helix was uncovered. The fact that the relatively free DNA generated in the absence of protein synthesis could be detected indicates that the vast excess of chromosomal protein present in the nucleus did not migrate to these bare regions. *It is therefore likely that the chromosome, as it exists in the interphase nucleus, is a very stable entity.* Consequently, it is not at all clear that non-histone proteins, or other proposed modulators of histone repression, can "self-assemble" with an intact interphase chromosome. Indeed, most successful attempts to "reorient" the transcriptive capacity of chromatin *in vitro* require extremely harsh conditions (PAUL et al., 1973). It is possible that the localized "loosening" effect DNA replication may have on chromatin establishes the proper environment for restructuring one or both daughter chromosomes for Hb-mRNA transcription. Clearly, more refined methods will be required for the demonstration of this point.

One important clue to the molecular basis for the observed coupling of differentiation and DNA replication is that the delay in the appearance of Hb occurs when DNA synthesis is only partially inhibited. Under these conditions the gross rate of DNA chain elongation is decreased (ENSMINGER and TAMM, 1970) and therefore certain experimentally testable predictions can be made about the changes in fine structure that may be occurring in the replicating regions of the chromosome. But perhaps the most important questions concern not the molecular basis for the coupling but the reason for its existence in the first place. A related problem deals with biological timing—in this case, why does Hb characteristically appear at about 35 hrs of incubation? As a background to all of these questions, it remains to be determined whether or not the genetic targets for the erythropoietic effects of BUdR are the same targets as those for the effects of FUdR.

Acknowledgements

I thank the many colleagues and associates who have contributed to the writing of this paper and the experiments described. This work was supported in part by the Helen Hay Whitney Foundation and in part by the National Science Foundation, Grant No. GB40148.

Note Added in Proof: M. GRODINE (personal communication) has shown that after BUdR is incorporated into 3 day, dividing erythroblasts, the rate of synthesis of Hb mRNA (as measured by protection from nuclease by a cDNA probe) is the same as that from unsubstituted cells. In contrast, Hb mRNA is not present in 3-day old embryos after Hb synthesis is prevented by prior treatment with BUdR at 25 hrs. These observations further emphasize the fact that BUdR is inhibiting a developmental program that is instituted at about 35 hrs, but is resistant to BUdR subsequently.

References

ABBOTT, J., MAYNE, R., HOLTZER, H.: Inhibition of cartilage development by BUdR. Develop. Biol. **28**, 430 (1972).

D'AMELIO, V.: The globins of adult and embryonic chick Hb. Biochim. Biophys. Acta **127**, 59 (1966).

AXEL, R., CEDAR, H., FELSENFELD, G.: Synthesis of globin RNA from duck reticulocyte chromatin *in vitro*. Proc. Natl. Acad. Sci. USA **70**, 2029 (1973).

BILLING, R. J., BONNER, J.: The structure of chromatin as revealed by DNase digestion studies. Biochim. Biophys. Acta **281**, 453 (1972).

BISCHOFF, R., HOLTZER, H.: Inhibition of myoblast fusion after one round of DNA synthesis in BUdR. J. Cell Biol. **44**, 134 (1970).

BRITTEN, R. J., DAVIDSON, E.: Genetic regulation in higher cells. Science **169**, 349 (1969).

CAMPBELL, G., WEINTRAUB, H., MAYALL, B., HOLTZER, H.: Primitive erythropoiesis in early chick embryogenesis. II. J. Cell Biol. **50**, 669 (1971).

COLEMAN, A. W., COLEMAN, J. R., KANKEL, D., WERNER, I.: The reversible control of animal cell differentiation by BUdR. Exp. Cell Res. **59**, 319 (1970).

ENSMINGER, W. D., TAMM, I.: The step in DNA synthesis blocked by Newcastle disease or mengovirus infection. Virology **40**, 152 (1970).

GEORGIEV, G.: The structure of transcriptional units in eukaryotic cells, In: MOSCONA, A. A., MONROY, A. (Eds.): Current Topics in Developmental Biology. New York: Academic Press 1972.

GILBERT, W., MAIZELS, N.: Sequences of controlling regions of the lactose operon of E. coli. Cold Spring Harbor Symp. Quant. Biol. **38** (1973).

GILL, F. M., SCHWARTZ, E.: Synthesis of globin chains in sickle-beta thalassaemia. J. Clin. Invest. **52**, 709 (1973).

GORDON, A. S.: The Regulation of Hematopoiesis. New York: Appleton-Century-Crofts 1971.

HAGOPIAN, H. K., LIPPKE, J., INGRAM, V.: Erythropoietic cell cultures from chick embryos. J. Cell Biol. **54**, 98 (1972).

HASHIMOTO, K., WILT, F.: The heterogeneity of chick Hb. Proc. Natl. Acad. Sci. **56**, 1477 (1966).

HOLTZER, H., SANGER, J. W., ISHIKAWA, H., STRAHS, K.: Selected aspects of myogenesis. Cold Spring Harbor Symp. Quant. Biol. **37**, 549 (1972).

HOLTZER, H., WEINTRAUB, H., MAYNE, R., MOCHAN, B.: DNA synthesis, the cell cycle, and differantiation. In: MOSCONA, A. A., MONROY, A. (Eds.): Current Topics in Developmental Biology. New York: Academic Press 1972.

ITZHAKI, R., COOPER, H. K.: Similarity of chromatin from different tissues. J. Mol. Biol. **73**, 199 (1973).

LEDER, P., AVIV, H., GIELEN, J., IKAWA, Y., PACKMAN, D., SWAN, D., ROSS, J.: The regulated expression of globin and immunoglobulin genes. Cold Spring Harbor Symp. Quant. Biol. **38** (1973).

LINS, S., RIGGS, A.: Lac operator analogues: BUdR substitution in the lac operator affects the rate of dissociation of the lac repressor. Proc. Natl. Acad. Sci. **69**, 2574 (1972).

LUCAS, A., JAMROZ, C.: Atlas of avian hematology. Washington, D.C.: U.S. Dept. of Agriculture Monograph 1966.

MANIATIS, T., PTASHNE, M.: Multiple repressor binding at the operators in bacteriophage lambda. Proc. Natl. Acad. Sci. **70**, 153 (1973).

MAYNE, R., ABBOTT, J., HOLTZER, H.: Requirement for cell proliferation for the effects of BUdR on cultures of chick chondrocytes. Exp. Cell Res. **77**, 255 (1973).

MAYNE, R., SANGER, J., HOLTZER, H.: Inhibition of mucopolysaccharide synthesis by BUdR in cultures of chick amnion cells. Develop. Biol. **25**, 547 (1971).

MIRSKY, A. E.: The structure of chromatin. Proc. Natl. Acad. Sci. **68**, 2945 (1971).

MIURA, Y., WILT, F. H.: The effects of BUdR on yolk sac erythropoiesis in the chick. J. Cell Biol. **48**, 523 (1971).

MOSS, B. A., JOYCE, W. G., INGRAM, V. M.: Histones in chick embryonic erythropoiesis. J. Biol. Chem. **248**, 1025 (1973).

OSTERTAG, N., CROZIER, T., KLUGE, N., MÉLDERIS, H., DUBE, S.: Action of BUdR on the induction of Hb synthesis in mouse leukemia cells resistant to BUdR. Nature New Biol. **243**, 203 (1973).

PAUL, J., HANISING, P., BIRNIE, G., HELL, A., YOUNG, B., GILMOUR, R. S., DREWIENKIEWICZ, G., WILLIAMSON, R.: The transcriptional Unit in eukaryotes. Cold Spring Harbor Symp. Quant. Biol. **38** (1973).

PAUL, J., HUNTER, J. A.: DNA synthesis is essential for increased Hb synthesis in response to erythropoietin. Nature **219**, 1362 (1968).

PEDERSON, T.: Chromatin structure and the cell cycle Proc. Natl. Acad. Sci. **69**, 2224 (1972).

RUBIN, G., SULSTON, J.: Physical linkage of 5s and 5.8s Genes (in press) (1973).

SCHWARTZ, E.: Heterozygous beta-thalassaemia: Balanced globin synthesis in bone marrow cells. Science **167**, 1513 (1970).

SHALEKAMPF, M., VAN GORR, D., SLINGERLAND, R.: Re-evaluation of the presence of multiple Hbs during ontogenesis of the chicken. J. Embryol. Exp. Morph. **28**, 681 (1972).

SMALL, J. V., DAVIES, H. G.: Erythropoiesis in the yolk sac of early chick embryos: An electronmicroscopic and micro-spectrophotometric study. Cell and Tissue **4**, 341 (1973).

STELLWAGEN, R., TOMKINS, G. M.: Differential effect of BUdR on the concentrations of specific enzymes in hepatoma cells in culture. Proc. Natl. Acad. Sci. **68**, 1147 (1971a).

STELLWAGEN, R., TOMKINS, G. M.: Preferential inhibition by BUdR of the synthesis of tyrosine aminotransferase in hepatoma cell cultures. J. Mol. Biol. **56**, 167 (1971b).

TERADA, M., CANTOR, L., METAFORA, S., RIFKIND, R., BANK, A., MARKS, P.: Globin mRNA activity in erythroid precursor cells and the effect of erythropoietin. Proc. Natl. Acad. Sci. **69**, 3575 (1972).

TURKINGTON, R., MAJUMDER, G., RIDDLE, M.: Inhibition of mammary gland differentiation *in vitro* by BUdR. J. Biol. Chem. **246**, 1814 (1971).

WEATHERALL, D. J., CLEGG, J. B.: The Thalassaemia Syndromes. Oxford: Blackwell 1972.

WEINTRAUB, H.: The assembly of newly-replicated DNA into chromatin. Cold Spring Harbor Symp. Quant. Biol. **38** (1973 a).

WEINTRAUB, H.: The size of the BUdR sensitive target for differantiation. Nature New Biol. **244**, 142 (1973 b).

WEINTRAUB, H., CAMPBELL, G., HOLTZER, H.: Primitive erythropoiesis in early chick embryogenesis. I. J. Cell. Biol. **50**, 65 (1971).

WEINTRAUB, H., CAMPBELL, G., HOLTZER, H.: Identification of a developmental program using BUdR. J. Mol. Biol. **70**, 337 (1972).

WEINTRAUB, H., CAMPBELL, G., HOLTZER, H.: Differentiation in the presence of BUdR is all-or-none. Nature New Biol. a244, 140 (1973).

WHITTAKER, R.: Segregation during ascidian embryogenesis of egg cytoplasmic information for tissue-specific enzyme development. Proc. Natl. Acad. Sci. **70**, 2096 (1973).

WILT, F. H.: Regulation of the initiation of chick embryo Hb synthesis. J. Mol. Biol. **12**, 331 (1965).

The Cell Cycle, Cell Lineage, and Neuronal Specificity

R. K. HUNT

Institute of Neurological Sciences, University of Pennsylvania School of Medicine, Philadelphia, PA, USA

I. Introduction

The first principle of neuronal differentiation is the generation of cellular diversity which, at least quantitatively, matches that found in the entire remainder of the embryo. More importantly, it is the characteristic *differences among neurons* (rather than properties which all neurons share to the exclusion of non-neural cells) that enables them to associate selectively and to assemble the synaptic circuits which are the substrates of macroscopic brain function. Neuronal differentiation must be viewed, therefore, not simply as a process by which "zygote" cells generate "nerve" cells, but rather as the systematic derivation from the zygote of a cell *population* with characteristically diverse properties and synaptic specificities. This review considers some aspects of neuronal specificity and the possible role(s) of the cell cycle and cell lineage in neuronal differentiation.

I should state at the outset that I would have preferred to summarize definitive experiments which delineate the succession of vectorial changes that occurs in the genesis of neuronal diversity, the degree to which certain of these exist as obligatory sequences, and the number and kinds of vectorial changes and intermediate states which can occur in a single cell within its own lifetime. At present, such experiments have not been done. Thus, I have adopted a more cautious approach, beginning with evidence that the invariant *qualitative* features of most neurons (their shape and orientation, transmitter pathways and synaptic associations) are intrinsically determined, aspects of the terminally differentiated state of distinct neuronal phenotypes which are potentially lineage-related. The bulk of the paper considers instances in which qualitative differences among neural precursor cells and orderly patterns of neural cell lineage have been described and how these are related to the specificity of the progeny neurons, the generation of specific phenotypes in the appropriate regions of the nervous system, and the deployment (among neurons of a single phenotype) of properties subserving position-dependent synaptic selectivity.

II. Programmed Features of Neuronal Diversity

That extraorganismic factors can play upon the developing brain is amply illustrated by the imprinting of young ducklings, infection of human infants with congenital rubella, and the effects of malnutrition or "deprived environments" on cortical maturation in young rats. Yet dramatic as these and other modifications are, the principal neural circuits are set down during pre-functional ontogeny and are not forged by experience out of equipotential neural nets (SPERRY, 1950; 1951). In fact, behaviorally *mal*adaptive circuits are never corrected by behavioral practice, and a frog or salamander with one eye rotated through 180° will show upside-down and back-to-front vision for the rest of its life (SPERRY, 1951). Similarly, tadpoles raised in darkness or with one eye covered by a skin graft develop normal visual circuits (SKARF, 1973); salamanders and lobsters develop their complex motor output circuits in the absence of sensory feedback (PIATT, 1949; HOLTZER and KAMRIN, 1956; DAVIS, 1973); frogs with translocated skin never adjust their abnormal cutaneosensory circuits by the experience of misdirected wiping (MINER, 1956).

Of the environmental modifications known, most involve only quantitative parameters. Despite the reduced number of cells and dendritic arborizations in the cortex of malnourished or environmentally deprived rats, for example, the stereotyped features of cell shape, orientation and layering, and the qualitative synaptic relations are normal. Where environmental factors have been shown to modify the qualitative features of neurons, they act selectively on a few susceptible cells: thus, the dramatic effects of structured visual experience, on neuronal properties in the kitten visual cortex (BLAKEMORE and COOPER, 1970), are actually confined to selection and fine-tuning of specific geniculocortical and intracortical synapses; they belie the presence of entirely normal retinal circuits, retinogeniculate and retinotectal connections, and a retinotopically organized projection from geniculate to cortex—some of whose elements had already formed highly specialized synaptic connections before the kitten's eyes opened (reviewed in JACOBSON, 1970).

The effects of most hormones and of afferent fibers and target neuron populations are also limited to quantitative parameters. For instance, prolactin-treated frogs and thyroidectomized rats show areas of neural hyperplasia and hypoplasia, hyper- and hypotrophy, increase or decrease in numbers of dendritic spines, or changes in total cell RNA or protein; but characteristic neuronal phenotypes are easily recognized and show normal qualitative features (HUNT and JACOBSON, 1971; unpubl. observations; HAMBURGH, 1973). Early eye removal in chick embryos leads to increased cell death and hypotrophic changes in the isthmo-optic nucleus and optic tectum, but the qualitative features of the surviving neurons are not markedly altered (COWAN and WENGER, 1968; KELLY and COWAN, 1972). The specific physiologic and cytochemical properties of neurons can likewise develop in the absence of afferent fibers: chronic X-irradiation of the cerebellum, which deprives the Purkinje cells of most of their imput, effects a withering of Purkinje cell dendrites and an increase in the *number* of climbing fiber afferents they individually entertain; yet their axonal projection, spontaneous firing rates, and receptors for GABA, seratonin, and norepinephrine all develop normally (WOOD-WARD et al., 1973). Finally, when previously isolated neuron populations are allowed to reintegrate into the nervous system (as when isolated spinal segments are allowed to reconnect with the neural axis), the neurons evince their original specificity in forming synaptic connections (PIATT, 1949; HOLTZER and KAMRIN, 1956; HUNT and JACOBSON, 1972a; 1973).

Documented cases in which exogenous molecules modify the qualitative features of post-mitotic neurons are few; e.g. corticosteroid induction of glutamine synthetase in chick retina; NGF induction of axonal proliferation in sympathetic neurons. And in all of these cases, the effector molecule acts on a limited number of neurons whose special susceptibility and specific response have been programmed into the cell. Thus, sympathetic neurons neither respond to corticosteroids, nor do they make glutamine synthetase in response to NGF. Both the *response* and the *responders* to a given effector molecule are predictable in advance. Thus, the capacity to proliferate neurites in response to NGF is best regarded as one of the unique set of properties which defines the phenotypic specificity of sympathetic neurons, and it is potentially lineage-related. The "exogenous control" is only relevant to our diversity analysis in that it alerts us to the fact that certain (potentially lineage-related) components of neuronal specificity may require appropriate exogenous agents in order to be expressed by the neuron or assayed by the investigator.

It is worth stressing, then, that there is no evidence for single effector molecules shaping diverse neurons into a common final shape, orientation, transmitter pathway, or set of synaptic relations[1]. Indeed, in the few cases where synaptic selectivity was thought to perhaps be induced in naive neurons through their terminal contacts, recent studies have forced a reconsideration of this view (SKLAR and HUNT, 1973; MARK, 1974). However, the viable possibility remains that *local* exogenous cues (e.g. resulting from intercellular communication, position in a gradient, etc.) might modify the specificity of post-mitotic neurons or the progression of precursor cells through a lineage. This must be examined, for each neural subsystem or neuronal phenotype, by transplantation or tissue culture methods.

Further evidence for the tight genetic and developmental constraints on neuronal differentiation come from autoradiographic studies indicating a highly invariant pattern of cell birthdates for the major neuronal phenotypes in thalamus, brain stem, retina, tectum, cortex, and cerebellum (cf. ALTMAN, 1967; ANGEVINE, 1970). Analysis of nascent neurons shows a similar invariance in migration patterns (SIDMAN and RAKIC, 1973), growth and axon formation (PRESTIGE, 1974),

---

[1] The one possible class of exceptions is provided by the metamorphic hormones of the amphibia and insects, which, although poorly understood, appear to directly or indirectly trigger cataclysmic changes throughout the developing brain. There is some evidence, however, than many of these changes involve a proliferative phase in early response to the hormone; moreover, the few cases in which direct qualitative modification of post-mitotic neuron populations has been analyzed in detail have revealed selective effects upon specific neuronal phenotypes in the area surveyed. Also, it remains to be ruled out that the variety of specific effects triggered by these hormones are in fact mediated by distinct intermediate messengers. In short, the metamorphic hormones may ultimately prove to be exceptions to the principles being developed here, but there is no compelling evidence at the moment that this is the case.

and ultrastructural differentiation (cf. HINDS and HINDS, 1974). Neurological mutants of the mouse are known in which most stereotyped features of neuronal diversity are normal but certain cell types cannot orient properly (DELONG and SIDMAN, 1970) or migrate in the normal way (SIDMAN and RAKIC, 1973). C. IDE and R. TOMPKINS (personal communication) have analyzed an autosomal recessive mutation in the Axolotyl, in which vestibular imput evokes *spastic* behavior: Physiologic studies revealed normal vestibular nerve and cell function, but severe abnormalities in the spastic cerebellum; and analysis of chimera made by embryonic grafting confirmed that the focus of the mutation was neuronal Abnormal physiologic response properties of retinal neurons or of thoracic motor neurons have been analyzed in various behavioral mutants of *Drosophila*, and gynandomorphs localized the focus of the mutations to the respective neural structures (IKEDA and KAPLAN, 1970; BENZER, 1973). That single gene mutations, with foci apparently in the nervous system, can perturb the invariant features of particular neurons or a small number of neuronal phenotypes suggests that many of the qualitative features of cell diversity in the nervous system are tightly controlled by specific neuronal genes.

To summarize: Most of the qualitative features of stereotyped diversity in the nervous system (cell shape and orientation, arborization patterns, transmitter pathways, postsynaptic receptors, and synaptic associations) arise in orderly spatiotemporal array under tight genetic control. In most instances, these are the very typological features which, along with location, have allowed the neurohistologist and histochemist to identify the major neurons that form the principal pathways and circuits of the nervous system, and to classify these neurons by name as Purkinje cells or granule cells, ganglion cells or retinal rods. Where they have been adequately studied (again, in the case of the major neural circuits), these features appear to be programmed components of the terminally differentiated state of distinct cell phenotypes, and they define the specificity of the final state in a developmental process whose intermediate states are potentially lineage-related. Exogenous and experiential influences rarely modify or shape these qualitative features; and where they do, the resulting modifications reflect a programmed susceptibility and a programmed response which are themselves additional components of the phenotypic specificity of the responding neurons.

III. Levels of Specificity and Boundary Criteria

The invariant qualitative features of terminally differentiated neurons are displayed, in all their complexity, in the frog retina, which contains 6 distinct neuronal phenotypes: rods and cones (photoreceptors); amocrine, bipolar, and horizontal cells (interneurons); and ganglion cells. Each phenotype has a characteristic shape and neurite pattern and (so far as it has been studied) appears to synthesize a characteristic set of neurotransmitters and postsynaptic receptors. The characteristic synaptic relationships among these neurons are locally confined, such that a cohort of adjacent photoreceptors connect with the few bipolar (and horizontal) cells immediately in front of them; these in turn converge onto the ganglion cell (and interact with amocrine cells) directly in front of them. The ganglion cell axons join to form the optic nerve and innervate the visual centers of the brain, including the midbrain optic tectum; there they form a point-to-point retinotopic pattern of connections with tectal neurons that reproduces a "map" of the retina across the tectal surface. In addition to the six basic cell phenotypes, several "subphenotypes" exist with subtle but characteristic anatomic, physiologic, and biochemical differences (HUNT and JACOBSON, 1974a). The comparative cytology of the amphibian retina, and some simple experiments on it, serve to illustrate three principles in the generation of neuronal diversity:

1. Different levels of specificity are seen when different neurons are compared. Thus, the ganglion cell differs from the rod in its position, ability to conduct action potentials, broadly arborized dendrites, long axon innervating the brain, connectivity with amocrine cells, ability to transduce specific chemical signals into electrotonic potentials, as well as its *inability* to synthesize rhodopsin, elaborate outer segments, or transduce light into electrotonic potentials. The four subphenotypes of ganglion cell (MATURANA et al., 1960) differ in their dendritic tree anatomy, physiologic response properties (e.g. to light-on vs. light-off), the precise pattern of synaptic connections they entertain with retinal interneurons, and the *depth* at which they terminate (and perhaps also the tectal cell type upon which they terminate) in the tectum. In contrast to these aspects of *phenotypic specificity*, individual ganglion cells at different positions in the retina differ in their affinities for different loci on the tectal surface, a property termed *locus specificity* (HUNT and JACOBSON, 1972b). Thus the genetic decisions needed to generate the diversity of ganglion cells and rods are superimposed upon (but likely to be very different from) those which generated the qualitative differences between Type I and Type II ganglion cells, and between ganglion cells of the nasal vs. temporal region of the retina.

2. Both locus specificity and phenotypic specificity are related to cell position in characteristic ways. The *development* of locus specificity, and perhaps even its long-term maintenance, are position-dependent processes (HUNT and JACOBSON, 1974a; YOON, 1973; see part VII.). The relation of phenotypic specificity to position is more complex. *By subsystem*, the differentiation of specific neuronal phenotypes is roughly position-dependent: thus, ganglion cells and rods differentiate in the retina and nowhere else. *Within subsystems*, however, phenotypic differentiation is not position-dependent, but rather the ability to assume a particular final position is a property of the particular phenotype. Thus, the six neuronal phenotypes of the frog retina are all generated from a single, monolayer neuroepithelium; and they will assume their normal proximodistal sequence of layers even if the presumptive retina is inverted in the mediolateral axis (EAKIN, 1943).

These observations are generally born out elsewhere in the developing nervous system as well. The differentiation of those neuronal phenotypes destined for cerebellum and brainstem, for midbrain tectum, for spinal cord, and for cerebrum is place-specific along the neuraxis; but within these subsystems, the positions at which the different phenotypes are generated (e.g. Purkinje cells vs. granule cells in cerebellum) bear little relation to the final positions they occupy. This distinction is crucial for distinguishing differentiations which could be induced by local signs (e.g. gradient value) from those in which the ability to *read* topographical local signs is a programmed feature of the differentiated state.

3. The specificity of the individual terminally differentiated neuron or neuronal phenotype is defined, to a large degree, by a unique *combination* of neuronal properties, drawn from a total set of shared neuronal properties, each of which is displayed two or more times across the nervous system. Thus, the ganglion cells are the only cells *in the retina* which conduct action potentials, exhibit long axons, deploy acetylcholinesterase, and show synaptic affinity for tectal neurons; but each of these properties appears, in concert with others which the ganglion cells lack, in a variety of other neuronal phenotypes. This concept of information-sharing in neuronal diversity extends to position (rods and cones occupy a common layer) and orientation (the long axis of amocrines and horizontals is parallel to, that of all other retinal neurons is perpendicular to, the retinal layers), and even to synaptic selectivity, which is built up by combinations of specifiers which can act independently (HUNT, 1974; JHAVERI and SCHNEIDER, 1974).

In summary, neurons differ from one another on different levels; the diversity is stereotyped, systematic, and related to cell position. Also, it is largely built out of unique combinations of standard (shared) neuronal properties in concert with what may be only a few truly unique biochemical properties (e.g. photopigments, membrane molecules that confer absolute synaptic affitity between two neuronal phenotypes). These principles have two important implications:

First, the diversity-generating mechanisms in neuronal differentiation must (1) work a standard set of genetic ingredients into the specific gene combinations that define the phenotypic specificity of Purkinje cells and rods; (2) avoid nonsense combinations, such as Purkinje cells which transduce light; (3) selectively discharge the Purkinje cell combination in the hindbrain area and the rod combination in the retina; and (4) superimpose individualized instructions for locus specificity upon single cells within an array of a single phenotype. Such a mechanism may squeeze a great deal of specificity out of a few structural genes. But this demands a complex system of regulatory genetics, which may require a long sequence of diverging intermediates to generate the many specific gene sets seen in the final state.

Second, these principles both illuminate and confound the task of establishing relevant *final-state boundaries* in the neuron population, by which to fractionate the final population into meaningful subpopulations, and with which to compare the sibship or *clonal boundaries* between their respective ancestoral cell groups. In fact, any of the systems of final-state boundaries shown in Table 1 represents a reasonable fractionation of neuronal diversity and could conform to clonal boundaries in the emergence of neural cell lineages. However, the principles (especially of position-dependence) developed here predict that *early* clonal boundaries, in the initial divergence of neural cell lineages, would be most likely to conform to geographical final-state boundaries (cerebellum vs. retina).

Late clonal boundaries, between precursor cells immediately antecedent to definitive neurons, would be most likely to conform to final-state boundaries based on phenotypic specificity, since the principal problem of regulation appears to be that of selecting particular combinations of shared neuronal properties for particular neuronal phenotypes. Possible, though less likely, are final-state

Table 1. Some possible final-state boundary systems for classifying neurons

Criterion	Neurons falling within same boundaries (select examples)	Neurons falling within other boundaries (select examples)
Geographical	Retinal ganglion cells, rods, cones amocrine cells, bipolars, etc.	Cerebellar Purkinje cells, basket cells, stellate cells, granule cells; Spinal motoneurons; cerebral pyramidal cells
Functional Association	Retinal ganglion cells, tectal and geniculate visual relay cells, striate cortical cells	Motor neurons, neurons of the eighth nerve, olfactory neurons
Single chemical properties	All dopaminergic neurons	Cholinergic neurons, adrenergic neurons
Single anatomic properties or physiologic properties	All neurons with long axon, including pyramidal cells, rubrospinal cells, retinal ganglion cells, Mauthner's cell	All neurons with short axon, including retinal bipolar cell, horizontal cell, cerebellar basket cell, spinal interneurons
Common locus specificity	Nasal retinal ganglion cells, caudal tectal cells, etc.	Temporal retinal ganglion cells, rostral tectal cells
Phenotypic specificity	Retinal ganglion cells	Rods or cones or bipolar cells

boundaries based on locus specificity or future functional associations; final-state boundaries based on single morphological, physiological, or chemical properties seem least likely of all. In fact, the evidence suggests that reproducible clonal boundaries *do* exist, which bear invariant relationships to particular final-state boundaries, and the relationships are precisely those predicted above.

IV. Qualitative Differences among Early Precursor Cells

Three lines of evidence indicate that early clonal boundaries have already been established in the neural plate, and that individual precursor cell groups are programmed to generate different *sets* of phenotypes characteristic of major neural subsystems. First, cell-marking experiments on the neural plate have identified localized cell groups destined to give rise to retina, or tectum, or hypothalamus, or brainstem (cf. C. O. JACOBSON, 1959). In addition, while these cell groups are invariant in position and bear a "rough and ready" relationship to the disposition of the final subsystems along the neuraxis, *local* proximity of precursor cell groups is not related to later anatomical disposition, functional relationships, or synaptic interconnections. Thus, the cells which are destined to give rise to retina are geographically separated, in the neural plate, from the ancestors of the tectum by a cell group destined to generate the neurons of the dorsal hypothalamus. Finally, single-marked cells do not move individually in the plate (although the whole sheet of cells moves during neurulation), even at the earliest stages; thus it is extremely unlikely that an early "visual system" clone segregates into the separate retinal and tectal cell groups, or that a single clone for a particular neuronal property (e.g. cholingergic receptor) arises at one locus and then disperses to take up positions in all those cell groups destined to generate subsystems in which some neurons bear this particular property.

The second line of evidence derives from transplantation of neural plate tissue. C. O. JACOBSON (1964) inverted the prospective hindbrain region of the plate ectoderm in *Amblystoma*, and found that the cranial nerve nuclei differentiated in inverted rostrocaudal sequence. Likewise, ectopic retinae containing all the normal neuronal phenotypes developed from transplants of the presumptive retinal cell group (FISCHER, 1969; refs. to early work in HARRISON, 1929). Retinal differentiation also occurred, in *Xenopus*, after CORNER (1963) excised *non*-retinal areas of the neural plate, and disrupted the contiguity of the plate ectoderm. Nevertheless, the presumptive retinal cell group successfully gave rise, at the usual time, to all the neuronal phenotypes of the retina—even though the microenvironment at the time of neuron origin was highly abnormal (e.g. opening onto the ventricle, immediately adjacent to a gaping wound, etc.). Similar observations have been made by a number of workers on slightly older neural tube tissue, which will generate its specified neuronal phenotypes when grafted to ectopic sites (DETWILER, 1936; PIATT, 1949; HOLTZER and KAMRIN, 1956).

The third line of evidence derives from studies on the development of isolated pieces of neural plate or neural tube *in vitro*. CORNER (1964), for example, cultured pieces of neural plate from *Xenopus* neurulae and found that sensory neurons differentiated in all major explants. However, montoneurons *never* differentiated from presumptive forebrain explants but always appeared after explantation of cell groups destined to give rise to midbrain or hindbrain. More recently, the presumptive retina of *Xenopus* (explanted at the time it begins to bud off the neural tube) was shown to give rise to an essentially normal larval retina *in vitro*, complete with all appropriate neuronal phenotypes and histotypic organization (E. FRANK et al., unpubl. observations). When reintroduced into a host orbit, these retinae will form functional synaptic connections with the visual centers of the brain (HUNT and JACOBSON, 1973 a).

The model for tissue culture studies, however, remains the classic work of STEFANELLI on the solitary Mauthner's (M) cell of the frog medulla. STEFANELLI (1950; 1951) explanted small pieces of neural plate or neural tube that either did or not contain the precursor cell group from which the M-cell derives *in situ*: one M-cell always appeared in the former, but never in the latter explants. Mid-late gastrula explants containing the M-cell's precursors sometimes generated an M-cell *in vitro* (other types of mid-late gastrula tissue never developed an M-cell), whereas no early-mid gastrula explants produced M-cells, including those containing the M-cell's normal ancestors. In addition, STEFANELLI observed the *embryo* after removal of the explant: gastrulae, from which the presumptive M-cell region was explanted and was seen to generate an M-cell *in vitro*, often developed an *additional* M-cell *in situ* from cells adjacent to the excised area. Early gastrulae could generate such an additional M-cell after removal of large pieces of the neural plate (i.e. from cells far away from the normal presumptive M-cell region), while late gastrulae could do so only after removal of small pieces. Post-gastrula embryos were never able to generate an additional M-cell *in situ* when the explant contained the normal M-cell precursors.

Such observations on the cell groups adjacent to excised regions of neural plate, taken with the evidence of local continuity of cell lines and absence of individual cell migration, indicate that the neural precursor cells of the late neural plate and neural tube are qualitatively different from those of the gastrula and early neural plate. At the cell populational level, at least, the latter can redirect their differentiation to generate additional phenotypes that would normally have arisen from adjacent cells. In contrast, the progeny of those early neural precursor cells, generated in the normal animal, lack this capacity to respond to excision of their neighbors.

To Summarize: Cell groups at different positions in the early neural ectoderm are qualitatively different at any one stage of development, and the clonal boundaries between these cell groups correspond to geographic final-state boundaries between major neural subsystems. The clonal boundaries conform to progressively narrower final-state boundaries as development proceeds, exactly as predicted for systems which generate cell diversity through lineage mechanisms. However, it must be stressed that the latter observation falls far short of proof. It remains to be determined in future experiments if what is true of cell groups is also true of single neural precursor cells. Also to be established is whether or not the progressive narrowing of final-state boundaries with time is inextricably tied to the emergence of successive generations of cells within each clone.

V. Late Neural Precursor Cells

The clonal boundaries between late precursor cells, the immediate ancestors of the definitive neurons, conform in many cases to final-state boundaries based on phenotypic specificity. In other cases, it is difficult to distinguish between final-state boundaries based on phenotypic specificity from those based on other criteria, such as locus specificity. Often times a cohort of neurons, which differ in phenotype but develop a common locus specificity, are generated at the same time and place from a single *small group* of precursor cells (e.g. in the frog optic tectum: STRAZNICKY and GAZE, 1972), and it is not known whether the *group* contains distinct precursors for each terminal phenotype.

In some such ambiguous cases, careful autoradiographic or genetic studies have been able to rule out a common clonal origin for diverse neurons with the same locus specificities. For example, JACOBSON (1970) speculated that the cohort of photoreceptors, interneurons, and ganglion cell that occupy one and the same retinal locus and that become connected to one another, might arise from a common precursor cell. However, HOLLYFIELD (1971) carefully followed the cells generated at a particular time point at the retinal margin and found that the interneurons ultimately synapsed with ganglion and photoreceptor cells which had been generated slightly earlier. HANSON et al. (cit. BENZER, 1973) approached another difficult system, the *Drosophila* eye, by making genetic mosaics for eye pigment mutations. While they confirmed that early clonal boundaries conformed to geographical final-state boundries, mosaic ommatidia were seen at the borders of mutant and wild-type regions, indicating that the different cell phenotypes of a single ommatidium do not share a common late precursor cell.

In a few systems, the autoradiographic studies have left open the possibility that neurons with a common locus specificity arise as a single clone. The best studied case is the butterfly optic lobe (NORDLANDER and EDWARDS, 1969a,b; see also SIDMAN and RAKIC, 1973, for a case in the mouse brainstem). When adult butterflies are sacrificed after a tritiated thymidine pulse at different larval stages, a spatiotemporal "gradient" of neuronal birthdates is seen in the optic lobe. As in the frog tectum, "columns" of neurons that traverse the entire depth of the lobe gray matter along a line normal to the surface are generated at about the same time in development. In this case, however, each "column" of neurons (whose fibers form a common bundle and project from the gray together) apparently derives from a single precursor cell.

Beginning on the first larval day, *neuroblasts* are generated from a small group of precursor cells near the larval neuropil. They migrate up to a position lateral to the protocerebrum and begin to undergo *symmetrical* divisions; the metaphase plate is perpendicular to the surface of the lobe anlagen and gives rise to two daughter cell neuroblasts, both of which may undergo further symmetrical divisions. By the third instar, some of the progeny neuroblasts begin to undergo *assymetrical* divisions, with the metaphase plate parallel to the surface of the lobe, generating another neuroblast (always the daughter closest to the surface) and a *ganglion mother cell*. The ganglion mother cell gives rise to two or three generations of cells, the last of which are *ganglion cells*, the post-mitotic neurons of the optic lobe. Eventually, the whole clone is displaced laterally along the lobe surface by new clones arising from more recently arrived neuroblasts.

These studies raise additional questions: Are symmetrically dividing neuroblasts identical to those that divide assymetrically? Are "first-generation" ganglion mother cells identical to those that divide to produce ganglion cells? They also suggest that the butterfly optic lobe may provide a model system for studying neuronal lineages. However, in regard to clonal origin for neurons of common locus specificity, the evidence must be considered as no more than strongly suggestive. In this regard, two notes of caution are necessary. First, there is no evidence that the various ganglion cells within a clone are of different phenotypes. Therefore, this system at best is unlikely to show clonal boundaries that conform to final-state boundaries, which are based on locus specificity and *cross* phenotypic boundaries. Second, it is conceivable that an occasional neuron from one clone crosses into another and joins the wrong bundle; and not without more refined techniques (cf. BENZER, 1973) could such deviations be detected.

In contrast to the difficult and ambiguous cased described above, many neural subsystems unequivocally evince late clonal boundaries that conform to final-state boundaries based on phenotypic specificity. From his regeneration studies, HOLTZER (1951) inferred that two or more qualitatively different precursor cell types exist in the ventricular zone of the urodele cord, one of which was specified for producing motor neurons. Likewise, the Purkinje cells and the other neuronal phenotypes of the cerebellum derive from separate cell lines. The Purkinje cells arise near the ventricle and migrate into the nascent cerebellum as post-mitotic cells. The other neurons arise later, from precursor cells which have migrated into an external granular layer, and there is circumstantial evidence that this layer contains distinct precursor cells for basket and granule cells (SIDMAN and RAKIC, 1973). TABER (1963) found that the neurons of sensory vs. motor cranial nerve nuclei arise at separate times from the rhombic lip, indicating their derivation from distinct precursor cells. In the mouse retina, HINDS and HINDS (1974) examined ganglion cell production by electron-microscopic reconstruction of "daughter cell pairs" in the retinal neuroepithelium. They found that when one member of the pair was a nascent ganglion cell, the other was usually a ganglion cell as well, or occasionally a neural precursor cell, but never a neuron of different phenotype.

In *Xenopus* retina, JACOBSON (1968b) found that the ganglion cells, interneurons, and photoreceptors of the central optic cup were generated at slightly different times between stages 28 and 32, the intervals being shorter than one cell cycle. BERGEY et al. (1973) have more recently treated the early optic cup with 5-bromodeoxyuridine, an agent which selectively suppresses differentiated functions without affecting cell viability or replication (see WEINTRAUB, this volume; DIENSTMAN and HOLTZER, this volume). Administered from stage 27^-, when the birth of the ganglion cells is imminent, the drug completely blocked the emergence of rods, cones, and interneurons but largely permitted the normal differentiation of the ganglion cells (see also, HUNT et al., 1974). Taken together, the labeling studies and 5-bromodeoxyuridine results indicate that *at least* two distinct precursor cell types exist in *Xenopus* retinal neuroepithelium, one of which is the immediate precursor for the ganglion cells of the central optic cup.

To Summarize: There is unambiguous evidence for the existence of late clonal boundaries, in many areas of the nervous system, which conform to final-state boundaries based on phenotypic specificity. There is some suggestive evidence for clonal origin of cells with common locus specificity, though in no case have such clonal boundaries been found to *cross* phenotypic boundaries.

VI. Invariance of Lineages and the Role of the Cell Cycle

An important prediciton for any system that used cell lineages to program stereotyped cellular diversity is that the final (terminally differentiated) state derives from an obligatory sequence of intermediate states, not all of which can occur in a single cell within its own lifetime. Thus, an irreducible number of cell generations must be spawned, which may be fewer than the number of generations spawned *in situ*, in order to generate the specificity and diversity of the final state. This prediction has never been rigorously tested in the developing nervous system. Nevertheless, there is a wealth of indirect evidence that late neural precursor cells are not able to express, within their own lifetime, the unique set of neuronal properties characteristic of their neuronal progeny; and that early neural precursor cells, which do not themselves divide to produce frank neurons, cannot acquire the capacity to do so within their own lifetime.

Thus, when SECOFF et al. (1972) explanted Drosophila neuroblasts (precursor cells similar to the butterfly neuroblasts described earlier), they found that clones of 8 or 16 cells were generated before frank ganglion cells appeared. Frequently, all 8 or 16 cells of the final clone, or 4 or 8 cells descended from one daughter of the first *in vitro* division, all began to display the properties of frank ganglion cells at about the same time. It seemed that the last few divisions culminating in neuron production were symmetrical, as they are *in vivo*; and in no case did neuroblasts themselves differentiate into neurons or give rise to neurons in one generation.

While clonal cell culture experiments have not yet been carried out on the precursor cells of the amphibian retina, this system also shows a remarkable refractoriness to experimental "telescoping" of its neural cell lineages. The first neurons are born at optic cup stages in all amphibia examined (stages 28–31 in *Xenopus*, about 1,5 days post-fertilization); their phenotypic properties are immediately expressed, and they shortly connect to form a functional but miniature larval eye (by stage 39 in Xenopus, 2,5 days post-fertilization), although many hundreds of thousands of neurons remain to be added to the retinal periphery during the larval growth period (JACOBSON, 1968b; HOLLYFIELD, 1971; STRAZNICKY and GAZE, 1971). HUNT et al. (unpubl. observations) recently showed that stage-22 eye primordia (primitive optic bulge whose cells are at least two generations proximal to the first retinal neurons) generate the same number of each neuronal phenotype of neuron, with roughly the same timecourse, when grafted to a stage 28–38 host as when grafted to a stage 22 (control) host, and when grafted onot the host flank vs. the host orbit. Moreover, half-eye primordia prepared at stage 25/26 generated about half the normal number of neurons of each phenotype, with roughly the normal timecourse, ruling out control of the time of neuron formation by "critical mass" (BERMAN and HUNT, 1975).

Xenoplastic transplantation of early eye primordia also showed intrinsic control of the number of cell generations between grafting and the emergence of retinal neurons, and of the time interval between these two events (HARRISON, 1929; FISCHER, 1969), through as many as four cell generations in some urodeles. One dramatic exception, however, is BALINSKY's (1958) observation that such transplantations *in early-mid gastrulae* did not lead to autonomous (donor-determined) differentiation of the retina. Instead, a host-dependent programming (or reprogramming), of the time and number of cell divisions of neuron formation, appeared to occur. Thus, grafted *Xenopus* eyes gave rise to a giant optic cup (2–4 times normal cell numbers) before neurons were generated, while urodele eyes generated neurons precociously in *Xenopus* hosts at a time when the optic cup constained less than half its normal complement of precursor cells. Balinsky's photomicrographs are particularly compelling, and in some cases show a *Xenopus* neuroepithelium so large that it had begun to subdivide into two optic cups. One interpretation of this intriguing result is that early "proliferative cell cycles" (HOLTZER, 1970) may occur, between the establishment of early (retinal vs. nonretinal) and late (ganglion cell vs. other neurons)clonal boundaries, that are dispensible to developmental programming and can be increased or decreased in number in response to the microenvironment of the embryo[2]. In contrast, slightly

---

[2] Such an interpretation must be considered tenative, however, until (1) both phenomena (host-dependent differentiation from early grafts, autonomous differentiation of later, neural plate grafts) can be reproduced in the same laboratory; and (2) it is definitively demonstrated that continued neural *induction* did not occur after transplantation of the early grafts.

later generations of retinal precursor cells (which show autonomous differentiation in Xenoplastic transplants) do not show this lability (FISCHER, 1969), suggesting that a "clock" has been activated which couples the final phases in retinal cell
differentiation to a specified number of cell generations.

A second strategy involves deliberate attempts to block passage of neural
precursor cells through the cell cycle, or to dissociate developmental programming from cell proliferation with special agents such as 5-bromodeoxyuridine.
Treatment of stage 24 ± 1 *Xenopus* with 5-bromodeoxyuridine (BrUdR) completely suppressed the therminal cytodifferentiation of all neuronal phenotypes in
the retina, and an abnormally large precursor cell population developed (BERGEY
et al., 1973). A similar suppression of neuronal differentiation has been in BrUdR-
treated chick retina, chick brain, and chick spinal cord (reviewed in HUNT et al.,
1975). Fluorodeoxyuridine (FUdR), cytosine arabinoside, and KC1 all blocked or
slowed DNA synthesis in neural precursor cells of *Xenopus* optic cup, and definitive neuronal properties were never seen in the retina for the duration of the block
(12–24 hrs); in contrast to BrUdR, however, the DNA-synthetic blocking agents
did not produce abnormally large numbers of neural precursor cells; in fact, the
number of precursor cells (as expected) failed to rise normally over the duration of
the block (HUNT et al., in preparation).

It appears, then, that the morphological features of neuronal phenotypic specificity do not appear when the lifetime of normal neural precursor cells is prolonged by DNA-synthetic blockade. Nor do neurons with "nonsense combinations" of shared neuronal properties appear in the presence of such a block. This
inference, while it certainly requires further documentation, is in no way incompatible with the report that *neuroblastoma* cells can be stimulated to adhere to
substratum and put out neurite-like processes in the presence of cytosine arabinoside (SCHUBERT and JACOB, 1970), which must be viewed as a peculiar aberration
of these deranged cells until proven otherwise. Whether neuronal properties can
penetrate in cells blocked in mitosis is not known, for conventional anti-mitotic
agents disrupt the neurotubules needed for axon and dendrite formation, and
because biochemical assays are generally limited to neurotransmitters and transmitter enzymes, which are poor markers of the terminally differentiated state.
Attainment of detectable levels of catecholamines, for example, occurs 7 days after
neuron origin in chick amocrine cells (HAUSCHILD, 1974) and *before* neuron
origin in sympathetic cells (COHEN, 1974). Indeed, the serendipitous relationship
between the time of neuron origin and the time of accumulation of catecholamines suggests that catecholamine biosynthesis may not be coordinately controlled
with other shared neuronal properties—although the atypical origin and development of sympathetic cells makes a generalization premature. Nevertheless, we
should be alerted to the possibility that not all of the properties which define the
phenotypic specificity of a neuron need be expressed at once, and that certain
biochemical features may in fact prove detectable (as the thresholds for our assays
go down) in antecedent precursor cell generations.

The role of cytokinesis is not well understood, though we have observed
apparent axons and dendrites on cells with 2 or 4 nuclei, following treatment of
Xenopus eye primordia with cytochalasin-B (BERGEY et al., unpubl. observations).
WHITTAKER (1973) blocked cytokinesis at the 16- or 32-cell stage in an ascidian

embryo (using cytochalasin-B) and found that the blastomeres from which the two pigmented brain cells (otolith and ocellus) normally derive became pigmented at the normal time. Thus, it appears that some neuronal information can be activated in the absence of cytokinesis. But can specified *combinations* of shared and unshared neuronal information be mobilized in the appropriate cells under these conditions? It is of great interest that WHITTAKER's pigmented blastomeres never developed the other properties of the otolith and ocellus, and many of the multinucleate retinal "neurons" could not, at least on simple histologic inspection, be recognized as one of the normal phenotypes of *Xenopus* retina. It may well be that cytokinesis specifically (and cell lineages generally) help to compartmentalize different cytoplasms for cytoplasmic-nuclear interactions. The ability of egg cytoplasm to activate DNA synthesis in brain cell nuclei (GURDON and WOODLAND, 1968) stresses the role of the cytoplasm in neuronal differentiation, as do the classic experiments of CARLSON (1952). He rotated the mitotic spindle, during the assymetric division of the grasshopper neuroblast, such that the chromosomes destined for one daughter (neuroblast) would up in the cytoplasm of the other (ganglion mother cell), and vice versa. CARLSON found that the respective cytoplasms determined the daughter cell phenotypes, and neurons were generated from the cell comprised of ganglion mother cell cytoplasm and neuroblast chromosomes.

To summarize: Although the evidence is fragmentary and largely indirect, we can say that neural precusor cells have thus far shown extreme reluctance to break from their lineage patterns. Early precursor cells, which do not normally generate frank neurons, do not do so when their cell cycle is prolonged or their microenvironment is manipulated; nor do such conditions cause late neural precursor cells to display definitive neuronal properties within their own lifetime.

VII. Coordinate Control of Developmental Programming

The relation of neuronal locus specificity to the cell cycle and to the lineages of origin of phenotypic specificity has been examined in only one system, the retinal ganglion cells of *Xenopus*. Recent studies have confirmed SPERRY's (1950, 1951) hypothesis that locus specificity is a position-dependent cytochemical property (probably localized to the cell surface), which enables the cell's axon to reach the appropriate locus in the retinotectal map. Locus specificities are patterned with respect to embryonic (AP and DV) axes set down in the early retina (HUNT and JACOBSON, 1972a, b) and aligned with the major axes of the embryo's body. For a time (stages 22–28) these retinal axes are stable but reversible: they survive *in vitro*, but *in vivo* rotation of the anlage leads to their replacement, in 2 to 6 hrs, by a new pair of (stable but reversible) axes, once again aligned with the embryo's body axes (HUNT and JACOBSON, 1973; 1974c). Between stages 28 and 31 (5 hrs), the AP and then the DV axis of the retina are fixed and *specified* as the permanent reference axes for cellular positional information (WOLPERT, 1971). Thereafter the eye will not undergo axial replacement, even if challenged by rotation into a stage 27/28 orbit and the patterning of ganglion cell locus specificities follows the original axial plan (JACOBSON, 1968a; JACOBSON and HUNT, 1973). Thus, the developmental program for spatial patterning of locus specificities is set down at the stages (28–31) when the first retinal neurons cease DNA synthesis and commence their terminal cytodifferentiation (JACOBSON, 1968b), but the program affects all the ganglion cells the retina will ever contain, including 70 to 100 thousand more which remain to be added to the retina (and integrated into the program) at larval stages.

There is evidence that cell lineage may play little or no formal role the process by which the late-arising ganglion cells become integrated into the original program, acquire positional information, and develop the proper locus specificity. HUNT and HOLTZER (in preparation) killed up to 75% of the precursor cells at the ciliary margin of the stage 38/40 retina; the survivors appeared to restore the continuity of the precursor cell ring (by undergoing "unscheduled" circumferential divisions) before resuming the normal pattern of radial growth. Nevertheless, a normal pattern of locus specificities, based on the original (stage 28–31) program, was seen in the adult retina, as assayed from the retinotectal maps. In contrast, killing the post-mitotic neurons in the central retina led to the development of scrambled maps. These results indicate that individual stem cells at the ciliary margin are not programmed to generate an obligatory sequence of qualitatively different (and unique) ganglion cells. Considered with other evidence showing cellular interactions within the retina (HUNT and JACOBSON, 1974a,d; HUNT, 1975), they suggest that nascent ganglion cells acquire their positional information and determine their appropriate locus specificities by interaction with the older, more centrally located cells.

On the other hand, axial specification and the initial establishment of the program at stages 28–31 do appear to be lineage-dependent. Axial specification can occur *in vitro* (HUNT and JACOBSON, 1973a), and its timing is intrinsically controlled within the eye primordium. Thus the time of specification always correlates with the stage of the optic cup and, in eyes grafted between embryos, could be neither accelerated nor delayed by altering the stage of the host (HUNT and JACOBSON, 1974b). HUNT et al. (1974) exploited the effects of BrUdR on the developing optic cup, to show that the time of specification was a programmed component of ganglion cell differentiation. When BrUdR (from stage 24) blocked the onset of differentiation of all retinal neurons, the eyes were still able to undergo replacement of their retinal axes at stage 32. Not the prolonged time needed for these eyes to reach stage 32, nor the attainment of unusually large size, nor the appearance of terminally differentiated non-neural cell types was sufficient to trigger axial specification. However, when some of the ganglion cells were allowed to commence their terminal cytodifferentiation (after BrUdR from stage 27⁻), axial specification occurred on schedule. An obligatory role for the interneurons and photoreceptors was thus eliminated in the control of axial specification. HUNT et al. (1975) inferred that the trigger for axial specification is localized to the gangliogenic precursor cells in the pre-stage 29 optic cup, and is activated at a predetermined point in their differentiation.

The importance of the precursor cell cycle is further underscored by observations on axial specification after treatment with FUdR (HUNT et al., in preparation). FUdR was injected at concentrations which (1) completely arrest retinal

cytodifferentiation and kill 100% of the retinal cells after 2 days without thymidine rescue, but which (2) permit recovery of functional eyes and histologically
normal retinae if exogenous thymidine is provided within 18 hrs or less. At
stage 32, the treated eyes were challenged to undergo replacement of their retinal
axes by grafting, in rotated orientation, into a normal stage 27/28 host orbit.
Surprisingly, when such a block was imposed from early stage 28 (arresting the
last generation of gangliogenic precursors in G_1 or S), AP-specification always
occurred and DV-specification almost always occurred by stage 32. Following the
axial replacement challenge and rescue with thymidine, these eyes went on to
generate completely rotated, or AP-rotated retinotectal maps. Similar results have
been obtained using cytosine arabinoside and KCl.

These experiments permit two important inferences: First, the actual cessation
of DNA synthesis and birth of definitive ganglion cells may not be necessary for
axial specification (a possibility also raised by the findings of JACOBSON et al.,
1974, on Rana pipiens). Second, the trigger for axial specification may be activated *several hours in advance* of the time at which the optic cup actually loses the
capacity to undergo axial replacement—early in the lifetime of the immediate
gangliogenic precursor cell or, prehaps, late in the lifetime of the penultimate
precursor. Once activated, the events of specification can proceed to completion
even if ganglion cell differentiation is halted. One important prediction of this
model has recently been met in preliminary experiments: early FUdR treatment
(from stage 20–24) *did* delay the time of axial specification by intervals approxi-
mating the duration of the block.

In summary, then, many more experiments will be needed before the role of
the cell cycle and cell lineage in the development of locus specificity is fully
known. Nevertheless, it appears that the optic cup of *Xenopus* has solved two of
its major pattern-formation problems, namely, coordination of spatial differentiation with cytodifferentiation and provision of a permanent blueprint for cellular
positional information in advance of the cells' needs to express position-dependent properties. This was accomplished by consigning control of axial specification to the *precursor cells* of those neurons whose need for positional information
is most pressing and immediate: the ganglion cells of the central retina.

VIII. Summary

A complete understanding of how the cell cycle and cell lineages function in
neurogenesis awaits full chemical characterization of neuronal diversity, as well as
analysis of the chemical continuities and discontinuities between each final-state
neuron and its antecedent generations of precursor cells. As an interim approach,
I have attempted to systematize neuronal diversity based on the morphological,
physiological, and (limited) chemical criteria currently available. Further, I have
endeavored to construct systems of *final-state boundaries* which fractionate the
final neuron population into arbitrary subpopulations based on biologically relevant distinctions and to determine, from the limited data on the cell cycle and cell
lineage in various developing neural subsystems, whether invariant *clonal boundaries* could be found that conformed to any of the final-state boundary systems.

The analysis led to the following observations: (1) Early clonal boundaries conform to geographic final-state boundaries separating major brain areas. (2) Late clonal boundaries, between precursor cells immediately antecedent to definitive neurons, most often conform to final-state boundaries based on phenotypic specificity. (3) In a few instances, evidence suggests late clonal boundaries which conform to final state boundaries based on locus specificity; but in no case were these proved, nor did they appear to *cross* phenotypic boundaries. In addition to the existence of such clonal boundaries, several other observations suggested a role for the cell cycle and cell lineage in generating neuronal diversity: (4) Phenotypic specificity in most (and perhaps all) neurons is intrinsically determined, and is largely defined by a unique *combination of shared neuronal properties* and may involve only a very few truly unique structural genes. (5) Within a replicating precursor cell group, qualitative changes in cell properties, including an early loss of the (regulative) capacity to redirect differentiation in response to extirpation of adjacent precursor cell groups, occur over a period of time. (6) Within a given precursor cell group, defined by early clonal boundaries, the emergence of phenotypic specificity and of final-state neurons takes place under tight spatiotemporal constraints that are intrinsically controlled within the cell group. (7) Under conditions prolonging their cell cycle or altering their microenvironment, early neural precursor cells have thus far failed to directly generate neurons; likewise late neural precursor cells have thus far failed to express, within their own lifetime, the phenotypic properties of their final-state progeny. (8) The coordinate development of phenotypic and locus specificity may be achieved by coupling control of pattern determination (e.g. axial specification) to progression of late precursor cells through a particular phenotypic lineage.

What emerges is a model of neuronal differentiation in which regulatory genetics are complex and sequential and structural genetics are combinatorial and largely redundant. Such a model certainly lends itself to lineage-dependent programming, but the evidence presented here merely testifies to the *plausibility* of a role for the cell cycle and cell lineage in neuronal differentiation. The exact nature of the role(s) is not known and may be as diverse as compartmentalizing cytoplasm in time and space for clean nuclear–cytoplasmic interactions, or coupling the activation of specific regulatory gene circuits to the replication of DNA. Likewise, it may be played out autonomously in individual replicating cells, or it may require cell interactions within each generation in the lineage in order to stabilize genetic decisions and permit movement through the lineage.

If lineages play any of these roles, however, the model predicts that the normal sequence of transitions that occur in the precursor cell groups for medulla or retina will not occur if progression through the cell cycle is blocked. Also, and more definitively, upon lifting such a block, they will only occur after new cell generations have appeared. Determining the minimum number of such cell generations intervening between two sequential transitions will help to define the differentiative capacity of single precursor cells within their own lifetimes. Detection and analysis of genetic mistakes and nonsense combinations of neuronal properties that may appear after pertubation of neural cell lineages, will provide insight into the complex system of regulatory mechanisms that serve to deposit a cell with the right combination of neuronal properties at the right time and place in the brain.

References

ALTMAN, J.: Post-natal growth and differentiation of the mammalian brain, with implications for a morphological theory of memory. In: QUARTON, G., MELNECHUK, T., SCHMITT, F. O. (Eds.): The Neurosciences. A Study Program. New York: Rockefeller University Press 1967.

ANGEVINE, J. B., JR.: Critical cellular events in the shaping of neural centers. In: SCHMITT, F. O. (Ed.): The Neurosciences: A Second Study Program. New York: Rockefeller University Press 1970.

BALINSKY, B. I.: On the factors controlling the size of the brain and eyes in anuran embryos. J. Exp. Zool. **139**, 403 (1958).

BENZER, S.: Genetic dissection of behavior. Sci. Am. **229** (6), 24 (1973).

BERGEY, G. K., HUNT, R. K., HOLTZER, H.: Selective effects of bromodeoxyuridine on developing *Xenopus laevis* retina. Anat. Rec. **175**, 271 (1973).

BERMAN, N. J., HUNT, R. K.: The visual projections to the optic tecta in *Xenopus* after partial extirpation of the embryonic eye. J. Comp. Neurol. (in press) (1975).

BLAKEMORE, C., COOPER, G. F.: Development of the brain depends on the visual environment. Nature **228**, 477 (1970).

CARLSON, J. G.: Microdissection studies of the dividing neuroblasts of the grasshopper, *Chortophaga viridifasciata* (De Gees). Chromosoma **5**, 199 (1952).

COHEN, A. M.: DNA synthesis and cell division in differentiated avian adrenergic neuroblasts. In: FAUCE, K., OLSON, L., ZOTTERMAN, Y. (Eds.): Dynamics of Regeneration and Growth of Neurons. Oxford: Pergamon Press 1974.

CORNER, M. A.: Development of the brain of *Xenopus laevis* after removal of parts of the neural plate. J. Exp. Zool. **153**, 301 (1963).

CORNER, M. A.: Localization of capacities for functional development in the neural plate of *Xenopus laevis*. J. Comp. Neurol. **123**, 243 (1964).

COWAN, W. M., WENGER, E.: Cell loss in the trochlear nucleus of the chick during normal development and after radical extirpation of the optic vesicle. J. Exp. Zool. **164**, 267 (1967).

DAVIS, W. J.: Development of locomotor patterns in the absence of peripheral sense organs and muscles. Proc. Natl. Acad. Sci. **70**, 956 (1973).

DELONG, G. R., SIDMAN, R. L.: Alignment defect of reaggregating cells in cultures of developing brain of reller mutant mice. Develop. Biol. **22**, 584 (1970).

DETWILER, S. R.: Neuroembryology: An Experimental Study. New York: MacMillan 1936.

EAKIN, R. M.: Determination and regulation of polarity in the retina of *Hyla regilla*. Univ. of Calif. Publ. Zool. **51**, 245 (1947).

FISCHER, A. F. TH.: Untersuchungen über die Entwicklung xenoplastischer Augenchimären bei Amphibien. Roux Archiv **163**, 81 (1969).

GAZE, R. M.: The formation of nerve connections. New York: Academic Press 1970.

GURDON, J. B., WOODLAND, H. R.: The cytoplasmic control of nuclear activity in animal development. Biol. Rev. **43**, 233 (1968).

HAMBURGER, V.: Specificity in neurogenesis. J. Cell Comp. Physiol. Suppl., 1, 60, 81 (1960).

HAMBURGH, M.: Hormones in development. New York: Academic Press 1973.

HARRISON, R. G.: Correlation in the development of the eye studied by means of heteroplastic transplantation. Roux Archiv **120**, 1 (1929).

HAUSCHILD, D. C.: Biogenic amine-containing neurons in chick embryo retina. Ph.D. Thesis, University of Pennsylvania (1974).

HERRICK, C. J.: The brain of the tiger salamander. Chicago: University of Chicago Press 1948.

HINDS, J., HINDS, P.: Early ganglion cell differentiation in the mouse retina: an electron microscopic analysis utilizing serial sections. Develop. Biol. **37**, 381 (1974).

HOLLYFIELD, J. G.: Differential growth of the neural retina in *Xenopus laevis* larvae. Develop. Biol. **24**, 264 (1971).

HOLTZER, H.: Reconstitution of the urodele spinal cord following unilateral ablation. I. Chronology of neuron regulation. J. Exp. Zool. **117**, 523 (1951).

HOLTZER, H.: Reconstitution of the urodele spinal cord following unilateral ablation. II. Regeneration of the longitudinal tracts and ectopic synaptic uniens of Mauthner's fiber. J. Exp. Zool. **119**, 263 (1952).

HOLTZER, H.: Proliferative and quantal cell cycles in myogenesis, chondrogenesis and erythrogenesis. Symp. Cell Biol. **9**, 69 (1970).

HOLTZER, H., KAMRIN, R. P.: Development of local coordination centers. Brachial centers in the salamander spinal cord. J. Exp. Zool. **132**, 391 (1956).

HUBEL, D. H., WIESEL, T. N.: Receptive fields of cells in striate cortex of very young, visually inexperienced kittens. J. Neurophysiol. **26**, 994 (1963).

HUNT, R. K.: Developmental programming for retinotectal patterns. In: Ciba Found. Symp. on Cell Patterning. (in press) (1975).

HUNT, R. K., BERGEY, G. K., HOLTZER, H.: 5-Bromodeoxyuridine: Localization of a developmental program in *Xenopus* optic cup. Develop. Biol. (in press) (1975).

HUNT, R. K., JACOBSON, M.: Neurogenesis in frogs after early larval treatment with somatotrophin or prolactin. Develop. Biol. **26**, 100 (1971).

HUNT, R. K., JACOBSON, M.: Development and stability of positional information in *Xenopus* retinal ganglion cells. Proc. Nat. Acad. Sci. USA **69**, 780 (1972a).

HUNT, R. K., JACOBSON, M.: Specification of positional information in retinal ganglion cells of *Xenopus*. Stability of the specified state. Proc. Nat. Acad. Sci. USA **69**, 2860 (1972b).

HUNT, R. K., JACOBSON, M.: Specification of positional information in retinal ganglion cells of *Xenopus*. Assays for analysis of the unspecified state. Proc. Nat. Acad. Sci. USA **70**, 503 (1973).

HUNT, R. K., JACOBSON, M.: Neuronal specificity revisited. Curr. Tops. Develop. Biol. **8**, 202 (1974a).

HUNT, R. K., JACOBSON, M.: Specification of positional information in retinal ganglion cells of *Xenopus*. Intraocular control of the time of specification. Proc. Nat. Acad. Sci. USA **71** (in press) (1974b).

HUNT, R. K., JACOBSON, M.: Rapid realignment of retinal axes in embryonic *Xenopus* eyes. J. Physiol. (Lond.), (in press) (1974c).

HUNT, R. K., JACOBSON, M.: Development of neuronal locus specificity in retinal ganglion cells of *Xenopus* after surgical eye transection or after fusion of whole embryonic eyes. Develop. Biol. (in press) (1974d).

IKEDA, K., KAPLAN, W. D.: Patterned neural activity of a mutant *Drosophila* melanogaster. Proc. Nat. Acad. Sci. USA **66**, 765 (1970).

JACOBSON, C. O.: The localization of the presumptive cerebral regions of the neural plate of the axolotl larva. J. Embryol. Exp. Morph. **7**, 1 (1959).

JACOBSON, C. O.: Motor nuclei, cranial nerve roots, and fibre pattern in the medulla oblongata after reversal experiments on the neural plate of Axolotl larvae. Zool. Bidr. Uppsala **36**, 73 (1964).

JACOBSON, M.: Development of neuronal specificity in retinal ganglion cells. Develop. Biol. **17**, 202 (1968a).

JACOBSON, M.: Cessation of DNA synthesis in retinal ganglion cells correlated with the time of specification of their central connections. Develop. Biol. **17**, 219 (1968b).

JACOBSON, M.: Developmental neurobiology. New York: Holt, Reinhart and Winston 1970.

JACOBSON, M., HIRSCH, H. V. B., DUDA, M., HUNT, R. K.: Dynamics of retinal specification in *Rana pipiens*. Submitted for review (1974).

JACOBSON, M., HUNT, R. K.: The origins of nerve cell specificity. Sci. Amer. 227 (2), 26 (1973).

JHAVERI, R. R., SCHNEIDER, G. E.: Retinal projections in syrian hamsters: Normal topography and alterations after partial tectum lesions at birth. Anat. Rec. **178**, 283 (1974).

KELLY, J. P., COWAN, W. M.: Studies on the development of the chick optic tectum. III. Effects of early eye removal. Brain Res. **42**, 263 (1972).

MARK, R. F.: Re-establishment of proper central-peripheral relations in the nervous system. In: Ciba Found. Symp. on Cell Patterning (in press) 1974.

MATURANA, H. R., LETTVIN, J. Y., McCULLOCH, W. S., PITTS, W. H.: Anatomy and physiology of vision in the frog (*Rana* pipiens). J. Gen. Physiol. Suppl. **43**, 129 (1960).

MINER, N.: Integumental specification of sensory fibers in the development of cutaneous local sign. J. Comp. Neurol. **105**, 161 (1956).

NORDLANDER, R. H., EDWARDS, J. S.: Postembryonic brian development in the monarch butterfly, *Danaus plexippus plexippus*. I. Cellular events during brain morphogenesis. Roux Archiv **162**, 197 (1969a).

NORDLANDER, R. H., EDWARDS, J. S.: II. Development of the optic lobes. Roux Archiv. **163**, 197 (1969b).

PIATT, J.: Heterotopic transplantation of brachial spinal cord segments followed by reimplantation into the orthotopic site. J. Exp. Zool. **111**, 1 (1949).

PRESTIGE, M.: Axon and cell numbers in the developing nervous system. Brit. Med. Bull. **30**, 107 (1974).

SCHUBERT, D., JACOB, F.: 5-bromodeoxyuridine-induced differentiation of a neuroblastoma. Proc. Nat. Acad. Sci. USA **67**, 247 (1970).

SEECOF, R. L., TPELITZ, R. L., GERSON, I., IKEDA, K., DONADY, J. J.: Differentiation of neuromuscular junctions of embryonic *Drosophila* cells. Proc. Nat. Acad. Sci. USA **69**, 566 (1972).

SIDMAN, R. L., RAKIC, P.: Neuronal migration, with special reference to the developing human brain: A review. Brain Res. **62**, 1 (1973).

SKARF, B.: Development of binocular single units in the optic tectum of frogs raised with disparate stimulation to the eyes. Brain Res. **51**, 352 (1973).

SKLAR, J. H., HUNT, R. K.: The acquisition of specificity in cutaneious sensory neurons. A reconsideration of the integumental specification hypothesis. Proc. Nat. Acad. Sci. USA **70**, 3684 (1973).

SPERRY, R. W.: Neuronal specificity. In: WEISS, P. (Ed.): Genetic Neurology. Chicago: University of Chicago Press 1950.

SPERRY, R. W.: Regulative factors in the orderly growth of neural circuits. Growth Symp. **10**, 63 (1951).

STEFANELLI, A.: Studies on the development of the Mauthner cell. In: WEISS, P. (Ed.): Genetic Neurology. Chicago: University of Chicago Press 1950.

STEFANELLI, A.: The Mauthnerian apparatus in the Ichthyopsida: Its nature and function and correlated problems of neurohistogenesis. Quant. Rev. Biol. **26**, 17 (1951).

STONE, L. S.: Polarization of the retina and the development of vision. J. Exp. Zool. **145**, 85 (1960).

STRAZNICKY, K., GAZE, R. M.: Growth of the retina in *Xenopus laevis*. An autoradiographic study. J. Embryol. Exp. Morph. **26**, 67 (1971).

STRAZNICKY, K., GAZE, R. M.: The development of the tectum in *Xenopus laevis*. An autoradiographic study. J. Embryol. Exp. Morph. **26**, 67 (1972).

TABER, E.: Histogenesis of brain stem neurons studied autoradiographically with thymidine-3 H in the mouse. Anat. Rec. **145**, 291 (1963).

WHITTAKER, J. R.: Segregation during ascidian embryogenesis of egg cytoplasmic information for tissue-specific enzyme development. Proc. Natl. Acad. Sci. **70**, 2096 (1973).

WOLPERT, L.: Positional information and pattern formation. Curr. Tops. Develop. Biol. **6**, 183 (1971).

WOODWARD, D. J., ALTMAN, J., HOFFER, B. J.: Electrophysiology and pharmacology of Purkinje cells in rat cerebellum degranulated by postnatal X-irradiation. Soc. Neurosci. Abstr. **3**, 368 (1973).

YOON, M.: Transposition of the visual projection from the nasal hemiretina onto the foreign rostral zone of the optic tectum in goldfish. Exp. Neurol. **37**, 451 (1973).

Neurogenesis and the Cell Cycle

CREIGHTON H. PHELPS and S. E. PFEIFFER

Departments of Anatomy and Microbiology, University of Connecticut Health Center, Farmington, CT, USA

I. Normal Development

A. Introduction

Despite the complexity of structure and function in the adult central nervous system, neurogenesis occurs in a fashion relatively parallel to other developing cell systems, with proliferation of cells taking place in an orderly programmed fashion. The complexity of the nervous system apparently derives from genetic and environmental influences which not only direct proliferation of the various cell types, but also control subsequent cell migrations, local overgrowth, selective cell death, differentiation of the various cell types, and the establishment of functionally appropriate connections between cells. This account is limited to the earlier stages of neurogenesis and the relationship of the cell cycle to the proliferation and differentiation of the various cell types. The excellent reviews of JACOBSON (1970), GAZE (1970), ANGEVINE (1970), SIDMAN (1970), and WATTERSON (1965) should be consulted for detailed consideration of other aspects of neurogenesis.

The nervous system originates from a longitudinal mid-dorsal thickening of the embryonic ectoderm called the neural plate. The neural plate, a layer of neuroepithelial cells, subsequently invaginates to form a neural groove. As a result of continued invagination and differential growth, the lateral margins of the neuroepithelium gradually become opposed and the neural groove becomes the neural tube. As the neural tube closes, some neuroepithelial cells detach dorsally to form the neural crest (for a complete description, see LANGMAN, 1969).

The cranial portion of the neural tube gradually expands to form the brain vesicles, while the caudal portion becomes the spinal cord. The cavities within the brain vesicles are termed the ventricles and are continuous with the central canal of the spinal cord. The majority of cells in the central nervous system originate from a restricted germinal zone called the primary germinal zone or ventricular layer (ANGEVINE et al., 1969), which lines the primitive ventricular cavity and the central canal of the spinal cord. In early development the ventricular layer is the only cell layer present, whereas later in development several other layers appear external to the ventricular layer.

B. Nuclear Migration during the Cell Cycle of Ventricular Layer Cells

The nuclei of the neuroepithelial cells in the ventricular layer are located at various distances from the ventricle. This led early investigators of neurogenesis to believe that there were several cell types present, arranged in a stratified manner (HIS, 1889, RAMÓN Y CAJAL, 1909). They further concluded that each of the cell types had different progeny which, after differentiation, became the adult neurons, neuroglia, and ependyma. Even though SCHAPER (1897) postulated that the neuroepithelium consisted of only one cell type which changed position during the cell cycle, it was not until SAUER (1935, 1936) described the mitotic activity within the neural tube that Schaper's original ideas were verified. Sauer confirmed that the neuroepithelium consisted of only one cell type, and that the apparent stratification was actually due to different positions of the nuclei within the cells during the various stages of the cell cycle. Cell division occurs only at the lumen. Between mitoses the cells extend peripheral processes to the outer surface of the neural

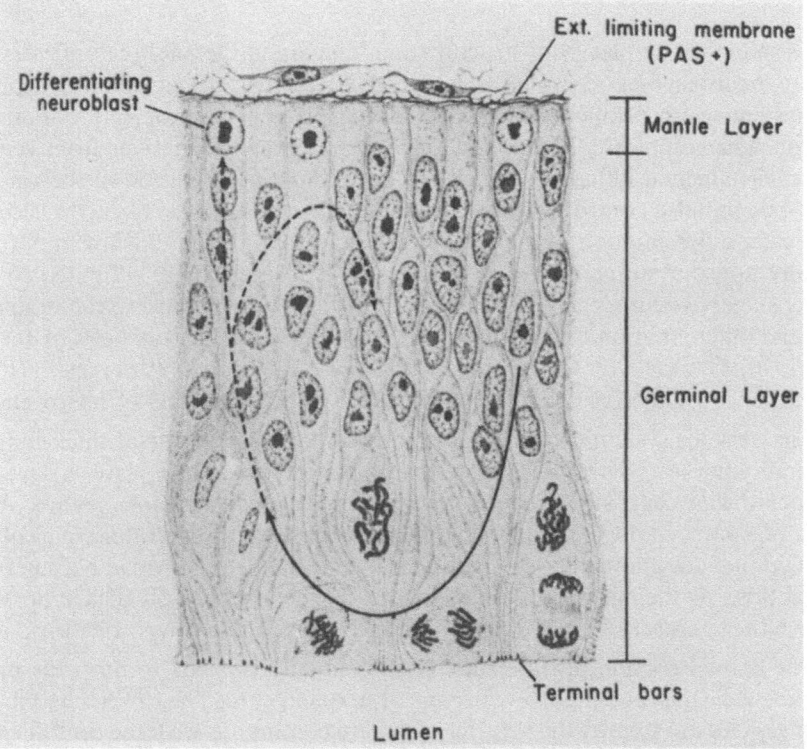

Fig. 1. Drawing of a transverse section through the wall of the neural tube. The nuclei of the neuroepithelial cells within the germinal layer (ventricular layer) migrate to the lumen (ventricle) where mitosis occurs and then back toward the periphery during G_1 and DNA synthesis. The mantle layer contains neuroblasts (young neurons) which formed in the germinal layer and then migrated peripherally. The neuroblasts no longer have the capacity to divide. (From LANGMAN et al., 1966)

tube, and the daughter nuclei migrate to the periphery. At the next mitosis, the cells retract their peripheral processes, and the nuclei again migrate toward the lumen of the neural tube, where the cells are held tightly together by terminal bars (Fig. 1). Sauer termed this phenomenon "interkinetic nuclear migration". These observations, made in the 1930's, were largely ignored until the late 1950's when several experimental techniques were employed to verify Sauer's observations.

1. Colchicine Blocking

The first experiments involved the use of colchicine to block mitosis in metaphase. If the concept of "interkinetic nuclear migration" is correct, and if all cells of the neuroepithelium move to the lumen to divide, then blockage of mitosis in metaphase should result in an accumulation of all dividing cells near the lumen. WATTERSON et al. (1956) observed that with continuous application of colchicine to the chick neural tube, there was a progressive accumulation of cells arrested in metaphase near the lumen. After 7.5 hrs exposure almost all of the neuroepithelial cells were blocked in metaphase. These experiments have been confirmed by KÄLLEN (1961, 1962) and LANGMAN et al. (1966).

2. Spectrophotometric Analyses

Further confirmation of SAUER's observations came with spectrophotometric analysis of the DNA content of cell nuclei at various positions within the wall of the neural tube. SAUER and CHITTENDEN (1959), after staining the nuclei with the Feulgen method for DNA, reported a gradation of DNA content of cell nuclei proportional both to nuclear size and position within the wall of the tube. Cells in telophase near the lumen were found to contain the diploid amount of DNA when compared to adult liver cells in the same species of animal. In contrast, cells in the outer portion of the tube contained quantities of DNA intermediate between diploid and tetraploid. Large late-prophase nuclei found in the inner third of the neuroepithelium contained tetraploid amounts of DNA, while small postmitotic nuclei in the inner third of the tube contained diploid amounts of DNA. The results of these experiments indicated that DNA synthesis begins very shortly after mitosis, continues as the nucleus migrates to the periphery, and is completed as it moves back toward the lumen to divide once again. The DNA synthetic (S) phase is much the longest phase in early embryos, occupying about half of the total cell cycle, while G_1 and G_2 are relatively short. However, as we shall discuss later, while the time spent in the S-phase remains relatively constant, G_1 gradually increases in length as development proceeds.

3. Tritiated Thymidine Labeling

A third, and perhaps the most useful, technique for studying the kinetics of cell proliferation during development of the nervous system is autoradiography, using H^3-thymidine, which is incorporated into DNA during the S-phase. When embryos are given a single injection of H^3-thymidine (pulse labeling), initially only those nuclei in the outer half of the germinal zone of the neural tube acquire label (SIDMAN et al., 1959). Since unincorporated H^3-thymidine is cleared from the body within 1 hr, those cells with label must have been in the DNA-synthetic phase of the cell cycles at the time of, or shortly after, the injection of labeled thymidine. More importantly, the label is retained through subsequent stages of development, thus allowing observation of cell migration and determination of the eventual final resting place of the cell and its progeny once they leave the germinal zone. 6 hrs after injection of H^3-thymidine, labeled cells were observed in the inner half of the neural tube and many were undergoing mitosis. Eventually some daughter cells (with half or less than half as much label) were observed outside the germinal zone, while other daughter cells remained within the germinal zone and repeated the cycle of migration and mitosis.

C. Developmental Changes in Cell Cycle Parameters

The above observations, which have been confirmed by numerous investigators in a variety of species, have not only allowed rather exact timing of the phases of the cell cycle, but also determination of changes in these parameters with progressive development. A common finding in studies of the timing of the cell cycle in the neural tube is an increase in the cell generation time with increasing age of the embryo. For example, in the optic tectum of the 3-day chick embryo, the neuroepithelial cell generation time is 8 hrs, while in the 6-day embryo the generation time is 15 hrs (WILSON, 1973). Similarly, in the rat brain, the generation time in the 12-day embryo is 11 hrs, while in the 18-day embryo it has increased to 19 hrs (v. WAECHTER and JAENSCH, 1972). As in other proliferating systems, the time spent in the S-phase and G_2-phase remains relatively constant despite dramatic changes in generation time. In the chick neural tube the mitotic phase shows small increases in duration as development proceeds (JELINIK, 1959; FUJITA, 1962; WILSON, 1973), but in the mouse neural tube no change in the length of mitosis was noted (KAUFMAN, 1968). The most substantial changes reported have been in the length of G_1. For example, in both the chick (FUJITA, 1962; WILSON, 1973) and mammalian (KAUFMAN, 1968; v. WAECHTER and JAENSCH, 1972) neural tube, increases in the time spent in G_1 were the most significant factor in the lengthening of the generation time. KAUFMAN (1968) suggested that the neuroepithelium, although morphologically homogeneous, may actually be mitotically heterogeneous, and that cells with short generation times differentiate early, leaving behind a proliferating population of cells with longer cell cycles. Although most of the cells in the nervous system are diploid and therefore arrested in G_1, certain populations of cells are thought to be polyploid and thus arrested in G_2.

D. Ploidy of Large Neurons

An important point to note is that large neurons in all parts of the nervous system all arise from the ventricular zone and not from the secondary germinal zones (see below). These include Purkinje cells of the cerebellum (MIALE and

SIDMAN, 1961), pyramidal cells of the cerebral cortex (ANGEVINE, 1965), ganglion cells of the retina (SIDMAN, 1961), mitral cells of the olfactory bulb (HINDS, 1968), anterior horn cells in the spinal cord (FUJITA, 1964), and large cells in the cochlear nuclei (TABER-PIERCE, 1967), habenular nuclei (ANGEVINE, 1970), and optic tectum (FUJITA, 1964). It is interesting that many of the large neurons are thought to be tetraploid (LAPHAM, 1968; HERMAN and LAPHAM, 1968; LENTZ and LAPHAM, 1969) and thus probably arrested in G_2. However, there has been some dispute as to whether these large neurons are really tetraploid, and in the cerebellum, MANN and YATES (1973) asserted that the polyploid cells in the PURKINJE cell layer are really large neuroglial cells and that claims of polyploidy in PURKINJE cells are not valid. Further refinements in measuring DNA in tissue sections should resolve this dispute. These large cells have never been seen to undergo mitosis.

E. Secondary Germinal Zones

The discussion thus far has been concerned with the primary germinal zone of the central nervous system, the ventricular layer. After a number of mitotic cycles, some cells begin to migrate from the ventricular layer to form a secondary layer at the periphery of the neural tube. This zone, termed the mantle zone, initially consists primarily of young neurons which no longer have the capacity to divide. Other cells, at least some of which retain their capacity for cell division, follow the neurons into the mantle zone and have been termed glioblasts. In still later stages of development, secondary germinal zones containing proliferating cells appear at various locations in the nervous system. These secondary zones originate from cells which have migrated from the ventricular zone and include (1) the subventricular layer which forms between the ventricular and mantle layers in the mammalian telencephalon (KERSHMAN, 1937) (2) an ill-defined hippocampal germinal zone (ANGEVINE, 1965), (3) the external granular layer in the cerebellum (RAMÓN Y CAJAL, 1890), and (4) an extensive zone in the hindbrain which gives rise to cells located in the medulla (TABER-PIERCE, 1967).

Smaller neurons and neuroglia arise both from the ventricular zone in earlier stages of development, and from the secondary germinal zones in later periods of development, including the postnatal period. These smaller cells are diploid and many of them (particularly the neuroglia; KORR et al., 1973) retain some capacity for cell division well into the postnatal period.

The external granular layer of the cerebellum (RAMÓN Y CAJAL, 1911) is the best characterized of the secondary germinal zones and is the site of origin of all interneurons of the cerebellar cortex. In the rat, the interneurons (basket cells, stellate cells, and granule cells) develop in a sequential fashion during the first three postnatal weeks (ALTMAN, 1969). The basket cells differentiate first, about five days postnatally, followed by the stellate cells at nine to ten days. Granule cell differentiation occurs throughout the early postnatal period, but the majority form during the second and third weeks. Basket cells and stellate cells are inhibitory in function while granule cells are excitatory. Axons from all three cell types establish synaptic connections with the already present Purkinje cells.

F. Experimental Modification of Normal Developmental Patterns

A number of external factors can influence the orderly proliferation and differentiation of cells in the nervous system. Among these are nutrition, hormonal
state, irradiation, and various chemicals and drugs. Few studies using $[H^3]$-
Thymidine have been carried out to determine how external factors affect cell
kinetics. Most studies have concentrated on changes in brain weight and brain
size or have measured total DNA content as an estimate of the number of cells
present.

Malnutrition in the early postnatal period curtails the normal increase in total
brain DNA and results in a reduced number of brain cells (Winick and Noble,
1966). At this time reduction was most pronounced in those areas of the brain
where cell proliferation is more active, such as in the external granular layer of the
cerebellum (Fish and Winick, 1969). Thus, rapidly proliferating regions of the
brain are affected more by malnutrition than areas where cell division is quiescent.

Thyroid hormone has long feen known to influence brain development. Animals that are made hyperthyroid or are hypothyroid at birth have reduced adult
brain weights (Eayrs and Taylor, 1951; Balaz et al., 1971). In an autoradiographic study of cell proliferation and differentiation in the rat cerebellar cortex,
Nicholson and Altman (1972) found that hypothyroidism prolongs proliferative
activity leading to greater final cell numbers in the adult. However, differentiation
was retarded. On the other hand, hyperthyroidism caused a premature cessation
of proliferation accompanied by earlier differentiation. Balaz and Cotterrell
(1972) found that cortisol treatment caused a decrease in proliferation in both the
cerebrum and cerebellum, resulting in a final deficit in cell number.

Altman has investigated the effects of X-irradiation on the external germinal
zone of the developing rat cerebellar cortex. The cells of the external germinal
layer can be almost totally eliminated in the newborn cerebellum by low level X-
irradiation. At these low levels, however, some cells remain undamaged and from
these the external germinal layer can be reconstituted. However, if the brain is
repeatedly irradiated on a daily basis, the external zone remains incomplete and a
cerebellar cortex may result which is devoid of all interneurous (Altman and
Anderson, 1972). Altman has speculated that given the knowledge of the sequential differentiation of the cerebellar interneurons (basket, stellate, and granule cells), and having the ability to control cell death and subsequent regrowth of
the external germinal zone, it should be possible not only to cause development of
a cerebellar cortex lacking in all interneurons, but also one devoid of a selected
type(s) of interneuron. However, there is no evidence to date that a cerebellar
cortex lacking a selected type(s) of neuron can be produced.

Similar studies of damage and repair of the cerebellar external granular layer
have been reported after treatment prenatally or postnatally with methylazoxy-
methanol (Shimada and Langman, 1970a) or postnatally with 5-fluorodeoxyu-
ridine (Shimada and Langman, 1970b). Although both types of treatment resulted in extensive destruction of the external granular layer, a few cells persisted
and from these a new layer was reconstituted. Prenatal reconstitution of the
external granular layer allowed subsequent normal development of the cerebellar

cortex, while postnatal treatment resulted in less complete recovery, with many cells failing to reach their final destination.

It would appear, then, that normal progressive development of the cerebellar cortex depends upon an intact proliferating cellular layer and that after irradiation or chemical injury, differentiation of the various interneurons can proceed only after the external granular layer has been reconstituted to a certain cell density. In summary, when germinal cell populations are damaged, further differentiation of neurons is impeded, either because the cells were never formed at all or because a terminal round of DNA synthesis was lacking before differentiation.

II. Tumor Cell Lines in Vitro

A. Neuroblastoma Cell Lines

Another approach to dissecting the varied but interrelated functions of the nervous system has been the development and analysis of biochemically differentiated clonal cell lines derived from tumors. Among the most thoroughly studied neoplasms is the murine neuroblastoma C 1300 (AUGUSTI-TOCCO and SATO, 1969; SCHUBERT et al., 1969; KLEBE and RUDDLE, 1969). This spontaneously occurring tumor was originally discovered in the abdominal cavity of an A/J mouse in 1940, and has been carried in vivo (where it grows as a relatively undifferentiated round cell tumor) through many passages.

Interest in this tumor has centered around the expression of a catalogue of biochemical and physiological functions exhibited by numerous clonal lines growing in cell culture. The expression of several of these functions is under experimental control.

When round C 1300 cells, either from the tumor or from suspension culture, are plated onto tissue culture dishes (similar to bacteriological petri dishes except for a chemical modification of the surface involving sulfonation to enhance sticking) processes up to 50 microns long are sent out within a few hours, and the cells become stainable with the Bodian silver stain. Microtubule protein has been isolated from neuroblastoma cells (OLMSTED et al., 1970) and may be involved in the maintenance of these neurites.

The cells have been shown to contain several enzymes involved in neurotransmitter synthesis and degradation, particularly choline acetyl transferase, tyrosine hydroxylase, and acetylcholinesterase (WILSON et al., 1972). AMANO et al. (1972) isolated from the parent tumor a great variety of clones distinguished by the relative activities of these enzymes. They can be grouped into three principle classes: cholinergic clones with high choline acetyl transferase but low tyrosine hydroxylase activities; adrenergic clones with the reverse enzyme activity levels; and clones with low or missing activities of both enzymes.

Many neuroblastoma clones have electrically excitable membranes (NELSON et al., 1969). While some clones are spontaneously active, others must be stimulated by current-passing electrodes or iontophoretic application of acetylcholine (HARRIS and DENNIS, 1970; NELSON et al., 1971). Although functional synapses have not been reported to develop between these neuroblastoma cells (electrical coupling has been reported), the altered acetylcholine sensitivity on the surface of cloned rat muscle cells when co-cultured with neuroblastoma suggests a rather specific trophic interaction (HARRIS and DENNIS, 1970; HARRIS et al., 1971).

Neuroblastoma cells have four glycosphingolipids considered to be characteristic of neural tissue, namely Gm_2, Gm_1, G_{D2}, G_{DIA} (DAWSON et al., 1971; YOGEESWARAN et al., 1973). At least one clonal line has low levels of Thy-1 surface antigen (SCHACHNER, 1973). Surface glycopeptides are different in clones that do not produce processes from clones that do have processes (GLICK et al., 1973), or which have process formation stimulated by the addition of bromodeoxyuridine (BROWN, 1971); at least one of the glycoproteins unique to cells with neurites appears by immunological criteria to be nervous system specific (AKESON and HERSCHMAN, 1974).

Other studies on these cell lines have emphasized RNA metabolism (AUGUSTI-TOCCO et al., 1973), 3', 5'-cyclic-monophosphate metabolism (GILMAN and NIRENBERG, 1971), and expression of neuronal properties in cell hybrids (MINNA et al., 1971).

Of special interest to our present study is the relationship between the progression of neuroblastoma cells through the cell cycle and the expression of one or more differentiated functions. Data from experiments performed with these cells have often been analyzed relative to a possible inverse relationship proliferation and "differentiation" (neurite outgrowth or expression of enzyme activity (Table 1). In studies such as these, two potential dangers should be carefully considered. First, are data that are derived from established tumor cell lines reliable? That is, do the defects in these cells extend beyond loss of proliferative controls to defective states of differentiation and their control? Secondly, do neuroblastoma cells actually differentiate in culture, or are they already end cells of a sort that can be induced to express, to changing degrees, various organ-specific functions?

Alternative answers to these questions have been well discussed by SCHUBERT et al. (1973); their conclusions are summarized here. In general, the fully differentiated, normal end cell is apparently incapable of further cell division (GROBSTEIN, 1955). Since the C1300 neuroblastoma cells are clearly not in a state of irreversible specialized gene expression, differentiation may be an unwise choice of description for neurite outgrowth and increases in specific activity of certain enzymes. At best, a state of transiently increased "morphological and biochemical complexity" exists. WEISS (1939) previously called such temporary changes in organ-specific cell properties "modulation"; such changes are dependent upon the continued presence of external stimulation (the one exception in Table 1 may be the long-term treatment with prostaglandin PGE_1). Contrary to this view, however, is the recognition of situations in which highly differentiated cells are, in fact, capable of returning to a proliferative state.

In addition, SCHUBERT et al. (1973) suggested that differentiation can be defined as the progression of a majority of the cells in a population from a state of relative simplicity, toward a specific goal of relative complexity, but not necessarily reaching "that final stage of ultimate functional complexity and specialization". Thus, C1300 neuroblastoma cells may continue to proliferate, continuously extending and retracting processes, for lack of the appropriate target cell with

Table 1. Summary of treatments of neuroblastoma C1300 clones in tissue culture and their effects on neurite outgrowth, certain enzymatic activities, and cell division

Agent	Effect of agent on			Ref.
	Neurite outgrowth	Enzyme and other activities	Cell division	
Dibutryl-3',5'-cyclic AMP, 1 mM	Induces neurite formation within 6 hours in some clones. Outgrowth complete by 24 hrs, and irreversible by 3 days. Blocked by cycloheximide (but not actinomycin D) and vinblastine sulfate.	Acetylcholinesterase activity markedly increased. Tyrosine hydroxylase: 39-fold increase at 0.5 mM dibutryl-3',5'-cyclic AMP. Catechol-0-methyl-transferase: no effect	Stopped by 2nd day. Effect irreversible by 3rd day.	(1–5)
Prostaglandin PGE$_1$	Substantial increase in neurite outgrowth by day 1; optimal by day 2. Causes much firmer cell attachment to dish. Effect irreversible after 3 days of treatment. Blocked by cycloheximide (but not actinomycin D) and vinblastine sulfate.	3',5'-cyclic AMP levels increased 3-fold after only 30 sec., 7-fold after 15 min. Effect potentiated by theophylline. Catechol-0-methyl transferase: no effect.	Reduced doubling time after 1 days; cell division stopped by 3rd day. Only slightly reversible after 3 days of treatment. Growth slowed when theophylline present.	(5, 6)
PGE$_2$	As effective as PGE$_1$ at 10 µg/ml		No data	
PGF$_2$α	No effect		No data	
Phosphodiesterase inhibitors papaverine	Maximum outgrowth after 24 hrs of treatment. Blocked by cycloheximide (but not actinomycin D) and vinblastine sulfate (which causes a reversal of previosly stimulated cells). Stimulation, but not maintenance blocked by cytochalasin B.	Tyrosine hydrolases enhanced levels	Maximum inhibition only after 2 days of treatment.	(3, 7)
theophylline	Small effect		No effect	(1)

Table 1 (continued)

Agent	Effect of agent on Neurite outgrowth	Enzyme and other activities	Cell division	Ref.
Sodium butyrate (0.5–1mM)	Not promoted	Tyrosine hydroxylase: 18-fold increase Catechol-o-methyl transferase: Increased 3-fold	Inhibited	(1, 3)
3′, 5′-cyclic AMP (0.5—1 mM)	Not promoted, or much less than by 3′, 5′-dibutyryl cyclic AMP	Acetylcholinsterase: increased activity	Inhibited	(1, 4)
5′-AMP (0.5 nM)	Not promoted		Inhibited	(2)
3′, 5′-cyclic GMP (0.5—1 mM)	Not promoted		Inhibited	(2)
5-Bromodeoxyuridine (BudR)	Neurite outgrowth promoted. Blocked by cycloheximide (but not actinomycin D), colchicine, and vinblastine sulfate. Stimulation occurs in virtual absence of DNA synthesis. Stimulation blocked by thymidine and deoxycytidine.	New surface glycopeptide appears.	DNA synthesis and cell division not effected	(8, 9)
Acetylcholine, 5 mM	Neurite outgrowth promoted	Acetylcholinesterase: 37-fold stimulation	No effect	(1)
Conditioned medium	Variable outgrowth promoted, depending on cells used to condition medium; glial cells somewhat better than others. Responsible compound dialyzable.		No effect	(11,12)
Lactic 0.003—0.01 M	Some promotion of neurite extention		Little or no effect	(11)

Solid substrate suspension versus monolayer culture or Petri dishes versus tissue culture dishes in serum-free medium)	Clone dependent promotion of neurite outgrowth. Petri dish surfaces promote neither firm cell attachments nor neurite outgrowth. Puromycin, cycloheximide, colchicine, and vinblastine sulfate block outgrowth. Actinomycin D at high concentrations (0.5—5 µg/ml) stimulates both attachment and neurite outgrowth in Petri dishes; but does not effect outgrowth in tissue culture dishes. Blocked by cycloheximide.	Acetycholinesterase: Specific activity increases. Microtubule formation. New nervous-system specific surface antigen(s) formed.	Cell division not inhibited directly by substrate shift, but inhibited due to absence of serum.	(11, 13–16)
High cell density at high (10%) serum concentrations	Some neurite outgrowth occurs when cell density is sufficiently high. Some dialyzable compound secreted by cells into medium may be involved (see "Lactic Acid Induction", this table).		May or may not be affected, depending upon cell density.	(11)
Low serum or no serum medium	Neurite outgrowth promoted. Not blocked by cycloheximide or actinomycin D, but blocked by vinblastin sulfate and colchicine. Blocking factor non-dializable. Mimiced by several serum proteins, including fetuin, and bovine Cohn fractions II plus III, and IV.	Acetylcholinesterase: Two step increase in specific activity. The first may be coupled to cell division inhibition, while the second may well be due to cell aging and death (21). Stimulation blocked by both actinomycin D and cyclohexamide (20). Tyrosine hydroxylase: No effect. Catechol-o-methyl transferase: No effect. Choline acetylase: No effect. Neurotubule protein: No effect. Pre-existing neurotubule protein presumed available for polymerization.	Inhibited markedly, but not necessarily in proportion and time sequence to neurite outgrowth.	(11, 17–20)

Table 1 (continued)

Agent	Effect of agent on Neurite outgrowth	Enzyme and other activities	Cell division	Ref.
X-ray	Some production of neurite outgrowth (increase from 8% to 27—39% of cells with neurites). Blocked by vinblastine sulfate.	Acetylcholinesterase: marked increase in specific activity. Tyrosine hydroxylase: No effect. Catechol-o-methyl transferase: 3-fold stimulation starting at 2 days after irradiation becoming maximal at 3 days. Cycloheximide and actinomycin D blocked this stimulation.	Inhibited	(3–5, 21)
6-Thioguanine	Promoted.			(22)
Inhibitors of DNA Synthesis cytosine arabinoside, FUdR	Promoted (20) or not promoted (12)	Acetylcholinesterase: 5-fold stimulation	Inhibited	(11, 19)
Mitomycin C thymidine	Did not promote neurite outgrowth in 10% serum, nor did they block neurite outgrowth in the absence of serum.			
Dopamine and 6-hydroxy-dopamine	Not promoted		Inhibited (in contrast) to cell lines CHO and BHK.	(23)
DL-glyceraldehyde	Not promoted		Inhibited	(6)

References for Table 1
1. FURMANSKI, et al., 1971.
2. PRASAD and HSIE, 1971.
3. PRASAD et al., 1972.
4. PRASAD and VERNADAKIS, 1972.
5. PRASAD and MANDEL, 1972.
6. PRASAD, 1972.
7. PRASAD and SHEPPARD, 1972.
8. SCHUBERT and JACOB, 1970.
9. BROWN, 1971.
10. HARKINS et al., 1972.
11. SCHUBERT et al., 1971a.
12. MONARD et al., 1973.
13. SCHUBERT et al., 1969.
14. AUGUSTI-TOCCO and SATO, 1969.
15. OLMSTED et al., 1970.
16. AKESON and HERSCHMANN, 1974.
17. SEEDS et al., 1970.
18. BLUME et al., 1970.
19. KATES et al., 1971.
20. SCHUBERT et al., 1971b.
21. PRASAD, 1971a.
22. PRASAD, GILMER, and KUMAR, 1973.
23. PRASAD, 1971b.

which to make a stable synaptic connection. In the absence of this end-stage specilization, the cells are unable to become stabilized with respect to division.

Most clones derived from C 1300 neuroblastoma can be induced to send out processes by a variety of agents; stimulation of organ-specific enzymatic activity often occurs as well. Can any definitive conclusion be reached from these studies regarding the relationship between cell cycle progression (proliferation) and the ability of these cells to "differentiate" (or perhaps more properly, "modulate" a differentiated function)? A summary of the types of agents used and the results obtained are presented in Table 1.

The general conclusion to be drawn from this body of results is that the cessation of cell division is not the key factor in the initiation of events leading to neurite outgrowth. When inhibition of cell division occurs along with neurite outgrowth upon administration of some agent, such inhibition generally occurs more slowly than does neurite outgrowth. In other cases, neurite outgrowth occurs in the absence of any demonstrable effect on cell division. It is important to note that this conclusion does not assert that neurite formation may not be antagonized to some extent by cell proliferation, particularly mitosis. In addition, SCHUBERT et al. (1971a) have provided convincing evidence that a particular number of cell divisions is not required in order for cells to produce neurites. Cells plated at varying initial densities did not send out neurites after a specific number of cell divisions, but rather when a characteristic population density was attained.

The factor that seems to play an important role in neurite outgrowth is the ability of the cells to attach to a solid substrate. Thus tissue culture dishes, which are treated to enhance cell adhesion, promote neurite outgrowth, while untreated Petri dishes support neither firm attachment nor neurite outgrowth [an exception appears to exist in the effect of relatively high concentrations of actinomycin D on cell attachment and neurite outgrowth in Petri dishes (SCHUBERT et al., 1971a), a phenomena marginally reminiscent of "superinduction" in certain hormone systems (TOMKINS et al., 1969)].

On the strength of these substrate experiments, the effects of other agents promoting neurite outgrowth have also been postulated to act via an ability to increase the interaction between the cells and the substrate (e.g., BUdR, SCHUBERT et al., 1971a). Thus the real value of these neuroblastoma clones for understanding the expression of nervous system specific function may lie in providing a system for studying cell-substrate interactions leading to new phenotypes.

B. Glioma Cell Lines

Only one example of the study of the expression of a glial differentiated function *in vitro* will be discussed here. PFEIFFER et al. (1970) showed that the regulation of an acidic protein unique to the nervous system ("S 100-protein"; MOORE and McGREGOR, 1965) in a rat glial line C_6 (BENDA et al., 1968) resided at least in part in interactions involving the cell surface and density-dependent inhibition of cell proliferation (STOKER and RUBIN, 1967). The evidence for this included: (1) the temporal relationship between the onset of "S 100-protein" accumulation and the beginning of the stationary phase of cell growth; (2) the requirement for homologous cell contact; (3) the absence of accumulation of "S 100-protein" in cells metabolically blocked in log phase; and, (4) the loss of the ability of the cells to accumulate "S 100-protein" when adapted to suspension culture, a change paralleled by the nearly complete loss of certain surface antigenicity (PFEIFFER et al., 1971); and (5) the return of the ability to accumulate this protein in monolayer cultures only after the cells have regained the ability to attain a high cell density stationary phase—a change paralled by the qualitative return of the surface antigenicity characteristic of monolayer cells.

Closer examination of the accumulation of "S 100-protein" during stationary phase revealed that in fact the so-called stationary phase cells continued to proliferate with a doubling time of 8 days (in contrast to the 18 hr doubling time of logarithmic phase cultures) until the cells reached a density fourfold greater than the initial stationary phase density. At that point, cell proliferation ceased and a gradual loss of cells from the monolayer began. Coincident with the complete cessation of proliferation, "S 100-protein" accumulation stopped as well. When cell density eventually returned to early stationary phase levels, due to cell loss, both cell proliferation and "S 100-protein" accumulation began anew (PFEIFFER et al., unpublished results). Further, following subculture of "inactive" long-term stationary phase cultures and their subsequent proliferation to a stationary phase density, these cells again accumulated "S 100-protein".

These latter data suggest that cell proliferation may be required not only prior to the initial differentiation (see, for example, TURKINGTON, 1968; HOLTZER et al., 1972), but also for the continued expression of differentiated function (CHACKO et al., 1969). Such a requirement could be due either to the renewal of some metabolic component depleted during extended stationary phase conditions or, alternatively, to the prevention of a gradual accumulation of cells in a non-functional stage of the cell cycle.

III. Conclusions
A. General

Cell differentiation can be envisioned as a two-step process. In the first step a stem cell A is changed to a cell B which is "determined", i.e. committed to a specific program of gene expression, but this program might not be expressed. Thus the cell has the potential to be an astrocyte, for example; but for some reason it does not necessarily realize that potential in a phenotypic sense. In the second step, cell type B is induced, either internally or externally, to actually express its organ specific function(s), thus becoming cell type C. Generally, the transition A→B is irreversible, while the transition B→C may or may not be irreversible.

The transition from A→B can be considered true development, while the second transition B→C represents a modulation or further expression of an inherent capacity. In the discussion above, the development of various precursors of neuronal and glial cell types from neural crest and germinal zones of the neural tube represents A→B transitions. In contrast, cellular migration, process formation, and the establishment of intercellular contacts among glia and neurons are examples of B→C type modulations of gene expression. Similarly, the induced extension of neurites in neuroblastoma cultures, and the stationary phase accumulation of "S100-protein" in C_6 glial cultures also represent B→C transitions. Note that the term gene expression is used in a very broad sense here, to include all possible controls on gene expression—from gene transcription to required post-translational modifications. In so far as they may have common elements, the study of both is relevant to an understanding of development.

B. Cell Division and Differentiation

1. Is there a requirement for cell division prior to differention? Transitions in phenotype of the A→B type are usually accompanied by cell proliferation. Central to the theme of the discussions in this volume is the question of whether or not cell division is an obligatory process in developmental transitions. In other words, do changes in regulatory processes controlling the expression of organ-specific events on the one hand, and those controlling cell cycle progression on the other, generally coincide only by chance, or do they coincide for mechanistic reasons (see RUTTER et al., 1973)?

Numerous hypotheses can be advanced to explain why a particular stable transition might require cell replication. For example: (a) the loss of a cytoplasmic gene coding for a negative control product could be caused by unequal daughter cell formation, thus allowing the expression of the target nuclear gene (the loss of a cytoplasmic gene is favored because of the evidence for the presence of a complete genome in differentiated somatic cell nuclei; e.g. GURDON and LASKEY, 1970); (b) a repressor, no longer synthesized, could be diluted out by cell division; (c) the concentration of a stable activator, whose synthesis was limited to a particular part of the cell cycle, could be increased to some threshold level by repeated passage of the cells through that phase of the cell cycle (a phase different from one in which resting cells were blocked); (d) differential gene expression could depend on a cellular response to an environmental signal operating via a surface receptor whose binding or cell recognition ability (DELONG, 1970; SIDMAN, 1970) varied as a function of the phase of the cell cycle. In each of these examples, the daughter cells differ in some fundamental and presumably stable way from that of the mother cell, thus fulfilling at least part of the criteria proposed by HOLTZER and co-workers (HOLTZER et al., 1972) for cells having traversed a so-called "quantal" cell cycle. As indicated above, if any of these mechanisms in fact operate, they are expected to do so for an A→B transition, thus allowing actual biochemical expression or further modulation to occur independently of any particular cell cycle, or frequently, as in neurons, in the absence of any cell cycling at all. Specifically, the induction of neurites and certain enzymatic activities in C1300 neuroblastoma is not "quantal" since any cell cycle can be induced; and cell density and surface adhesion appear to be more important than the extent of cell proliferation (i.e., the number of cell cycles) or lack of it. In the case of neurogenesis *in vivo*, sufficient evidence is not available to either substantiate or discard the concept of special cell cycles in which crucial developmental events occur.

2. Are cell proliferation and expression of differentiated function incompatable? Since many differentiated cells are non-proliferative or at least nearly so, Grobstein (1955) suggested that a basic incompatibility may exist between cell proliferation and differentiation. In the extreme form this "dictum" is not true, since cell types do exist that express their organ-specific function while proliferating (e.g., erythroid and cartilage cells; Campbell et al., 1971; Coon, 1966). In only a few systems is there a complete loss of proliferation upon final differentiation (e.g., neurogenesis, Sidman, 1970; and myogenesis, Stockdale and Holtzer, 1961).

Specifically for one type of cell discussed in this chapter, C 1300 neuroblastoma cells are able to proliferate and express organ-specific functions at the same time. On the other hand, rat C_6 glioma cells in culture appear to require stationary phase conditions to express their ability to produce "S 100 protein" (but see below).

3. Is some cell proliferation required to *maintain* a differentiated state? There are at least a couple of examples in which it appears that either a very slow rate of cell cycle progression, or an occasional round of proliferation at normal rate, is required to maintain the continued expression of a differentiated function. One such example was discussed above for the C_6 glioma cells, in which a very slow (8-day) doubling time apparently must be maintained for the continued accumulation of "S 100 protein". As suggested earlier, this could be due to the necessity to avoid a gradual accumulation of cells in an inactive part of the cell cycle, or to allow some proliferation-dependent repair. At present we know of no evidence for proliferation-dependent maintenance of organ-specific function in the normal nervous system; indeed, many of the cells in the adult nervous system appear incapable of proliferation.

At present, it appears that the available data relevant to neurogenesis do not permit serious analysis of the detailed interaction between cell cycle progression and differentiation. However, techniques that are now available, as well as the development of new techniques, promise to make the nervous system, in spite of its complexity, an ideal source of material for further exploration of this question.

Acknowledgements

Supported in part by grants from the American Cancer Society No. VC 124-B and the National Institutes of Health No. NS 10861-03.

References

Akeson, R., Herschman, H. R.: Modulation of cell surface antigens of a murine neuroblastoma. Proc. Natl. Acad. Sci. **71**, 187–191 (1974).

Altman, J.: Autoradiographic and histological studies of postnatal neurogenesis. III. Dating the time of production and onset of differentiation of cerebellar microneurons. J. Comp. Neurol. **136**, 269–294 (1969).

ALTMAN, J., ANDERSON, W. J.: Experimental reorganization of the cerebellar cortex. I. Morphological effects of elimination of all microneurons with prolonged X-irradiation started at birth. J. Comp. Neurol. **146**, 355–406 (1972).

AMANO, T., RICHELSON, E., NIRENBERG, M.: Neurotransmitter synthesis by neuroblastoma clones. Proc. Natl. Acad. Sci. **69**, 258–263 (1972).

ANGEVINE, JR., J. B.: Time of neuron origin in the hippocampal region: an autoradiographic study in the mouse. Exp. Neurol. Suppl. **2**, 1–70 (1965).

ANGEVINE, JR., J. B.: Time of neuron origin in the diencephalon of the mouse. J. Comp. Neurol. **139**, 129–187 (1970a).

ANGEVINE, JR., J. B.: Critial cellular events in the shaping of neural centers. In: The Neurosciences—Second Study Program (Ed. F. O. SCHMITT), 62–72. New York: Rockefeller Univ. Press 1970b.

ANGEVINE, JR., J. B., BODIAN, D., COULOMBRE, A. J., EDDS, JR., M. V., HAMBURGER, V., JACOBSON, M., LYSER, K. M., PRESTIGE, M. C., SIDMAN, R. L., VARON, S., WEISS, P.: Embryonic vertebrate central nervous system: revised terminology. Anat. Rec. **166**, 257–262 (1970).

AUGUSTI-TOCCO, G., CASOLA, L., PARISI, E., ROMANO, M., ZUCCO, F.: Biochemical characterization of a clonal line of neuroblastoma. In: SATO, G. (Ed.): Tissue Culture of the Nervous System. New York: Plenum Press 1973.

AUGUSTI-TOCCO, G., SATO, G.: Establishment of functional clonal lines of neurons from mouse neuroblastoma. Proc. Natl. Acad. Sci. **64**, 311–315 (1969).

BALAZ, R., COTTERRELL, M.: Effect of hormonal state on cell number and functional maturation of the brain. Nature **236**, 348–350 (1972).

BALAZ, R., KOVACS, S., COCKS, W. A., JOHNSON, A. L., EAYRS, J. T.: Effect of thyroid hormone on the biochemical maturation of rat brain: postnatal cell formation. Brain Res. **25**, 555–570 (1971).

BENDA, P., LIGHTBODY, J., SATO, G. H., LEVINE, L., SWEET, W.: Differentiated rat glial cell strain in tissue culture. Science **161**, 370–371 (1968).

BLUME, A., GILBERT, F., WILSON, S., FARBER, J., ROSENBERG, R., NIRENBERG, M.: Regulation of acetylcholinesterase in neuroblastoma cells. Proc. Natl. Acad. Sci. **67**, 786–792 (1970).

BROWN, J. C.: Surface glycoprotein characteristic of the differentiated state of neuroblastoma C-1300 cells. Exptl. Cell Res. **69**, 440–442 (1971).

CAMPBELL, G. LeM., WEINTRAUB, H., MAYALL, B. H., HOLTZER, H.: Primitive Erythropoiesis in early chick embryogenesis. II. Correlation between hemoglobin synthesis and mitotic history. J. Cell. Biol. **50**, 669–681 (1971).

CHACKO, S., ABBOTT, J., HOLTZER, S., HOLTZER, H.: The loss of phenotypic traits by differentiated cells. VI. Behavior of the progeny of a single chondrocyte. J. Exptl. Med. **130**, 417–442 (1969).

COON, H.: Clonal stability and phenotypic expression of chick cartilage cells in vitro. Proc. Natl. Acad. Sci. **55**, 66–73 (1966).

DAWSON, G., KEMP, S. F., STOOLMILLER, A. C., DORFMAN, A.: Biosynthesis of glycosphingolipids by mouse neuroblastoma (NB41A), rat glia (RGC-6) human glia (CHB4) in cell culture. Biochem. Biophys. Res. Commun. **44**, 687–694 (1971).

DE LONG, G. R.: Histogenesis of mouse isocortex and hippocampus in reaggregating cell cultures. Develop. Biol. **22**, 563–583 (1970).

DETWILER, S. R.: Observations upon the migration of neural crest cells, and upon the development of the spinal ganglia and vertebral arches in Amblystoma. Am. J. Anat. **61**, 63–94 (1937).

EAYRS, J. T., TAYLOR, S. H.: The effect of thyroid deficiency induced by methyl thiouracil on the maturation of the central nervous system. J. Anat. (Lond.) **85**, 350–358 (1951).

FISH, I., WINICK, M.: Effect of malnutrition on regional growth of the developing brain. Exp. Neurol. **25**, 534–540 (1969).

FUJITA, S.: Kinetics of cell proliferation. Exp. Cell Res. **28**, 52–60 (1962).

FUJITA, S.: Analysis of neuron differentiation in the central nervous system by tritiated thymidine autoradiography. J. Comp. Neurol. **122**, 311–328 (1964).

FURMANSKI, P., SILVERMAN, D. J., LUBIN, M.: Expression of differentiated functions in mouse neuroblastoma mediated by 3′, 5′-Cyclic AmP. Nature **233**, 413–415 (1971).

Gaze,R.M.: The formation of nervous connections. London-New York: Academic Press 1970.

Gilman,A.G., Nirenberg,M.: Regulation of adenosine 3', 5'-cyclic monophosphate metabolism in cultured neuroblastoma cells. Nature **234**, 356–358 (1971).

Glick,M., Kimhi,C.Y., Littauer,U.Z.: Glycopeptides from surface membranes of neuroblastoma cells. Proc. Natl. Acad. Sci. **70**, 1682–1687 (1973).

Grobstein,C.: Tissue disaggregation in relation to determination and stability of cell type. Ann. N. Y. Acad. Sci. **60**, 1095–1107 (1955).

Gurdon,J.B., Laskey,R.A.: The transplantation of nuclei from single cultured cells into enucleate frogs' eggs. J. Embryol. Exp. Morph. **24**, 227–248 (1970).

Harkins,J., Arsenault,M., Schlesinger,K., Kates,J.: Induction of neuronal functions: Acetylcholine-induced acetylcholinesterase activity in mouse neuroblastoma cells. Proc. Natl. Acad. Sci. **69**, 3161–3164 (1972).

Harris,A.J., Dennis,M.J.: Acetylcholine sensitivity and distribution on mouse neuroblastoma cells. Science **167**, 1253–1255 (1970).

Harris,A.J., Heinemann,S., Schubert,D., Tarikas,H.: Trophic interaction between cloned tissue culture lines of nerve and muscle. Nature **231**, 296–301 (1971).

Harrison,R.G.: Neuroblast versus sheath cell in the development of peripheral nerves. J. Comp. Neurol. **37**, 123–205 (1924).

Herman,C.J., Lapham,L.W.: Neuronal polyploidy and nuclear volumes in the cat central nervous system. Brain Res. **15**, 35–48 (1969).

Hinds,J.W.: Autoradiographic study of histogenesis in the mouse olfactory bulb. I. Time of origin of neurons and neuroglia. J. Comp. Neurol. **134**, 287–304 (1968).

His,W.: Die Neuroblasten und deren Entstehung im embryonalen Mark. Abhandl. Kgl. Sächs. Ges. Wissensch. Math. Phys. Kl. **15**, 313–372 (1889).

Holtzer,H., Weintraub,H., Mayne,B., Mochan,B.: The cell cycle, cell lineages, and cell differentiation. In: Moscona,A.A., Monroy,A. (Eds.): Current Topics in Developmental Biology, pp.229–256. New York: Academic Press 1972.

Jacobson,M.: Developmental neurobiology. New York: Holt, Rinehart and Winston 1970.

Jelinik,R.: Proliferace v centrálním nervovem kurecich zarodku. I. Doba travání mitosy v germinální zone michy od 2. do 6. dne żarodecného vývoje Csk. Morfologie **7**, 163–173 (1959).

Källen,B.: Studies on cell proliferation in the brain of chick embryos with special reference to the mesencephalon. Z. Anat. Entwicklungsgeschichte **122**, 388–401 (1961).

Källen,B.: Mitotic patterning in the central nervous system of chick embryos studied by a colchicine method. Z. Anat. Entwicklungsgeschichte **123**, 309–319 (1962).

Kates,J.R., Winterton,R., Schlesinger,K.: Induction of acetylcholinesterase activity in mouse neuroblastoma tissue culture cells. Nature **229**, 345–347 (1971).

Kaufman,S.L.: Lengthening of the generation cycle during embryonic differentiation of the mouse neural tube. Exp. Cell Res. **49**, 420–424 (1968).

Kershman,J.: The medulloblast and the medulloblastoma; a study of human embryos. Arch. Neurol. Psychiat. (Chicago) **40**, 937–967 (1938).

Klebe,R.J., Ruddle,F.H.: Neuroblastoma: Cell culture analysis of a differentiating stem cell system. J. Cell Biol. **43**, 69 A (1969).

Korr,H., Schultze,B., Maurer,W.: Autoradiographic investigations of glial proliferation in the brain of adult mice. I. The DNA synthesis phase of neuroglia and endothelial cells. J. Comp. Neurol. **150**, 169–176 (1973).

Langman,J.: Medical embryology (Sec. Ed.). Baltimore: Williams and Wilkins 1969.

Langman,J., Guerrant,R., Freeman,B.: Behavior of neuroepithelial cells. J. Comp. Neurol. **127**, 399–412 (1966).

Lapham,L.W.: Tetraploid DNA content of Purkinje neurons of human cerebellar cortex. Science. **159**, 310–312 (1968).

Lentz,R.D., Lapham,L.W.: A quantitative cytochemical study of the DNA content of neurons of the rat cerebellar cortex. J. Neurochem. **16**, 379–384 (1969).

Mann,D.M.A., Yates,P.O.: Polyploidy in the human nervous system. J. Neurol. Sci. **18**, 183–196 (1973).

MIALE, I. L., SIDMAN, R. L.: An autoradiographic analysis of histogenesis in the mouse cerebellum. Exp. Neurol. **4**, 277–296 (1961).

MINNA, J., NELSON, P., PEACOCK, J., GLASER, D., NIRENBERG, M.: Genes for neuronal properties expressed in neuroblastoma X L cell hybrids. Proc. Natl. Acad. Sci. **68**, 234–239 (1971).

MONARD, D., SOLOMON, F., RENTSCH, M., GYSIN, R.: Glia-induced morphological differentiation in neuroblastoma cells. Proc. Natl. Acad. Sci. **70**, 1894–1897 (1973).

NELSON, P., PEACOCK, J. H., AMANO, T.: Responses of neuroblastoma cells to iontophoretically applied acetylcholine. J. Cell. Physiol. **77**, 353–362 (1971).

NELSON, P., RUFFNER, B. W., NIRENBERG, M.: Neuronal tumor cells with excitable membranes grown *in vitro*. Proc. Natl. Acad. Sci. **64**, 1004–1010 (1969).

NICHOLSON, J. L., ALTMAN, J.: The effects of early hypo- and hyperthyroidism on the development of rat cerebellar cortex. I. Cell proliferation and differentiation. Brain Res. **44**, 13–23 (1972).

OLMSTED, J. B., CARLSON, K., KLEBE, R., RUDDLE, F., ROSENBAUM, J.: Isolation of microtubule protein from cultured mouse neuroblastoma cells. Proc. Natl. Acad. Sci. **65**, 129–136 (1970).

PFEIFFER, S. E., HERSCHMAN, H. R., LIGHTBODY, J., SATO, G.: Synthesis by a clonal line of rat glial cells of a protein unique to the nervous system. J. Cell Physiol. **75**, 329–340 (1970).

PFEIFFER, S. E., HERSCHMAN, H. R., LIGHTBODY, J. E., SATO, G., LEVINE, L.: Modification of cell surface antigens as a function of culture conditions. J. Cell Physiol. **78**, 145–152 (1971).

PRASAD, K. N.: X-ray-induced morphological differentiation of mouse neuroblastoma cells *in vitro*. Nature **234**, 471–473 (1971a).

PRASAD, K. N.: Effect of dopamine and 6-hydroxydopamine on mouse neuroblastoma cells *in vitro*. Cancer Res. **31**, 1457–1460 (1971b).

PRASAD, K. N.: Morphological differentiation induced by prostaglandin in mouse neuroblastoma cells in culture. Nature New Biol. **236**, 49–52 (1972).

PRASAD, K. N., GILMER, K., KUMAR, S.: Morphologically "differentiated" mouse neuroblastoma cells induced by noncyclic AMP agents: Levels of cyclic AMP, nucleic acid and protein. Proc. Soc. Exp. Med. Biol. **143**, 1168–1171 (1973).

PRASAD, K. N., HSIE, A. W.: Morphologic differentiation of mouse neuroblastoma cells induced *in vitro* by dibutyryl adenosine 3′:5′ cyclic monophosphate. Nature **233**, 141–142 (1971).

PRASAD, K. N., MANDAL, B.: Catechol-o-methyl-transferase activity in dibutyryl cyclic AMP, prostaglandin and X-ray-induced differentiated neuroblastoma cell cultures. Exp. Cell Res. **74**, 532–534 (1972).

PRASAD, K. N., SHEPPARD, J. R.: Inhibitors of cyclic-nucleotide phosphodiesterase induced morphological differentiation of mouse neuroblastoma cell cultures. Exp. Cell Res. **73**, 436–440 (1972).

PRASAD, K. N., VERNADAKIS, A.: Morphological and biochemical study of X-ray and dibutyryl cyclic AMP-induced differentiated neuroblastoma cells. Exp. Cell Res. **70**, 27–32 (1972).

PRASAD, K. N., WAYMIRE, J. C., WEINER, N.: A further study on the morphology and biochemistry of X-ray and dibutyryl cyclic AMP-induced differentiated neuroblastoma cells in culture. Exp. Cell Res. **74**, 110–114 (1972).

RAMÓN Y CAJAL, S.: A propos de certains éléments bipolaires du cervelet avec quelques détails nouveaux sur l'évolution des fibres cérébelleuses. Int. Monatsschrift. Anat. Physiol. (Leipzig) **7**, 12–31 (1890).

RAMÓN Y CAJAL, S.: Histologie du Systeme Nerveux de l'Homme et des Vertébrés, 2 vols. (1909–1911). L. AZOULAY (trans.). Madrid: Reprinted by Instituto Ramón y Cajal del C.S.I.C. 1952–1955.

RUTTER, W. J., PICTET, R. L., MORRIS, P. W.: Toward molecular mechanism of developmental processes. Ann. Rev. Biochem. **42**, 601–646 (1973).

SAUER, F. C.: Mitosis in the neural tube. J. Comp. Neurol. **62**, 377–405 (1935).

SAUER, F. C.: The interkinetic migration of embryonic epithelial nuclei. J. Morphol. **60**, 1–11 (1936).

SAUER, M. E., CHUTTENDEN, A. C.: Deoxyribonucleic acid content of cell nuclei in the neural tube of the chick embryo: evidence for intermitotic migration of nuclei. Exp. Cell Res. **16**, 1–6 (1959).

SCHACHNER, M.: Serologically demonstrable cell surface specificities on mouse neuroblastoma C 1300. Nature New Biol. **243**, 117–119 (1973).

SCHAPER, A.: The earliest differentiation in the central nervous system of vertebrates. Science **5**, 430–431 (1897).

SCHUBERT, D., JACOB, F.: 5-Bromodeoxyuridine-induced differentiation of a neuroblastoma. Proc. Natl. Acad. Sci. **67**, 247–254 (1970).

SCHUBERT, D., HARRIS, A. J., HEINEMANN, S., KIDOKORO, Y., PATRICK, J., STEINBACH, J. H.: Differentiation and interaction of clonal lines of nerve and muscle. In: G. SATO (Ed.): Tissue Culture of the Nervous System. New York: Plenum Press 1973.

SCHUBERT, D., HUMPHREYS, S., BARONI, C., COHN, M.: *In vitro* differentiation of a mouse neuroblastoma. Proc. Natl. Acad. Sci. **64**, 316–323 (1969).

SCHUBERT, D., HUMPHREYS, S., VITRY, F. D., JACOB, F.: Induced differentiation of a neuroblastoma. Develop. Biol. **25**, 514–546 (1971 a).

SCHUBERT, D., TARIKAS, H., HARRIS, A. J., HEINEMANN, S.: Induction of acetylcholine esterase activity in a mouse neuroblastoma. Nature **233**, 79–80 (1971 b).

SEEDS, N. W., GILMAN, A. G., AMANO, T., NIRENBERG, M.: Regulation of axon formation by clonal lines of a neural tumor. Prod. Natl. Acad. Sci. **66**, 160–167 (1970).

SHIMADA, M., LANGMAN, J.: Repair of the external granular layer of the hamster cerebellum after prenatal and postnatal administration of methylazoxy methanol. Teratology **3**, 119–134 (1970 a).

SHIMADA, M., LANGMAN, J.: Repair of the external granular layer after postnatal treatment with 5-fluorodeoxyuridine. Am. J. Anat. **129**, 247–260 (1970 b).

SIDMAN, R. L.: Histogenesis of the mouse retina studied with tritiated thymidine. In: SMELSER, G. K. (Ed.): The Structure of the Eye, 487–505. New York: Academic Press 1961.

SIDMAN, R. L.: Autoradiographic methods and principles for study of the nervous system with Thymidine-H^3. In: W. J. H. NAUTA, S. O. EBBESON (Eds.): Contemporary Research Methods in Neuroanatomy, pp. 252–273. Berlin-Heidelberg-New York: Springer 1970 a.

SIDMAN, R. L.: Cell proliferation, migration, and interaction in the developing mammalian central nervous system. In: SCHMITT, F. O. (Ed.): The Neurosciences, pp. 100–107. New York: Rockefeller Univ. Press 1970 b.

SIDMAN, R. L., MIALE, I. L., FEDER, N.: Cell proliferation and migration in the primitive ependymal zone: an autoradiographic study of histogenesis in the nervous system. Exp. Neurol. **1**, 322–333 (1959).

STOCKDALE, F. E., HOLTZER, H.: DNA synthesis and myogenesis. Exp. Cell Res. **24**, 508–520 (1961).

STOKER, M., RUBIN, H.: Density dependent inhibition of cell growth in culture. Nature **215**, 171–172 (1967).

TABER-PIERCE, E.: Histogenesis of the dorsal and ventral cochlear nuclei in the mouse. An autoradiographic study. J. Comp. Neurol. **131**, 27–54 (1967).

TOMKINS, G. M., GELEHRTER, T., GRANNER, D., MARIN, D. W., SAMUELS, H., THOMPSON, E. B.: Control of specific gene expression in higher organisms. Science **166**, 1474–1478 (1969).

TURKINGTON, R. W.: Hormone-dependent differentiation of mammary gland *in vitro*. In: A. A. MOSCONA, A. MONROY (Eds.): Current Topics in Developmental Biology, 210–212. New York: Academic Press 1968.

v. WAECHTER, R., JAENSCH, B.: Generation times of the matrix cells during embryonic brain development: an autoradiographic study in rats. Brain Res. **46**, 235–250 (1972).

WATTERSON, R. L.: Structure and mitotic behavior of the early neural tube. In: DEHAAN, R. L., URSPRUNG, H. (Eds.): Organogenesis, pp. 129–159. New York: Holt, Rinehart und Winston 1965.

WATTERSON, R. L., VENEZIANO, P., BARTHA, A.: Absence of a true germinal zone in neural tubes of young chick embryos as demonstrated by the colchicine technique. Anat. Rec. **124**, 379 (1956).

WEISS, P.: Principles of development. New York: Holt 1939.

WILSON, D. B.: Chronological changes in the cell cycle of chick neuropithelial cells. J. Embryol. Exp. Morph. **29**, 745–751 (1973).

WILSON, S., SCHRIER, B. K., FARBER, J. L., THOMPSON, E. J., ROSENBERG, BLUME, A. J., NIREN-BERG, M. W.: Markers for gene expression in cultured cells from the nervous system. J. Biol. Chem. **247**, 3159–3169 (1972).

WINICK, M., NOBLE, A.: Cellular response in rat during malnutrition at various ages. J. Nutr. **89**, 300–306 (1966).

YNTEMA, C. L., HAMMOND, W. S.: The development of the autonomic nervous system. Biol. Rev. **22**, 344–359 (1947).

YOGEESWARAN, G., MURRAY, R. K., PEARSON, M. L., SNAWAL, B. D., MCMORRIS, F. A., RUD-DLE, F. H.: Glycosphingolipids of clonal lines of mouse neuroblastoma and neuroblastoma X L cell hybrids. J. Biol. Chem. **248**, 1231–1239 (1973).

The Cell Cycle and Cell Differentiation in the Drosophila Ovary

ROBERT C. KING

Department of Biological Sciences, Northwestern University, Evanston, IL, USA

I. Introduction

Females belonging to certain evolutionarily advanced groups of insects (such as the Diptera, Lepidoptera, and the Hymenoptera) are characterized by ovarioles that contain egg chambers in which the oocyte is one member of a cluster of interconnected cells (BROWN and KING, 1964; KING and AGGARWAL, 1965; CASSIDY and KING, 1972). The other cells of the cluster, the "nurse cells", grow and simultaneously transfer their cytoplasm to the oocyte by a system of intercellular canals. In *Drosophila melanogaster*, as in most Diptera belonging to the superfamily Muscoida, each egg chamber consists of an oocyte, its fifteen "nurse" cells, and a surrounding monolayer of follicle cells (see Fig. 1 B). The egg and its fifteen interconnected nurse cells are fourth generation descendants of a single cell, the germinal "cystoblast". The interconnected cells formed by the division of a cystoblast are called "cystocytes". It is within the apical portion of the ovariole (the germarium) that the consecutive divisions occur which produce the sixteen cell cluster, and it is here also that the cluster becomes enveloped by profollicle cells. However, it is within the more posterior part of the ovariole (the vitellarium) that the major growth of the egg chamber occurs (see Fig. 1 A). Here, under optimal conditions it takes only three days for the cytoplasmic volume of the oocyte to increase ninety thousand times (KING, 1970, his Fig. II-19).

In *Drosophila melanogaster* a considerable body of information is now available concerning the details of how reproduction is accomplished and regulated in the female, and many insights are given into the genetic control of oogenesis through studies of the aberrant forms of ovarian development characteristically seen in females possessing certain mutant genes that markedly influence their fertility (KING, 1970, his Table I-1). Of special interest are mutations which affect the earliest stages in the differentiation of the oocyte. As we shall see, the microscopically visible changes in chromosome behavior that signal the differentiation of the oocyte are only the terminal steps in a series of qualitative alterations in the synthetic activities that have taken place over a considerable period of time, and which involve several generations of cells.

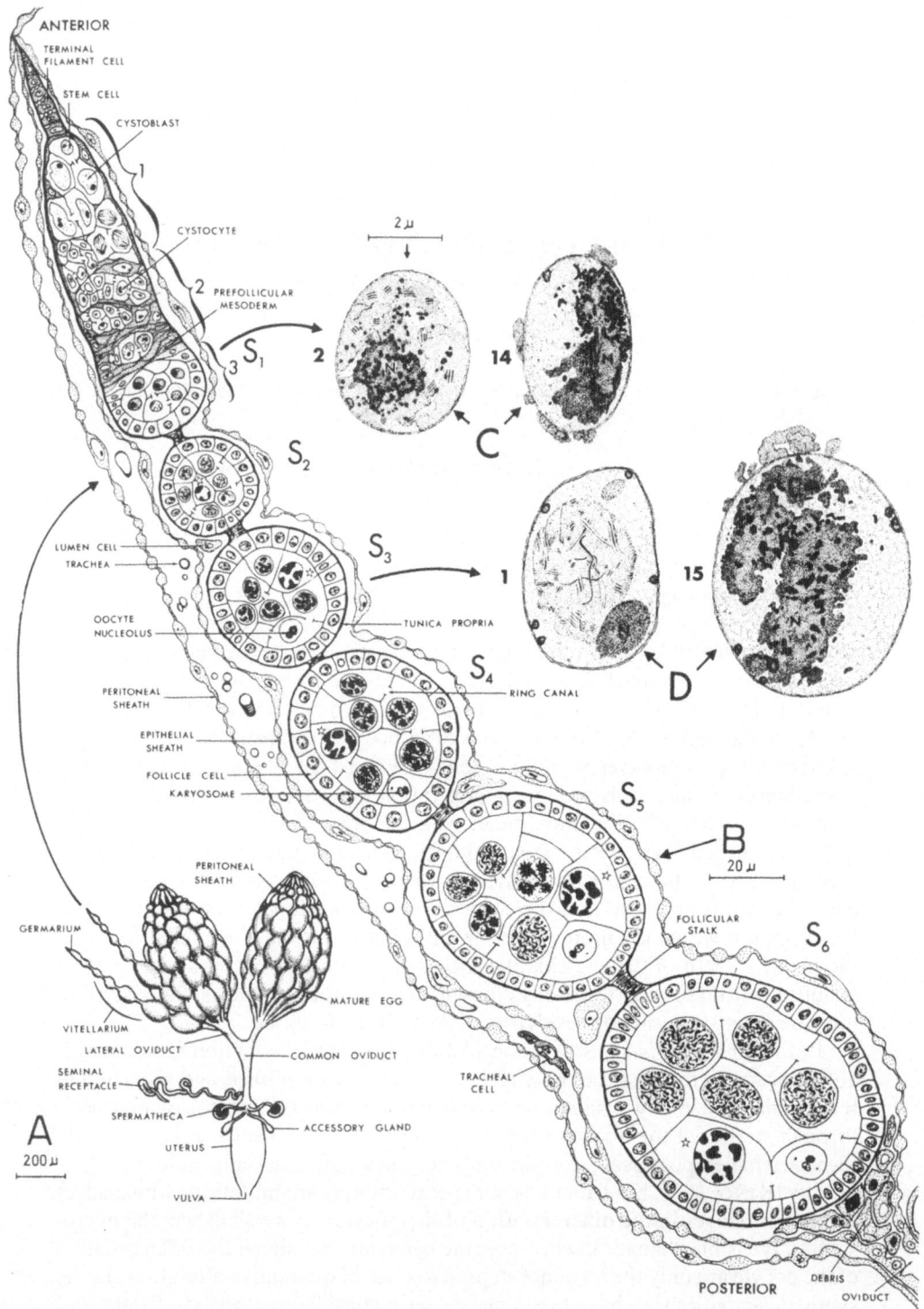

Fig. 1 A—D

II. The Production of Branching Chains of Cystocytes

BROWN and KING (1964) have demonstrated that the *Drosophila* oocyte and its fifteen nurse cells are joined by large canals, each of which is enclosed by a ring-shaped plasmalemmal thickening (see Fig. 2). They followed the pattern of interconnections after making serial reconstructions of sections viewed with the light microscope and found that the oocyte and one nurse cell were always connected to each other and to three other nurse cells as well. Two of the nurse cells were each connected to 3 other cells, 4 nurse cells were each connected to 2 cells, and 8 were connected to only one other cell (see Fig. 3, C_4). BROWN and KING suggested that the lineage of the cells could be determined by the pattern of their interconnections. Since the number of "ring canals" was equal to the number of spindles formed during the four consecutive divisions, we suggested that the interconnected cells arose as the result of incomplete cytokinesis.

KOCH and KING (1966) made an electron microscopic study of serially-sectioned germaria to identify cells in various developmental stages in the production of egg chambers. They found an apical, mitotically-active area in the germarium which contained in its most anterior region one or two single cells followed posteriorly by clusters of 2, 4, 8, and 16 interconnected cells. Furthermore, the volumes of the cystocytes in this developmental sequence varied in a characteristic fashion. The single cells were the largest and the cystocytes of 16-cell clusters were the smallest cells in the series. To explain these findings, we postulated that each germarium contained one or two stem line oogonia. Each divides, forming two daughter cells, and these separate. The apical one continues to behave as a stem cell; whereas the other cell functions as a cystoblast. The cystoblast and the cells derived from it do not double their birth sizes before entering the next division cycle, as does a stem cell, and the sister cystocytes do not separate, but form permanent, interconnecting ring canals. Therefore, as the divisions continue

---

Fig. 1. (A) A dorsal view of the internal reproductive system of an adult female *Drosophila melanogaster*. Two ovarioles have been pulled loose from the left ovary. Sperm are stored in the ventral receptacle (which is drawn uncoiled) and in the spermathecae. The uterus is drawn expanded, as it would be when it contains a mature egg. (B) A diagram of a single ovariole and its investing membranes. The nurse cell-oocyte complexes are representatives of the first 6 of the 14 stages of oogenesis. Within the vitellarium, sectioned egg chambers are drawn so as to show the morphology of the nuclei of the oocyte and 6 of the 15 nurse cells. The distribution of the nucleolar material is drawn in the starred nurse cell nucleus, whereas the other 5 nurse nuclei show the appearance of the chromosomal material. The distributions of both DNA and nucleolar RNA are shown for the oocyte and follicle cells. The oocyte nucleolus begins to fragment during stages 4 through 6. (C) Drawings illustrating the morphology of a pro-oocyte nucleus (left) and a nucleus in an adjacent pro-nurse cell (right) from germarial region 3. Synaptonemal complexes are present in the nuclei of 2 pro-oocytes. The nuclei of 14 pro-nurse cells lack synaptonemal complexes and contain more nucleolar material (N). A cloud of particles adheres to the surface of these nuclei. (D) The oocyte nucleus (left) and a nucleus from an adjacent nurse cell (right) from a stage-3 chamber in the vitellarium. The magnification is the same as in (C). Nucleolar material is far more abundant in each of the 15 nurse cell nuclei than in the oocyte nucleus. The synatptonemal complexes have congregated into a central area which under the light microscope appears as a Feulgen positive mass adjacent to the nucleolus. (From KOCH et al., 1967)

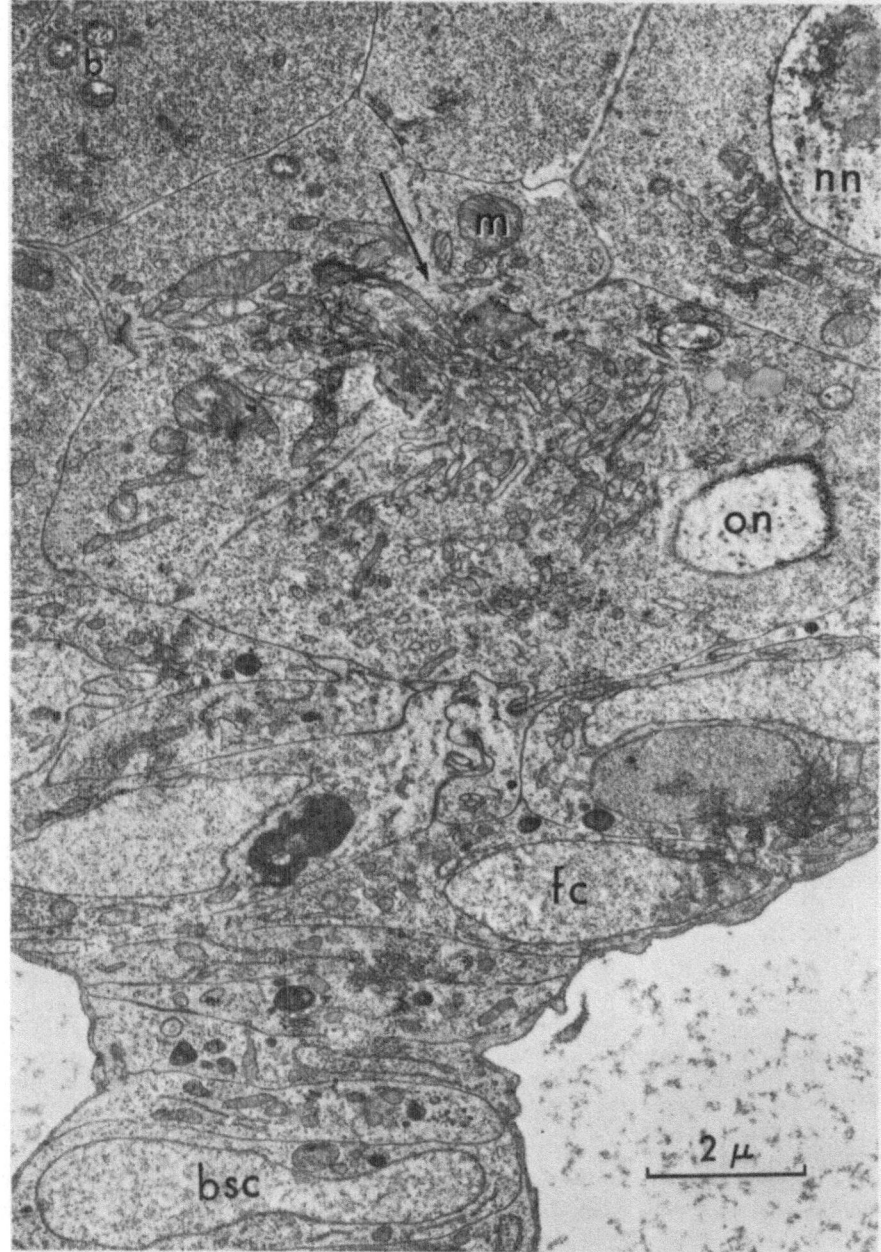

Fig. 2. An electron micrograph of the ring canal (see arrow) connecting the oocyte with an adjacent nurse cell in a young egg chamber. Note that mitochondria were passing through the canal at the time the chamber was fixed. (*b*) Symbiotic bacteria; (*bsc*) a nucleus of a basal stalk cell; (*fc*) follicle cell nucleus; (*m*) mitochondrion; (*on*) oocyte nucleus; (*nn*) nurse cell nucleus. (From KING et al., 1968)

these cystocytes become smaller and smaller and the ring canal system becomes more complex (see Fig. 3).

To explain the generation of branching chains of cystocytes we suggested (KOCH et al., 1967) that the cleavage program of the cystocytes was generated by the migratory behavior of their centrioles. If during the interval between mitoses, a mother and a newly formed daughter centriole always moved in opposite directions roughly one quarter the circumference of the cell and if the direction of movement was always perpendicular to the long axis of the previous spindle, then the cleavage furrows would always develop at right angles to the previous furrow. In a given division cycle this behavior would result in one sister cell receiving all of the previously formed canals, the other none. After three or more divisions the interconnected cystocytes would be arranged in branching chains (see Fig. 3; C_3, C_4).

III. The Hormonal Control of Oogonial Transformation

KING et al. (1968) have made an electron microscopic study of the metamorphosis of the *Drosophila* ovary. It is during the pupal period that the ovarioles differentiate. Two hours after puparium formation the pre-pupal ovary consists of an ellipsoidal cellular mass surrounded by an outer, acellular membrane (see Fig. 4 A). The differentiation of ovarioles has already begun, since terminal filaments are already evident. By the 24th hour after puparium formation the germarial region of the ovariole has differentiated (see Fig. 4 B). The cytological changes which occur between the second and 24th hour suggest that the large and small central cells are segregated into ovarioles by apical cells which multiply, grow and push basally until they reach the population of basal cells. Tunica propria is laid down at the interface between this advancing population of apically-derived, interstitial cells and the adjacent central cells. The result is a germarium, laterally enclosed in tunica propria, but naked basally. The more basally disposed germarial cells are smaller, and some are connected by ring canals (see arrows, Fig. 4 B). Since ring canals are formed between cells that are descendants of cystoblasts, it follows that between the second and 24th hour after puparium formation cystoblasts have formed, and the early cystocyte divisions are well underway in the 24 hrs ovary.

It is clear that oogonia divide during the larval period, and that by the time of puparium formation sufficient cells have been formed to provide each differentiating germarium with several oogonia. However, the cystocyte divisions do not begin until early in the pupal period. Thus the signal resulting in the transformation of oogonia into cystoblasts coincides with and presumably depends upon the same hormonal stimulus that triggers metamorphosis.

BODENSTEIN (1938) demonstrated by ligation experiments that puparium formation depends upon a hormonal factor liberated from the anterior part of the *Drosophila melanogaster* larva about 12 hrs before pupation. This timing was confirmed by BECKER (1962) who found that the pupation hormone is secreted 3.5 to 5 hrs before puparium formation. VOGT (1942) proved that secretions of the ring gland were responsible not only for puparium formation, but for various

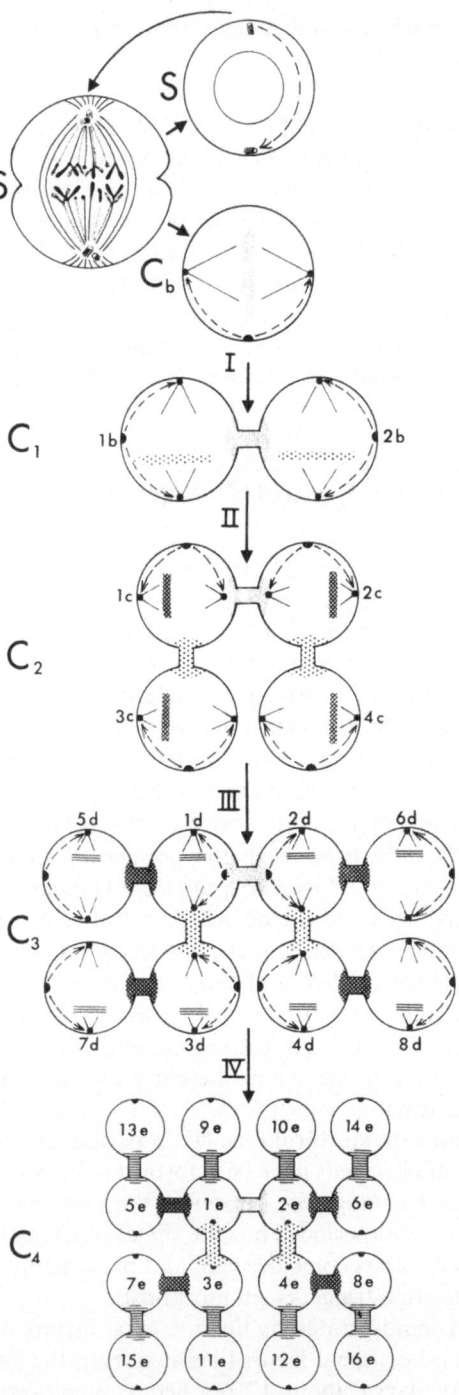

Fig. 3

metamorphic changes accompanying pupation, such as the breakdown of the larval fat body and the differentiation of ovarioles in the ovary and of the ommatidia in the optic imaginal disc.

In *Drosophila melanogaster* the ring gland lies above the brain, straddling the aorta (see Fig. 5A and B). The gland is a compound structure containing the prothoracic glands, the corpus allatum, and the corpus cardiacum. VOGT subdivided the ring gland into specific fragments which she then implanted into isolated larval abdomens. She was able to show (VOGT, 1943) that the prothoracic gland fragments alone produced the hormone that stimulated the development of the imaginal discs (during the larval period) and their differentiation (during the pupal period).

The hormone of the insect prothoracic gland, ecdysone, has been shown to be a derivative of cholesterol (KARLSON and HOFFMEISTER, 1963). *Drosophila* cannot synthesize cholesterol, and therefore it must be obtained in the diet (SANG and KING, 1961). In *Calliphora erythrocephala* ecdysone is secreted in the greatest quantities just before the last larval molt and throughout most of the prepupal period (SHAAYA and KARLSON, 1965). AGGARWAL and KING (1969) have made an electron microscopic study of the ring gland of *Drosophila melanogaster*, and they found that smooth surfaced endoplasmic reticulum (er) is the most common subcellular component in prothoracic gland cells. They therefore assigned to the smooth er the function of sequestering cholesterol and transforming it to ecdysone. They calculated that the volume of smooth er per prothoracic gland cell increases by 10 times during the 4 hour prepupal period. Late in metamorphosis the corpus allatum and corpus cardiacum move posteriorly and ventrally relative to the brain (contrast Fig. 5C and D), and the prothoracic gland degenerates.

When the *Drosophila* ovary is first subdivided into ovarioles, each newly formed germarium contains typical oogonia that undergo complete cleavage. The pupal transformation of oogonia to cystoblasts proceeds in a polarized fashion, extending anteriorly until all but a few apical oogonia are affected. If we assume that all oogonial cells are transformed during the pupal period, then whether such

---

Fig. 3. A diagrammatic model showing steps in the production of a cluster of 16 interconnected cystocytes. In this drawing the cells are represented by circles lying in a single plane, and the ring canals have been lengthened for clarity. The area of each circle is proportional to the volume of the cell. The stem cell (S) divides into 2 daughters; one behaves like its parent; the other differentiates into a cystoblast (C_b) which by a series of 4 divisions (I–IV) produces 16 interconnected cystocytes. (C_1) First, (C_2) second, (C_3) third, (C_4) fourth generation cystocytes. The upper and lower portion of the diagram represents the anterior and posterior portions of germarial region 1, respectively. The original stem cell is shown at early anaphase. Each parent-daughter pair of centrioles is attached to a part of the membrane by astral rays. The daughter centriole is drawn slightly smaller than its parent. The daughter stem cell receives one pair of centrioles. One remains in place while the other moves to the opposite pole. The movement is represented by a broken arrow. In the daughter cystoblast and all cystocytes, the initial position of the original centriole pair is represented by a solid half-circle; whereas their final positions are represented by solid circles. The position of each of the future cleavage furrows is drawn as a strip of defined texture. The canal derived from the contractile ring of the furrow is treated in a similar fashion. See text for further discussion. (From KOCH et al., 1967)

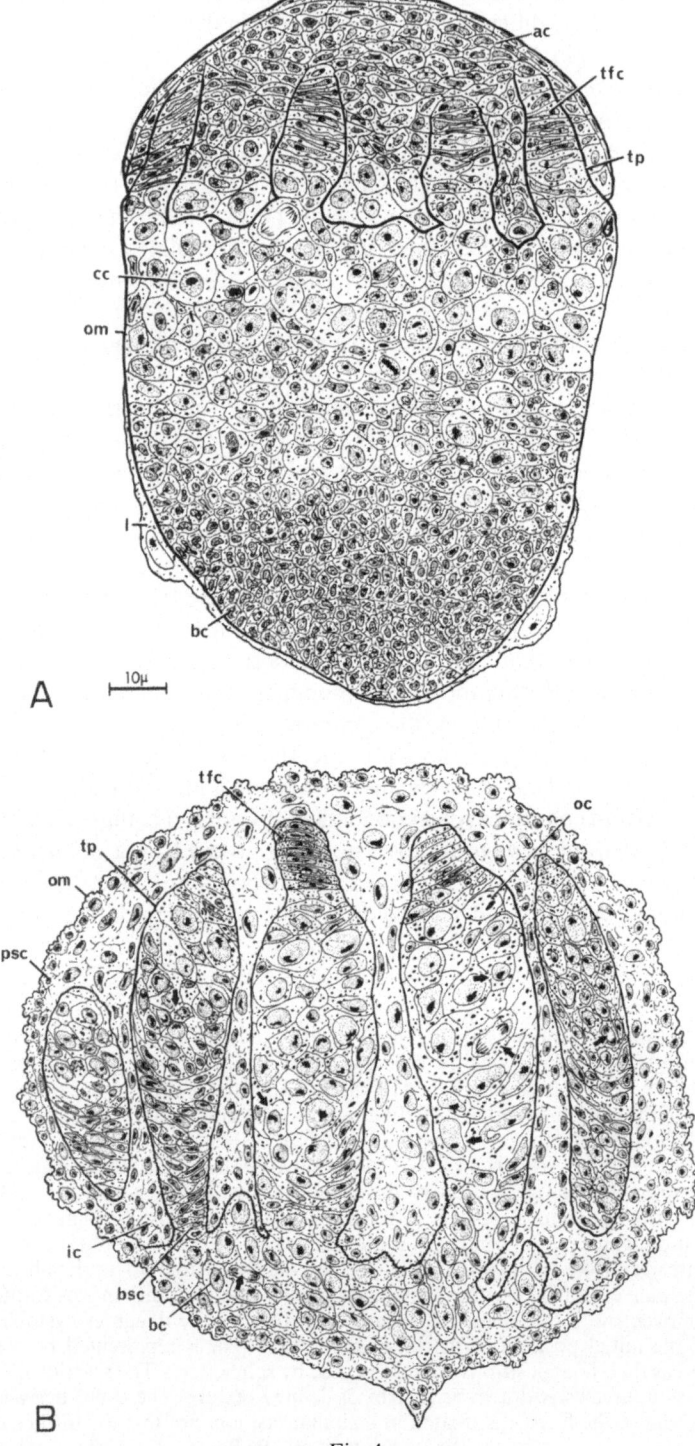

Fig. 4

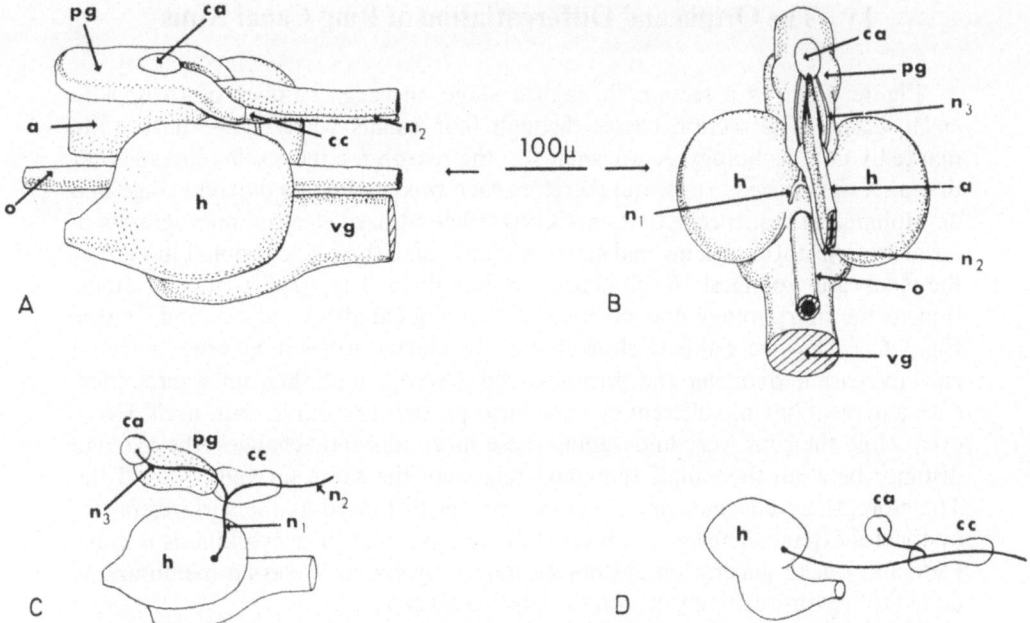

Fig. 5. A-D. The larval ring gland of *Drosophila melanogaster* and its associated organs, viewed from the side (A) and from above (B). The positions of the brain and the endocrine organs are contrasted for the larva and adult in diagrams (C) and (D), respectively. (*a*) Aorta; (*ca*) corpus allatum; (*cc*) corpus cardiacum; (*h*) hemisphere of brain; (*n₁*) afferent nerve to *cc*; (*n₂*) efferent nerve from *cc*; (*n₃*) nerve from *cc* to *ca*; (*o*) oesophagus; (*pg*) prothoracic gland; (*vg*) ventral ganglion. (From EHRIE and KING, 1971)

"potential cystoblasts" undergo complete or incomplete cytokinesis depends upon their positions relative to the apex of the germarium.

The observations described in the preceding paragraphs suggest that once germ cells receive a pulse of ecdysone from the prepupal prothoracic gland they are programmed to undergo incomplete cytokinesis. Furthermore, this behavior is transmitted to daughter cells, since cystocyte divisions continue in the adult after the prothoracic glands have degenerated and the ecdysone level has dropped to zero. The fact that stem cells still undergo complete cytokinesis in the apex of the germaria of pupae or adults suggests that an interaction between follicle cells in the region of the terminal filament (see Fig. 1 B) and adjacent germ cells prevents them from forming ring canals.

Fig. 4. (A) The *Drosophila melanogaster* prepupal ovary 2 hrs after puparium formation. See text for description. (*ac*) Apical cell; (*bc*) basal cell; (*cc*) central cell; (*l*) lamellocyte; (*om*) outer membrane; (*tfc*) terminal filament cell; (*tp*) tunica propria. (B) The pupal ovary 24 hrs after puparium formation. See text for description. (*bc*) Basal cell; (*bsc*) basal stalk cell; (*ic*) interstitial cell; (*oc*) oogonial cell; (*om*) outer membrane; (*psc*) peritoneal sheath cell; (*tfc*) terminal filament cell; (*tp*) tunica propria. Arrows point to mitotic figures and to ring canals. (From KING et al., 1968)

IV. The Origin and Differentiation of Ring Canal Rims

Figure 6 shows a section through a stage one egg chamber of *Drosophila melanogaster*. The section passes through four canals whose rims differ quite markedly in morphology. As we shall see, the reason for these differences is that the canal rims varied in age, and therefore each was caught at a different stage in a developmental sequence. KOCH and KING (1969) utilized electron micrographs of serial sections through a normal germarium to make three dimensional models of the oldest and youngest 16-cell clusters it contained. They paid particular attention to the morphology and position of the ring canals in cells 1e and 2e (see Fig. 3, C_4). They were able to show that as the cluster grows in volume the canal rims increase in diameter and thickness and develop along their inner circumference a deposit having different cytochemical properties from the rim itself. However, while the rims were undergoing these morphogenetic changes, the average distance between the canals remained relatively the same for cells 1e and 2e. Therefore, these plasmalemmal regions are not disturbed as the surfaces of the cystocytes expand. The above observations suggest that once cytokinesis is complete and fourth generation cystocytes start to grow, new plasma membrane is added to regions relatively distant from the canal rims.

In *Drosophila virilis* the pattern of interconnections among the 16 sister cystocytes is the same as that seen in *Drosophila melanogaster*. When consecutive horizontal sections through ring canals of *Drosophila virilis* were photographed, traced and reconstructed, KINDERMAN and KING (1973) found that the rim was composed of several overlapping leaves. The leaf arrangements resembled those first described for canal rims in germarial cystocytes of *Habrobracon juglandis* by CASSIDY and KING (1969). In this wasp each rim is composed of eight leaves arranged in four pairs, and it appears that the rims can dilate as the cystocytes grow by the sliding of certain leaves past one another (contrast Fig. 7A and B). In both *Habrobracon juglandis* and *Drosophila virilis* each leaf is composed of a monolayer of microtubules (see Fig. 7C) that have a mean outer diameter of 20 mμ. Each microtubule is separated from its neighbor by a space about 10 mμ wide.

In *Bombyx mori*, sister spermatocytes are connected by a canal system similar to the ones observed in clusters of ovarian cystocytes (KING and AKAI, 1971a). The midbody that forms at the site of future canals between sister spermatocytes consists of a disc of electron-dense material in which spindle microtubules are embedded. Immediately beneath the plasma membrane of the cleavage furrow is an area of dense cytoplasm 30-40 mμ thick called the contractile ring. It is composed of an ordered array of filaments (each 4-7 mμ in diameter) that are aligned circumferentially along the equator of the cell (see Fig. 8). AKAI and I suggested that the outer tubules of the midbodies are somehow crosslinked to the component fibrils of the contractile ring, and they are therefore prevented from constricting the orifice further. Subsequently the midbody dissolves and the "stabilized" contractile ring serves as the canal rim. Presumably the contractile ring also serves as the organelle about which the ring canal is elaborated in *Drosophila melanogaster*.

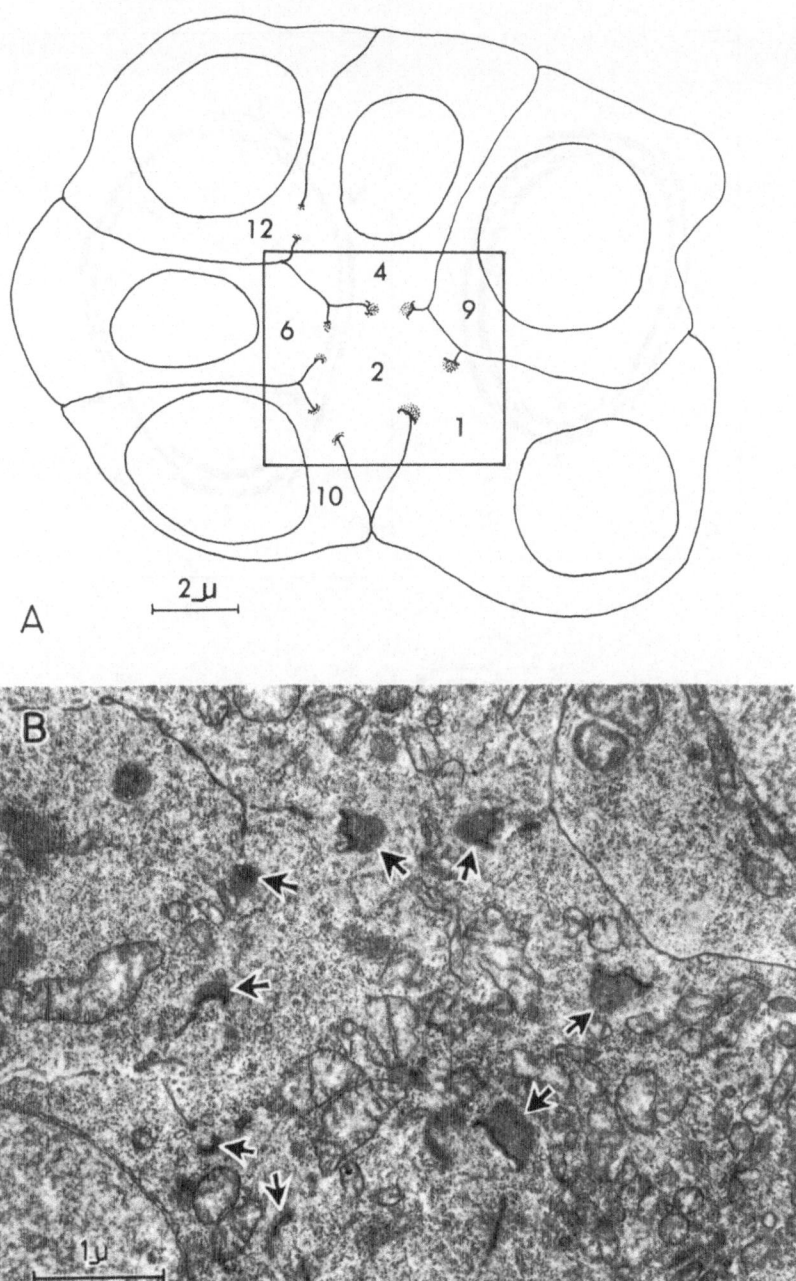

Fig. 6. (A) An outline of a magnified section through a stage 1 *Drosophila* egg chamber. The interconnected fourth generation cystocytes are numbered according to the system shown in Fig. 3, C_4. The area within the square is included in the electron micrograph in (B). (B) This micrograph shows the canals connecting cell 2 with cells 1, 4, 6, and 10. These canals were formed during the first, second, third, and fourth cystocyte divisions, respectively. The arrows point to the canal rims. (From KOCH and KING, 1969)

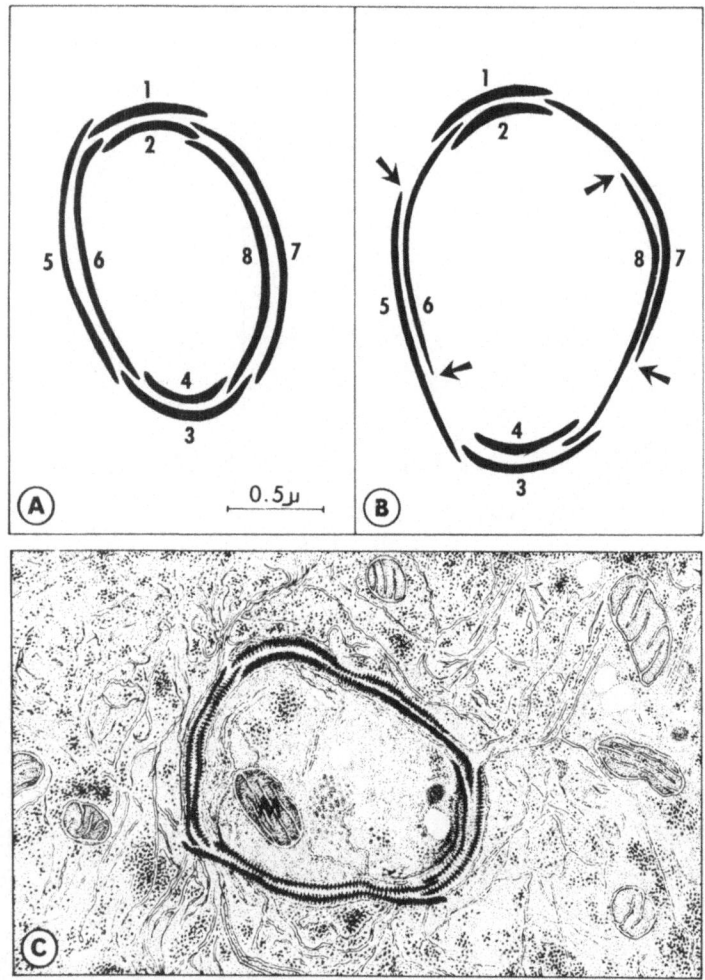

Fig. 7. (A) An outline drawing of a horizontally sectioned ring canal rim of *Habrobracon* showing the orientation of the eight leaves that form the ring. (B) An outline drawing of a rim that is dilated relative to the canal shown in (A). The dilation has been accompanied by a change in the orientation of the long, narrow leaves (see arrows). (C) A drawing illustrating the striated appearance of the leaves which is due to a monolayer of between 40 and 70 short, parallel microtubules embedded in each leaf. (From CASSIDY and KING, 1969)

V. Factors Controlling the Cessation of Mitosis and the Differentiation of Cystocytes

In the *Drosophila* ovariole the 16 cystocytes in each chamber are the mitotic products of a single cell. Mitosis should insure the nuclear equivalence of all 16 cells, and the particulate cytoplasmic components should be distributed initially between all sister cells in a random fashion. Yet two specific cells undergo a different type of development than that of the remaining 14. These cells are

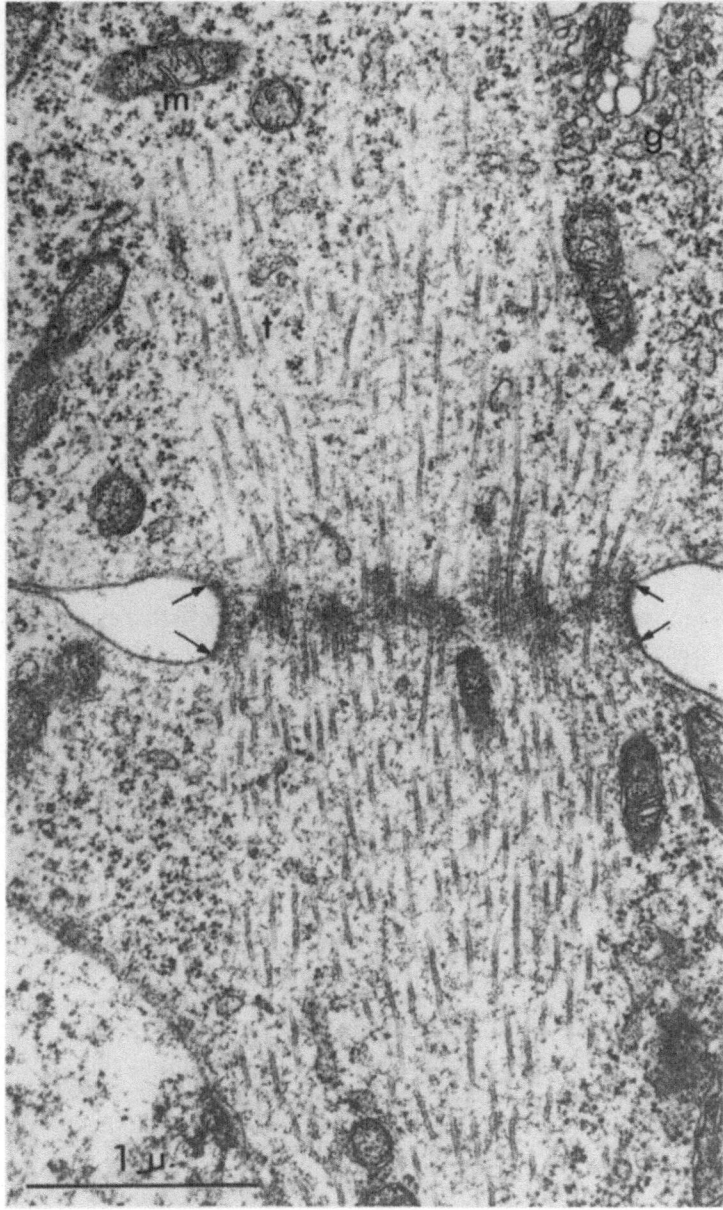

Fig. 8. A midbody forming between two sister spermatocytes in *Bombyx mori*. The midbody contains dense material which is laid down in the region where microtubules extending from each spermatocyte overlap. The arrows show the edges of the contractile ring. (*g*) Golgi material; (*m*) mitochondrion; (*t*) spindle microtubule. (From KING and AKAI, 1971a)

cystocytes 1 e and 2e. KOCH et al. (1967) showed that synaptonemal complexes formed in the nuclei of these cells shortly after they were produced in the germarium by the fourth cystocyte division (see Fig. 1 C). The construction of synaptonemal complexes demonstrates that homologous chromosomes are forming synaptic bivalents and hence that cells 1 e and 2e are entering the zygotene stage of meiotic prophase. We therefore named these cells "pro-oocytes" and referred to the 14 sister cells which lacked synaptonemal complexes as "pro-nurse" cells. Subsequent studies have shown that cells 1 e and 2e differentiate into pro-oocytes and 3 e-16 e as pro-nurse cells in *Drosophila virilis* (KINDERMAN and KING, 1973).

What is the source of the cue that initiates this differential behavior? Since the pro-oocytes contain a unique number of canals, it seems reasonable to suggest that in *Drosophila* the canal rims somehow direct the differentiation of cystocytes to the oocyte pathway when four are present.

In the germaria of *Drosophila melanogaster* mitotic figures are found to be grouped into clusters of 2, 4, and 8 (BUCHER, 1957; JOHNSON and KING, 1972; GRELL, 1973), and this observation demonstrates that interconnected cystocytes divide simultaneously. In *Xenopus laevis* somatic cell nuclei can be induced to undergo DNA replication by exposing them to egg cytoplasm (GRAHAM et al., 1966). This finding demonstrates that the test cytoplasm contains factors that stimulate mitosis. Perhaps in *Drosophila* the system of canals between cystocytes allows passage of soluble "mitogens" which initiate division in all cells in a given cluster. Thus sister cystocytes should show an "all or none" type of mitotic activity, and their number (N) after their final division should be given by $N = 2^M$, where M equals the number of consecutive mitoses preceding meiosis. The values of N observed in a number of higher insects follow the 2^M rule. For example, the final number of cystocytes per cluster during oogenesis is 8, 16, and 32 in the case of *Bombyx mori*, *Drosophila melanogaster*, and *Habrobracon juglandis*, respectively.

These observations support the hypotheses that the ring canal system insures the mitotic synchrony of sister cystocytes during oogenesis. Since the division of all sister cystocytes seems to be controlled as a unit, it is reasonable to suggest that a cue which causes one or more of the interconnected cells to differentiate will generate a sequence of reactions that terminates the mitotic activity of all other cells in the cluster. If the formation of two cells with four ring canals during the fourth cystocyte division is the cue for the differentiation of the pro-oocytes, their differentiation may terminate further mitoses among the other 14 sister cells.

VI. The Influence of the Fes Mutation upon the Division and Differentiation of Cystocytes

It is clear from the foregoing discussion that female gametocytes undergo, just before meiosis, a unique series of mitoses that are followed by incomplete cytokinesis. This conclusion leads to the prediction that genes exist which are activated only in cystocytes and that the products of these genes function to arrest subsequent cleavage furrows in a permanent, partially open condition. Mutant genes of

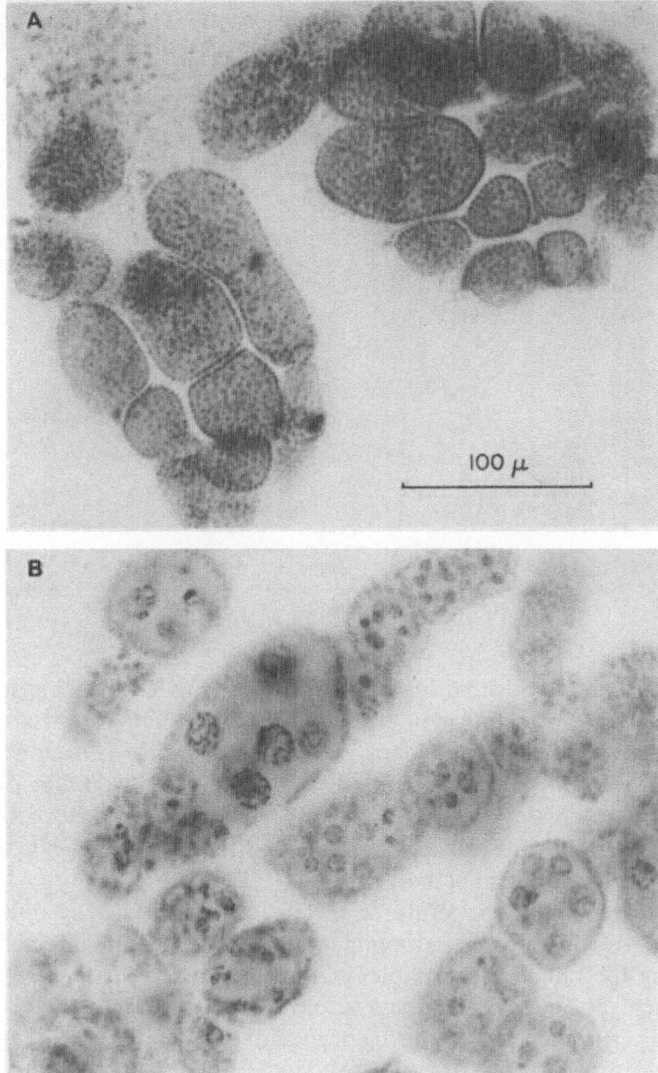

Fig. 9. A and B. Feulgen-stained, ovarian wholemounts from 2-day-old females homozygous for the *fes* gene. (A) Only tumors are seen in the ovaries of the female reared at 25° C. (B) Only chambers containing nurse-like cells are seen in the ovary from the 18° C series. (From KING, 1969)

this class might be expected to produce defective molecules, with the result that all or some fraction of the cystocyte cleavages would be complete. One would expect that such mutations should generate aberrations in gametogenesis and cause complete or partial sterility. The *fes* mutation is a case in point.

Females of *Drosophila melanogaster* homozygous for the autosomal recessive gene *female sterile (fes)* are rendered sterile because they produce "ovarian tumors" instead of eggs (see KING, 1969, for review). Each tumor in the vitellarium

Table 1. Catalogue of distribution of ring canals among 124 cystocytes found in 13 chambers from *fes* females reared at 18° C. All cystocytes were nurse-like except for 3 cells (*), which resembled oocytes. (From unpublished data of P. A. SMITH and P. MURPHY)

Chamber	Total cells	Total cells with × canals			
		1	2	3	4
1	15	6	5	4	0
2	12	6	3	2	1*
3	12	5	4	3	0
4	11	4	5	2	0
5	11	5	3	3	0
6	11	6	2	2	1*
7	8	4	3	0	1*
8	8	4	2	2	0
9	8	4	2	2	0
10	7	3	3	1	0
11	7	3	3	1	0
12	7	3	3	1	0
13	7	2	5	0	0
	124	55	43	23	3

is composed of hundreds to thousands of cells (see Fig. 9 A) which resemble cysto-cytes in that they are similar in size, are mitotically active, and are sometimes interconnected. Ovaries of *fes* genotype become "tumorous" even when they are transplanted into the abdomens of wild type females, and wild type ovaries trans-planted into *fes* females do not become tumorous (KING and BODENSTEIN, 1965). Therefore, there is no evidence for diffusible tumorigenic agents as initiating factors in the development of these tumors.

When reared at low temperatures, *fes* females will produce chambers in which most cystocytes, instead of continuing to divide, differentiate into cells which, on the basis of their nuclear morphology, resemble nurse cells (see Fig. 9 B). About 2% of such *fes* "nurse" chambers also contain an oocyte that is capable of undergoing vitellogenesis. We have concluded from an analysis of egg chambers containing subnormal and supernormal numbers of nurse cells, that vitellogenesis can proceed with as few as 7 and as many as 30 nurse cells. The frequency of yolky oocytes is 80 times higher in *fes* flies reared at 18° C that at 25° C (KING et al., 1961).

The three-dimensional morphology of 13 *fes* chambers containing nurselike cells was determined by an analysis of serial sections, and the data are presented in Table 1. Note that all oocyte-deficient chambers lack cystocytes possessing 4 ring canals and that all cystocytes with 4 canals differentiated into oocytes. These findings support the hypotheses, developed on p. 98, that wild-type cystocytes possessing less than 4 ring canal rims are programmed to develop as nurse cells, whereas those with 4 become oocytes.

Since *fes* cystocytes frequently continue dividing rather than differentiating, it is likely that the primary effect of the mutation is an alteration of the pattern of the cystocyte divisions in the germarium. Recently the three dimensional interrelations of the cells in a *fes* germarium was worked out, utilizing electron micro-graphs taken of serial, ultra-thin sections (JOHNSON and KING, 1972). This germarium contained a large number of unconnected cells and clusters made up of only a few interconnected cystocytes. In these cases the rims of the *fes* ring canals remained thin and delicate with little or no internal coating.

In order to study the dynamics of cell division in the germarium, the mitotic figures in hundreds of *fes* and wild type germaria were observed in Feulgen-stained wholemounts. The low frequency of metaphase figures observed made it desirable to investigate germaria in which the divisions produced over a period of several hours were accumulated. The injection of flies with the mitotic poison, colchicine, produced the desired effect. All cystocyte divisions were found to take place in the anterior third of each wild-type germarium. Metaphases occured throughout the *fes* germarium, and the number of isolated metaphase figures observed was 15-20 times higher than in wild type. Clusters of two metaphases were twice as abundant in *fes* as in wild type, and clusters of four were equally abundant. Clusters of 8 were seen about six times more often in wild type than in *fes*, but clusters of 3, 5, 6, 7, 9, 10, and 11 metaphases (which are never observed in wild type germaria) do occur in *fes*. More than twice as many dividing cells were found in the average *fes* germarium than in wild type. Similar frequencies of metaphases were seen in *fes* at both 25° C and 18° C. Johnson and I concluded that the average *fes* cystocyte undergoes one additional cycle of division before leaving the germarium.

The abnormal patterns of interconnections found in *fes* clusters suggest that the primary effect of the *fes* mutation is the elimination of some of the cells within the cluster. This elimination occurs at random and affects any cell in a cluster independently of its neighbors, causing asymmetric deviations from the normal pattern of connections. As a consequence of this abnormality, the remaining cells may not stop dividing after the normal number of division cycles. There are two simple ways in which a cell may be eliminated from a cluster. Either some cells fail to divide, or the nuclear divisions are normal, but cytokinesis is complete. Since the first hypothesis is rendered untenable by the observation that more dividing cells are found in mutant than in wild type germaria, the second must be the correct one.

We conclude that while all wild-type cystocytes undergo incomplete cytokinesis, a significant fraction of *fes* cystocytes undergo complete cytokinesis. An algebraic model developed from this hypothesis predicts the relative frequencies of single cells and clusters containing between 2 and 8 cells and enables us to calculate q, the probability that cystocytes will undergo complete cytokinesis. The predicted frequencies were not significantly different from those observed, and the hypothesis was also consistent with the observed rate of division found in the colchicine-treated ovaries and with the patterns of cystocyte interconnections found in the reconstructed *fes* germarium. Germaria from *fes* females reared at 18° C and 25° C gave q values of 0.356 and 0.403, respectively. We concluded that the product of the *fes* + gene is required for the formation of a stable canal system and suggested that the product of the mutant gene is defective in this regard and thermolabile. The *fes* + substance may function to prevent the complete closing down of the contractile ring during cystocyte cytokinesis.

The ovarioles of *fes* females reared in the cold contain many chambers in which cystocytes have stopped dividing and have differentiated into nurse cells, even though the clusters lack pro-oocytes (see Fig. 9 B). The number of nurse cells per chamber varies widely. For example, in a Feulgen-stained wholemount of 41 such chambers from a *fes* ovary, we observed an average of 13 nurse nuclei per chamber, with a range between 1 and 39 (KING et al., 1961). In such cases the stimulus to cease dividing could not come from the 4 canal cells, since presumably none were present, and the cue cannot be related to canal rims at all in situations where a lone, mutant cystocyte has differentiated into a nurse cell. It follows that in the *Drosophila* female, a cystocyte can complete its differentiation into an endopolyploid, nurse-like cell in the absence of canal rims.

In the normal germarium, cystocytes do not double their volumes between divisions, and consequently the mass of each individual cell is reduced with each division. The average cell in a newly formed, 16-cell cluster is only one-fifth the volume of the original cystoblast. In *fes* the frequency of clusters of nurse cells in the vitellarium is about ten times greater for flies reared at 18° C than at 25° C (KING et al., 1961). We know, however, that in the *fes* germarium mitoses occur with equal frequencies at both temperatures (see Table I, JOHNSON and KING, 1972). Perhaps lowering the temperature slows down the growth between divisions, so that after a few divisions the cystocytes fall below the "critical minimum mass" that is required for further division. If the cystocyte is prevented from dividing, it then enters the nurse cell developmental pathway. The concept that cell division can be prevented by reducing the size of the cell below a critical mass has considerable experimental support. For example, PRESCOTT (1956) was able to abolish cytokinesis in *Amoeba proteus* by preventing the cell from reaching a critical volume. This was done by amputating pieces of cytoplasm from the animal at periodic intervals.

VII. Hereditary Ovarian Tumors of Fused Females

Females of *Drosophila melanogaster* homozygous for the sex-linked gene *fused* (*fu*) also produce "ovarian tumors". However, the ovaries of newly eclosed females appear normal, and tumors only develop as the flies age. The rate at which such tumors are generated varies with the allele studied (SMITH and KING, 1966). For example, there is a sixfold difference between the rates at which ovarian tumors form in the ovaries of fu^1 and of fu^{59} homozygotes. When a fly carrying a *fu* allele characterized by a low tumor incidence is crossed with a fly carrying an allele characterized by a high incidence, the compound, F_1 female generally shows a low incidence of ovarian tumors. If fu^1 or fu^{59} ovaries are implanted into an abdomen of a fly homozygous for the normal *fu* allele, the implanted ovaries develop tumors at the rate characteristic of each mutant allele (SMITH et al., 1965). Females homozygous for fu^1, when reared at 18° C, develop tumors at a rate 10% that of females reared at 25° C, and the reduced tumor frequency in ovaries from *fu* flies reared in the cold is not due to a general retardation in the rate of ovarian development.

The *fused* gene is an example of a pleiotropic mutation. In fact, it was named for its effect upon wing venation. In all the *fu* alleles so far studied, the alteration in wing vein morphology is accompanied by the production of ovarian tumors. How the product of the *fu* gene controls such seemingly independent developmental processes is not understood.

By appropriate crosses, SMITH and KING (1966) were able to produce females containing a double dose of both the *fu* and *fes* genes. Such flies showed the *fused* wing phenotype and the *fes* ovarian phenotype. The ovaries of females homozygous for *fu* and heterozygous for *fes* have a higher incidence of tumors than do *fu/fu; +/+* females of the same age. Thus, the "recessive" *fes* gene can be detected in the heterozygous condition by its interaction with *fu*.

I suggested earlier that the product of the + allele of the *fes* gene may function to prevent the constriction of the contractile ring during cystocyte cytokinesis and that the product of the mutant allele may be thermolabile. In case of *fu*, since the tumor incidence is reduced by lowering the environmental temperature, one could also argue that the *fu* + substance is thermolabile. However, the mutant wing phenotype is not modified by lowering the environmental temperature. In terms of the differentiation of cystocytes the product of the *fes* gene must be required before that of the *fu* gene, since tumors develop prior to eclosion in *fes* but sometime after eclosion in *fu*. It remains to be seen whether the *fu* gene also increases the probability that cystocytes will undergo complete cytokinesis.

VIII. The Fate of the Pro-Oocytes

The nuclei of cystocytes 1e and 2e both form synaptonemal complexes during the period that the 16-cell cluster moves through the germarium. However, chambers in the vitellarium contain 15 nurse cells and 1 oocyte (contrast Figs. 1C and

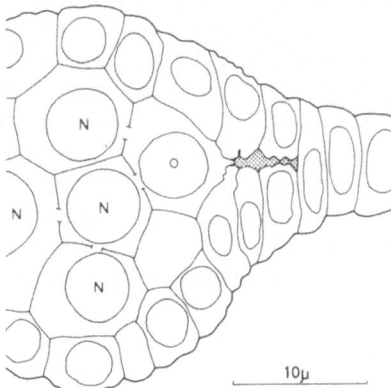

Fig. 10. A drawing of a stage-2 chamber and the stalk that connects it to the next egg chamber in the vitellarium. See Fig. 1B (S_2 and S_3) for orientation purposes. The canal that connects the oocyte (*O*) with the most anterior stalk cell is stippled. (*N*) nurse cell nucleus. The remaining nuclei belong to follicle cells. (From KING, 1972)

D). These observations demonstrate that a 4 canal cystocyte can enter the developmental pathway leading to oocyte differentiation and subsequently switch to the alternative pathway which characterizes the nurse cell. What is the source of the stimulus that causes the other pro-oocyte to continue developing as an oocyte? As the 16-ell cluster moves posteriorly through the germarium and is enveloped by follicle cells, one or the other of the 4 canal cells eventually comes to lie opposite those follicle cells making up the stalk that connects the germarium to the vitellarium (see Fig. 10). This observation suggests that the continued differentiation of the oocyte requires an interaction with adjacent stalk cells. The nurse cells begin to transfer their cytoplasm to the oocyte via the system of ring canals once the chamber enters the vitellarium. WOODRUFF and TELFER (1973) have shown in *Hyalophora cecropia* that a potential difference develops between the oocyte and the nurse cells which is of sufficient magnitude to effect the transfer of charged particles from nurse cells to oocyte by electrophoresis. Perhaps one function of the stalk cells is to generate this potential difference.

IX. The Control of Chromosomal Replication and Transcription during Gametogenesis

Some insights into the complexity of the systems that control chromosomal behavior can be obtained by contrasting the mitotic and meiotic divisions which occur during gametogenesis in males and females of *Drosophila melanogaster*. Four cystocyte divisions take place in both sexes. In the male, all cystocytes undergo the two divisions of meiosis, and thus generate bundles containing 64 spermatids (see Fig. 4 in TOKUYASU et al., 1972). In the female fruitfly, only one of the 16 cystocytes (the oocyte) completes the meiotic divisions. The mechanism that suppresses the meiotic divisions in certain cystocytes presumably arose during the evolution of the higher insects to allow for the production of the nurse cells. In these cells DNA replication continues but cytokinesis is suppressed, a situation that is the reverse of that characterizing the second meiotic division.

In both sexes all fourth generation cystocytes replicate their DNA to generate the 4C amount, and DNA replication is then switched off, except for the nurse cells. As these develop in the vitellarium, 7 or 8 more doublings occur (see Table II-1 in KING, 1970). The provision of the oocyte with 15 highly polyploid nurse cells serves to multiply the rDNA available for transcription thousands of times, and the cistrons that specify the ribosomal proteins, tRNAs, and the enzymes functioning in translation are multiplied concurrently. Although methods for the selective extra-chromosomal replication of oocyte rDNA have evolved in some insect species (GALL et al., 1969), this has not taken place in *Drosophila* (MOHAN and RITOSSA, 1970).

Studies on the mechanism of action of steroid hormones in birds and mammals (O'MALLEY et al., 1970; STEGGLES et al., 1971) indicate that the cells of target tissues contain specific receptor proteins that bind to the hormones. The receptor protein–hormone complex then enters the nucleus, attaches to the chromatin, and stimulates the transcription of mRNAs for the proteins whose synthesis is known to be induced by the hormones. If the synthetic activities of the chromosomes of insect cystocytes are controlled by proteins of the sort mentioned above, then nuclei bathed by a common cytoplasm should engage in similar synthetic activities. However, in the case of interconnected ovarian cystocytes, the oocyte behaves in a completely different fashion from its sister nurse cells. A premeiotic replication of DNA occurs within the nuclei of oocytes before they enter the vitellarium (CHANDLEY, 1966), and the DNA concentration is maintained at the 4C value throughout the remainder of oogenesis. Thus DNA replication is switched off in the oocyte before the chamber enters the vitellarium; whereas it continues in the sister nurse cells. Oocytes have also been shown to be much less active in transcription than are their sister nurse cells (HUGHES and BERRY, 1970).

In the developing egg chamber of *Drosophila virilis* clusters of microfibrils adhere to the oocyte nucleus which eventually develops a coating of amorphous material 1 μ thick that is dense enough to exclude perinuclear mitochondria and ribosomes. The nuclei of the 15 sister nurse cells never develop such a coating (KINDERMAN and KING, 1973). We suggested that the nuclear coating prevents the transfer of large molecules to the oocyte nucleoplasm and so insulates the oocyte chromosomes from the influences of those compounds that stimulate replication and transcription in the sister nurse cells. We have presented cytological evidence suggesting that protein microfibrils continue to be deposited on the canal rims and that later masses of clustered microfibrils break off each rim and are carried away in the cytoplasmic stream that flows into the oocyte. We suggested the ring canal rim is composed of old equatorial plasmalemma that retains an affinity for contractile ring microfibrils. Subsequent to the arrested cleavage, additional microfibrils bind to the canal rims and also attach to the outer surface of the oocyte nucleus. Perhaps the receptor molecules necessary for the subsequent binding of microfibrils pass from the rims to the oocyte nucleus; but this is merely one of many plausible interpretations.

X. Discussion and Conclusions

The cycle of differentiation of the germ cells postulated for the *Drosophila* female is diagrammed in Fig. 11. Oogonia (O) undergo complete cytokinesis until metamorphosis. Ecdysone secreted by the prepupal prothoracic gland transforms oogonia to cystoblasts (C_b). Genes (such as *fes* +) that arrest cytokinesis are switched on in all cystoblasts, except those adjacent to the follicle cells (FC) in the apex of the germarium. Cystocyte divisions continue until two cells (1e and 2e) receive 4 ring canals. The canal rims then generate signals of unknown nature that stimulate the nuclei of these cells to enter meiotic prophase and become pro-oocytes (PO). Subsequent signals from these differentiated nuclei pass to all interconnected cells and switch off those genes controlling mitosis. Those cystocytes with fewer than 4 canal rims (cells 3–16e) differentiate into nurse cells (NC). Upon reaching the posterior region of the germarium the pro-oocyte that comes to lie adjacent to follicle cells (FC) of the stalk which connects the germarium to the

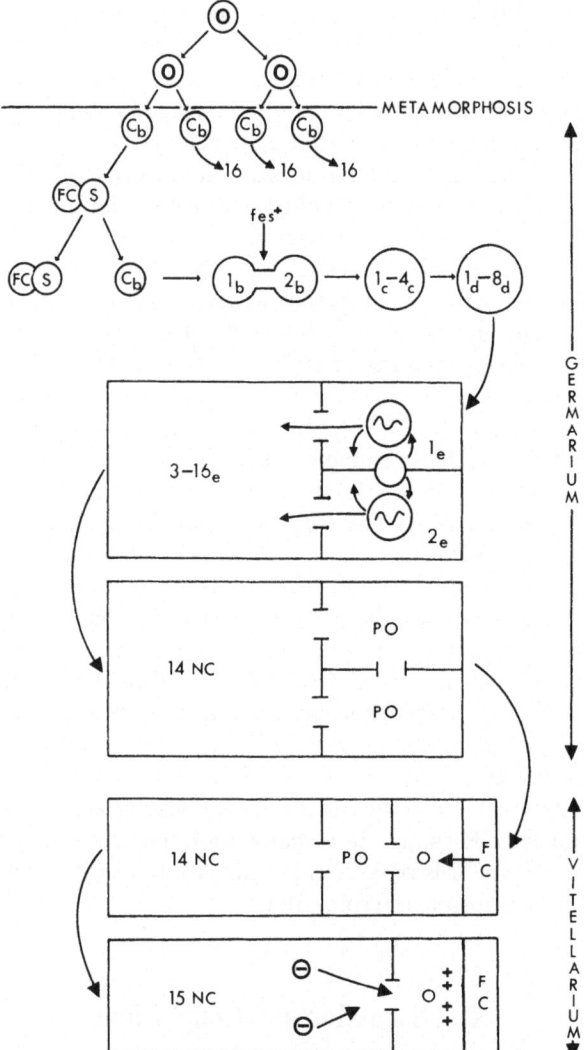

Fig. 11. A diagram outlining the proposed cycle of differentiation of germ cells in the *Droso-phila* ovary. See text for further discussion

vitellarium receives the stimuli necessary for its continued differentiation as an oocyte (o). Further DNA replication is suppressed. The other pro-oocyte switches to the alternative pathway and becomes the fifteenth nurse cell. In the vitellarium the 15 nurse cells continue to replicate their DNA, and cytoplasmic particles flow from these toward the oocyte in response to a potential gradient.

In the male, branching chains of cystocytes are also produced, and again the final number is 16. It follows that in *Drosophila* both male and female gametocytes undergo a specialized type of mitotic division, prior to meiosis, which involves incomplete cytokinesis. Four cystocyte divisions invariably take place in fruitflies of both sexes, and the number of consecutive cystocyte divisions is species-specific in other insects as well. For example, in *Bombyx mori* it is 3 and 6 for the cystocytes of females and males, respectively (KING and AKAI, 1971b). Since clusters of interconnected, sister germ cells have been reported in the ovaries and testes of animals belonging to 4 orders of holometabolous insects and 5 orders of placental mammals (see literature reviews in CASSIDY and KING, 1969; and DYM and FAWCETT, 1971), it is clear that the phenomenon of arrested cleavage preceding meiosis is widespread and presumably of great antiquity.

TOKUYASU et al. (1972) have demonstrated that in *Drosophila melanogaster* the component cells in a bundle of 64 spermatids are interconnected and that they differentiate in synchrony. It follows that cleavages are also arrested during the meiotic divisions and that the cytoplasmic channels generated between spermatids facilitate the synchronous development of all the cells in the bundle.

The study of the arrested cleavages and subsequent differentiation of spermatogonia and spermatocytes provides insights into a mechanism for the synchronization of mitoses and of the subsequent differentiation of sister cells. Future studies may demonstrate that arrested cleavages are not restricted to germ cells and that during embryogenesis defined numbers of developmentally synchronized cells are generated in this way. Indeed, ring canals have been observed in a recent study of the epithelial cells of the midgut of *Calliphora* embryos (VAN DER STARRE-VAN DER MOLEN et al., 1973, their Fig. 7).

The study of the differentiation of ovarian cystocytes provides insights to the cues that may switch genetically identical cells into different developmental pathways. The division cycle I have described (p. 90) generates daughter cells which differ in that one sister receives all, the other none of the previously formed canals. If complete cytokinesis follows, the sister cells will still differ in terms of the distribution of "cleavage plaques" on their surfaces. Such areas, unlike the adjacent plasmalemma, presumably contain the receptor molecules to which the microfibrils of the contractile ring attach. Thus, depending on the number of consecutive divisions each clone undergoes, varying numbers of cells containing different numbers of cleavage plaques will be generated. If, for example, the average clone undergoes 4 divisions, 16 cells would be produced, and of these 2 would each have 4 plaques, 2 would have 3, 4 would have 2, and 8 would have 1 plaque each. If such cortical islands can generate stimuli that derepress different genes in the adjacent nuclei, then one would expect that the cells from the same clone that differ in the number of cleavage plaques each contains might enter different developmental pathways.

References

AGGARWAL, S. K., KING, R. C.: A comparative study of the ring glands from wild type and *lgl* mutant *Drosophila melanogaster*. J. Morphol. **129**, 171–200 (1969).

BECKER, H. J.: Die Puffs der Speicheldrüsenchromosomen von *Drosophila melanogaster*. II. Die Auslösung der Puffbildung, ihre Spezifität und ihre Beziehung zur Funktion der Ringdrüse. Chromosoma **13**, 341–384 (1962).

BODENSTEIN, D.: Untersuchungen zum Metamorphoseproblem. I. Kombinierte Schnürungs- und Transplantations-Experimente an *Drosophila*. Wilhelm Roux Arch. Entwicklungs- mech. Org. **137**, 474–505 (1938).

BROWN, E. H., KING, R. C.: Studies on the events resulting in the formation of an egg chamber in *Drosophila melanogaster*. Growth **28**, 41–81 (1964).

BUCHER, N.: Experimentelle Untersuchungen über die Beziehungen zwischen Keimzellen und somatischen Zellen im Ovar von *Drosophila melanogaster*. Rev. Suisse Zool. **64**, 91–188 (1957).

CASSIDY, J. D., KING, R. C.: The dilatable ring canals of the ovarian cystocytes of *Habrobracon juglandis*. Biol. Bull. **137**, 429–437 (1969).

CASSIDY, J. D., KING, R. C.: Ovarian development in *Habrobracon juglandis* (Ashmead) (Hy- menoptera: Braconidae). I. The origin and differentiation of the oocyte-nurse cell com- plex. Biol. Bull. **143**, 483–505 (1972).

CHANDLEY, A. C.: Studies on oogenesis in *Drosophila melanogaster* with ^{3}H-thymidine label. Exp. Cell Res. **44**, 201–215 (1966).

DYM, M., FAWCETT, D. W.: Further observations on the number of spermatogonia, spermato- cytes, and spermatids connected by intercellular bridges in the mammalian testis. Biol. Reprod. **4**, 195–215 (1971).

EHRIE, M. G., KING, R. C.: The anatomy of the larval ring gland of *Drosophila melanogaster* and its associated organs. Drosophila Inform. Serv. **47**, 79–80 (1971).

GALL, J. G., MACGREGOR, H. C., KIDSTON, M. E.: Gene amplification in the oocytes of Dytiscid water beetles. Chromosoma **26**, 169–187 (1969).

GRAHAM, C. F., ARMS, K., GURDON, J. B.: The induction of DNA synthesis by frog egg cyto- plasm. Develop. Biol. **14**, 349–381 (1966).

GRELL, R. F.: Recombination and DNA replication in the *Drosophila melanogaster* oocyte. Genetics **73**, 87–108 (1973).

HUGHES, M., BERRY, S. J.: The synthesis and secretion of ribosomes by nurse cells of *Anther- aea polyphemus*. Develop. Biol. **23**, 651–664 (1970).

JOHNSON, J. H., KING, R. C.: Studies on *fes*, a mutation affecting cystocyte cytokinesis, in *Drosophila melanogaster*. Biol. Bull. **143**, 525–547 (1972).

KARLSON, P., HOFFMEISTER, H.: Zur Biogenese des Ecdysons. I. Umwandlung von Cholesterin in Ecdyson. Z. Physiol. Chem. **331**, 298–300 (1963).

KINDERMAN, N. B., KING, R. C.: Oogenesis in *Drosophila virilis*. I. Interactions between the ring canal rims and the nucleus of the oocyte. Biol. Bull. **144**, 331–354 (1973).

KING, R. C.: The hereditary ovarian tumors of *Drosophila melanogaster*. Natl. Cancer Inst. Monograph. **No. 31**, pp. 323–345 (1969).

KING, R. C.: Ovarian Development in *Drosophila melanogaster*. p. 227. New York: Academic Press 1970.

KING, R. C.: Drosophila oogenesis and its genetic control. In: BIGGERS, J. D., SCHUETZ, A. W. (Eds.): Oogenesis, pp. 253–275. Baltimore: University Park Press 1972.

KING, R. C., AGGARWAL, S. K.: Oogenesis in *Hyalophora cecropia*. Growth **29**, 17–83 (1965).

KING, R. C., AGGARWAL, S. K., AGGARWAL, U.: The development of the female *Drosophila* reproductive system. J. Morphol. **124**, 143–166 (1968).

KING, R. C., AKAI, H.: Spermatogenesis in *Bombyx mori*. I. The canal system joining sister spermatocytes. J. Morphol. **134**, 47–56 (1971a).

KING, R. C., AKAI, H.: Spermatogenesis in *Bombyx mori*. II. The ultrastructure of synapsed bivalents. J. Morphol. **134**, 181–194 (1971b).

KING, R. C., BODENSTEIN, D.: The transplantation of ovaries between genetically sterile and wild type *Drosophila melanogaster*. Z. Naturforsch. **20b**, 292–297 (1965).

KING, R. C., KOCH, E. A., CASSENS, G. A.: The effect of temperature upon the hereditary ovarian tumors of the *fes* mutant of *Drosophila melanogaster*. Growth **25**, 45–65 (1961).

KOCH, E. A., KING, R. C.: The origin and early differentiation of the egg chamber of *Droso- phila melanogaster*. J. Morphol. **119**, 283–304 (1966).

KOCH, E. A., KING, R. C.: Further studies on the ring canal system of the ovarian cystocytes of *Drosophila melanogaster*. Z. Zellforsch. **102**, 129–152 (1969).

KOCH, E. A., SMITH, P. A., KING, R. C.: The division and differentiation of *Drosophila* cysto- cytes. J. Morphol. **121**, 55–70 (1967).

MOHAN,J., RITOSSA,R.M.: Regulation of ribosomal RNA synthesis and its bearing on the *bobbed* phenotype in *Drosophila melanogaster*. Develop. Biol. **22**, 495–512 (1970).

O'MALLEY,B.W., SHERMAN,M.R., TOFT,D.O.: Progesterone "receptors" in the cytoplasm and nucleus of chick oviduct target tissue. Proc. Natl. Acad. Sci. **67**, 501–508 (1970).

PRESCOTT,D.M.: Relation between cell growth and cell division. III. Changes in nuclear volume and growth rate and prevention of cell division in *Amoeba proteus* resulting from cytoplasmic amputation. Exp. Cell Res. **11**, 94–98 (1956).

SANG,J.H., KING,R.C.: Nutritional requirements of auxenically cultured *Drosophila melanogaster* adults. J. Exp. Biol. **38**, 793–809 (1961).

SHAAYA,E., KARLSON,P.: Der Ecdysontiter während der Insektenentwicklung. II. Die postembryonale Entwicklung der Schmeißfliege *Calliphora erythrocephala*. Meig. J. Insect Physiol. **11**, 65–69 (1965).

SMITH,P.A., BODENSTEIN,D., KING,R.C.: Autonomy of *fu* and *fu*^{59} ovarian implants with respect to rate of tumor production. J. Exp. Zool. **159**, 333–336 (1965).

SMITH,P.A., KING,R.C.: Studies on *fused*, a mutant gene producing ovarian tumors in *Drosophila melanogaster*. J. Natl. Cancer Inst. **36**, 445–463 (1966).

STARRE-VAN DER MOLEN,L.G. VAN DER, PLANQUÉ-HUIDEKOPER,B., DE PRIESTER,W.: Embryogenesis of *Calliphora erythrocephala*. III. Ultrastructure of the midgut epithelial cells during late embryonic development. Z. Zellforsch. **144**, 117–138 (1973).

STEGGLES,A.W., SPELSBERG,T.C., GLASSER,S.R., O'MALLEY,B.W.: Soluble complexes between steroid hormones and target-tissue receptors bind specifically to target tissue chromatin. Proc. Natl. Acad. Sci. **68**, 1479–1482 (1971).

TOKUYASU,K.T., PEACOCK,W.J., HARDY,R.W.: Dynamics of spermiogenesis in *Drosophila melanogaster*. I. Individuation process. Z. Zellforsch. **124**, 479–506 (1972).

VOGT,M.: Induktion von Metamorphoseprozessen durch implantierte Ringdrüsen bei *Drosophila*. Wilhelm Roux' Arch. Enwicklungsmech. Org. **142**, 131–182 (1942).

VOGT,M.: Hormonale Auslösung früher Entwicklungsprozesse in der Augenantennen-Anlage von *Drosophila*. Naturwissenschaften **31**, 200–201 (1943).

WOODRUFF,R.I., TELFER,W.H.: Polarized intercellular bridges in ovarian follicles of the cecropia moth. J. Cell Biol. **58**, 172–188 (1973).

The Cell Cycle and Cellular Differentiation in Insects

PETER LAWRENCE

Medical Research Council, Laboratory of Molecular Biology, Cambridge, England

I. Introduction

When one event regularly follows another it is common practice to assume a causal relationship between the two. Such a view may be of little heuristic value unless the mechanism is specified. The study of the interrelation between the cell cycle and differentiation is a case in point. Because differentiation always follows some cell division, a theory of any usefulness must depend on experiments that pinpoint a determinative *element* in the preceding cell cycle.

There is one class of divisions where this element is conspicuous, for the mitosis itself is asymmetric. In such divisions cell Type A usually gives rise to cells of Type A and B, or B and C. Here the cytoplasm is unequally divided and this may play a causal role in generating the two different daughter cells. Asymmetric division may determine the fate of a daughter cell in another way: by placing the daughter cells in different environments. With symmetric divisions (equal cytokinesis) the evidence for determinative elements is indirect and their existence has become as much a matter for dogma as for experimental test. We shall look into these problems using insects as examples. Further, the role of the cell cycle in initiation and propagation of determination is far from clear; some relevant cases are considered in the last section of the chapter.

II. Asymmetric Divisions

In these differential divisions the cleavage plane is asymmetrically placed so that the daughter cells receive different amounts of cytoplasm. Moreover, the division plane is often peculiarly oriented so that the daughter cells come to lie in different regions of the tissue. One example is the development of scales in the wings of Lepidoptera; here spaced-out epidermal cells transform into scale mother cells and undergo an oblique division to give rise to a small internal cell that degenerates, and a large external cell that divides to form the scale and socket cell (STOSSBERG, 1938) (Fig. 1).

Although the differential divisions presumed to give rise to *Drosophila* bristles have not been observed, there is circumstantial evidence pointing to the import-

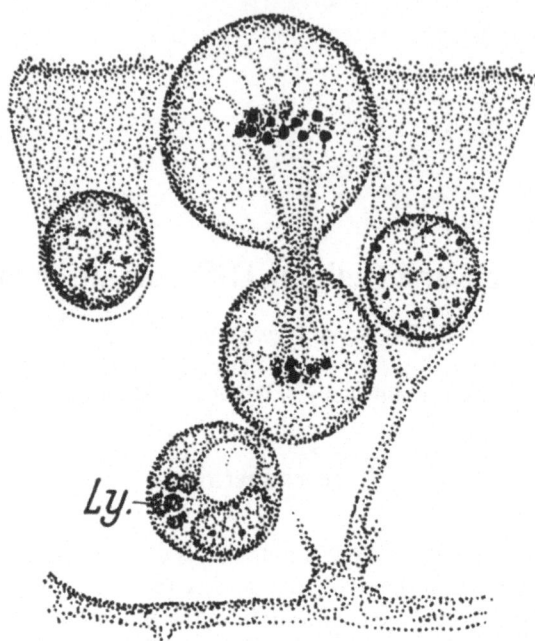

Fig. 1. The first asymmetric division of the scale mother cell. The upper and larger daughter subsequently divides to give rise to the scale and socket cell, the lower daughter degenerates. Ly = "lymphocyte". (From STOSSBERG, 1938)

ance of the orientation of the cleavage plane. In the mutant *Hairless* two polytene cells develop, but instead of one forming the socket and one the shaft of the bristle as happens in the wild type, in the mutant two socket cells are formed. In the wild type the two cells lie at different levels in the epithelium, only the presumptive socket cell having contact with the cuticular surface; but in *Hairless* flies both cells are at the same level and both have contact with the external surface. LEES and WADDINGTON (1942) suggested that the cleavage plane may be orthogonal to the surface in the mutant but obliquely oriented in the wild type.

In an important series of experiments CARLSON (1952) demonstrated conclusively that it is the differential partitioning of the *cytoplasm* rather than the nucleus which is determinative. Insect neuroblasts normally divide unequally to give a smaller daughter which becomes a ganglion mother cell, and a large daughter which remains as a neuroblast and undergoes further differential divisions. CARLSON was able to keep neuroblasts of the grasshopper *Chortophaga* in hanging drop culture. Using a needle, he rotated the spindle with the chromosomes of dividing neuroblasts through 180 degrees, so that each daughter cell received the chromosomes that would have entered the other. The operation had no effect on the fate of the daughter cells. This experiment points to the unequal cytokinesis as being essential to the generation of diversity. We still need to know what initiates the division and how the position of the cleavage plane is controlled. In *Oncopeltus* asymmetric divisions can be detected before cytokinesis begins; unlike cells in normal epidermal divisions, which maintain the structure of the sheet, the bulk of the hair mother cell retracts from the surface before the obliquely oriented division occurs. Moreover, the period of time intervening between the end of DNA synthesis and metaphase is' different for each of the three kinds of asymmetric divisions involved in the development of an *Oncopeltus* hair (LAWRENCE, 1968). The unusual orientation of differential divisions presumably depends on the earlier relocation of cellular organelles such as centrioles (WHITTEN, 1973). Clearly, therefore, the cell about to undergo asymmetric division is already differentiating before the mitosis occurs.

It is suspected that a number of differential divisions can follow each other in a strictly defined sequence to generate small epidermal organs that consist of several cell types. Certainly this is the case in the scales of the moth *Ephestia* (STOSSBERG, 1938), the hairs of the hemipteran bug *Oncopeltus* (LAWRENCE, 1966a), and the fly *Calliphora* (Peters, 1965) (3–4 different cell types), but the precise lineage has not been followed for more complex epidermal sensillae such as chemosensory scales, where each organ may consist of 12 or more cells (CLEVER, 1958; LAWRENCE, 1966a).

The development of the daughter cells of a differential division may be mutually dependent. CLEVER (1960) showed that the maturation of the scale and socket cells in *Ephestia* depends on the presence of nerve cells; when these were selectively killed with methylene blue, the scale and socket cells failed to become polytene. Such experiments show that, in principle, differences between sister cells need not necessarily descend directly from the previous mitosis, but can be built up by means of local interactions. In spite of this, the idea of asymmetrical divisions has a strong appeal to the theoretician; HENKE (1946) for example proposed an elaborate theory to explain aspects of the scale pattern in Lepidoptera. He suggested that the pupal epidermis consists of cells containing a standard quantity of a growth substance essential for DNA synthesis. In an initial differential division these cells would give rise to scale mother cells and epidermal cells. The amount of growth substance could be unequally partitioned between the two daughter cells. In some areas of the wing the epidermal mother cell would receive a larger share of the growth substance and give rise to a larger clone of epidermal cells (say 16), while in this same area the ploidy of the scale-forming cell would be low (say, 4n). In other areas where the scale ploidy is high (32n) the number of intervening epidermal cells would be low (2/scale). This theory, while appealing to numerologists, seems to have been ruled out by measurements of cell number and ploidy (SMOLKA, 1958; LAWRENCE, 1973). Likewise, it had generally been thought (KÜHN, 1965) that where organs consist of precise numbers of cells arranged in small groups, each of these groups descends from a single mother cell. Ommatidia were thought to develop from a single progenitor, which underwent a strictly determined sequence of cell divisions to give rise to the component cells of the ommatidium (about 20 depending on the species). BERNARD (1937) even went some way towards describing the lineage of the ommatidium of the ant *Formicina*. Nevertheless, analysis of genetic mosaics of single ommatidia of *Drosophila* (HANSON et al., 1972) and *Oncopeltus* (SHELTON and LAWRENCE, 1973) has shown that ommatidia in these species do not develop from a single primordial cell. In these mosaics near the boundary between large areas of mutant and wild-type eye, the ommatidia are genetically mosaic, sometimes only containing one cell of the minority genotype. It may therefore be that the constituent cells of the ommatidia are determined only after the cell divisions have ceased. This example illustrates the danger of generalizing too far from limited evidence. In fact, precise mosaic development and differential divisions may be limited to organs consisting of only a very small number of cell types.

III. Symmetric Divisions

These are cell divisions where the two progeny are identical to each other. Sometimes the daughter cells subsequently become different from the parent cell. The question is whether this differentiation is in any way dependent on events occurring in the preceding cell cycle. The basic idea of a "quantal" mitosis (OKA-ZAKI and HOLTZER, 1966) is that there is an event in the preceding cell cycle which is *the* essential component in the process leading to differentiation of the daughters. This view can lead to the idea that differentiation cannot occur without such a mitosis, and finally by a somewhat circular argument to the belief that, if cell function changes independently of a preceding mitosis or round of DNA synthesis, it is not "true" differentiation (HOLTZER et al., 1972).

What experimental test would demonstrate a "quantal" mitosis? First we need a useful definition of differentiation; one definition would be that a particular protein not formed by the mother cell is produced by the daughter cells. This is a good criterion in principle, but in practice it seems too restrictive. Cytologists can easily distinguish cell types on the grounds of differing structure and behavior, even where there is no evidence that these cell types contain different proteins. In my opinion, these criteria can be just as valid. In insects the secretion of a cuticle that is quite different in structure and pigmentation (Fig. 2) would seem to be an adequate indicator for differentiation, even in the absence of specific biochemical analysis.

We also need a way of proving causal relationships: for example if the differentiation is induced by a hormone, it should be possible to show that the hormone is only effective during some restricted phase in the critical cell cycle. Another approach is to use an agent that inhibits differentiation but not division, and to show that it will inhibit subsequent differentiation only when present at a particular and early stage in the cell cycle (e.g. bromodeoxyuridine; see HOLTZER and WEINTRAUB, this volume).

IV. Juvenile Hormone and Cell Division

Juvenile hormone would seem to be an ideal agent for testing the "quantal" mitosis hypothesis. Often pupal or adult epidermal cells secrete a very different cuticle from their larval progenitors (Fig. 2); analysis of gels suggests that they contain different proteins (e.g. ANDERSON, 1973). When an insect epidermal cell divides, the daughter cells normally remain side by side (LAWRENCE, 1968) and

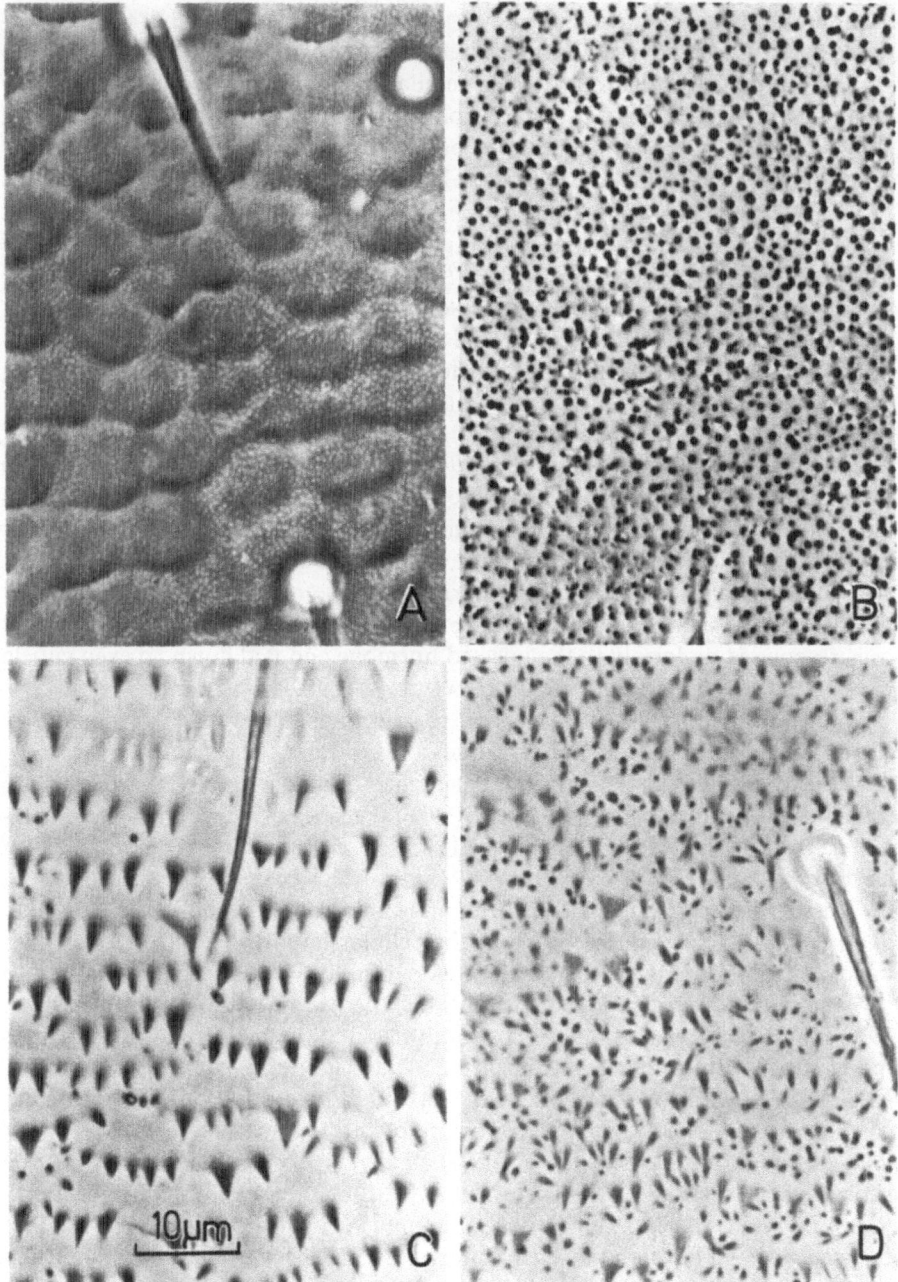

Fig. 2 A–D. Phase-contrast photographs of larval (A) and adult (B) cuticle from the same region of *Oncopeltus fasciatus*. The larval cuticle is thick, darkly pigmented and smooth; the adult cuticle, thin, transparent and covered with small tubercles. Anterior is at the top of the page. (C) and (D) are two intermediate cuticles produced as a result of injecting juvenile hormone late in the molt cycle. (From LAWRENCE, 1969)

each secretes its own patch of cuticle (WIGGLESWORTH, 1940) so that the differentiated state of sister cells can be monitored. WIGGLESWORTH (1940) in fact noted a general temporal correlation between cellular divisions and response to juvenile hormone introduced during the last molt in *Rhodnius*. KRISHNAKUMARAN et al. (1967) first tested experimentally the possibility of a causal link between cell division and cellular metamorphosis. They suggested that "in Saturniid moths DNA synthesis seems to be an early obligatory step in the differentiation of most cells" and asked: "Is it possible that a genome... cannot be switched to a new kind of function without replication ...?" Even their own experiments provided some exceptions: while most of the epidermal cells underwent DNA synthesis during the larval-pupal molt, some did not. Moreover, in *Rhodnius* (WIGGLESWORTH, 1940, 1963) epidermal and bristle-forming cells can make larval and then adult structures without intervening mitosis, and probably without DNA synthesis. *Tenebrio* epidermal cells can also secrete pupal, then adult, cuticle without intervening mitoses (WIGGLESWORTH, 1948; CAVENEY, 1973). In *Oncopeltus* it was possible, by injecting a juvenile hormone analogue, to partially inhibit metamorphosis late in the last molt. At that stage all but a few cell divisions had already occurred in the absence of juvenile hormone. If there were increased sensitivity of individual dividing cells to juvenile hormone, those remaining divisions should give rise to sparse pairs of cells secreting a larval cuticle. In fact, all the epidermal cells responded uniformly to the injected hormone and laid down an even cuticle of intermediate larval/adult structure (LAWRENCE, 1969) (Fig. 2).

Galleria epidermis is also sensitive to injections of juvenile hormone after many cell divisions have occurred. Large doses injected 7 days after the previous larval ecdysis can still result in the emergence of a supernumerary larval stage (SEHNAL, 1971). In *Tenebrio*, CAVENEY (1970) provides evidence that one epidermal cell, during the deposition of a single cuticle, can secrete first larval and then adult structure.

These experiments rule out a general and simple relation between cell division and response to juvenile hormone at the cellular level. Nevertheless, some evidence for local correlation between DNA synthesis and increased sensitivity to juvenile hormone is mentioned, but not documented in SCHNEIDERMAN (1969a; 1972). After simultaneous injection of both tritiated thymidine and juvenile hormone into *Antheraea* pupae, the resulting adults had pupal patches which "coincided with patches of cells that showed maximum incorporation of the isotope". This experiment would seem to demonstrate a correlation between cell division and juvenile hormone sensitivity, although the experiments previously quoted show that cell division is not a necessary condition for response to the hormone. SEHNAL (1969; SEHNAL and NOVAK, 1969), working with *Galleria*, noted that injected juvenile hormone failed to affect regions where epidermal mitosis had already occurred; he suggested that mitoses in the absence of juvenile hormone could promote metamorphosis in the dividing cells and confer resistance to later injection of hormone. SEHNAL (1969) suggested that this resistance might spread from the dividing cells into nearby cells.

The evidence linking wound healing with increased sensitivity to juvenile hormone is much more convincing. One assay for the hormone is based on the observation that a small wound dramatically increases local sensitivity to the hormone (SCHNEIDERMAN and GILBERT, 1958). This observation has been made on several insects: *Rhodnius* (WIGGLESWORTH, 1936), *Galleria* (SCHNEIDERMAN et al., 1965; DE WILDE et al., 1968), *Tenebrio* (CAVENEY, 1970), and *Oncopeltus* (WILLIS and LAWRENCE, 1971). The cell division often associated with wound healing is credited with responsibility for increased response to juvenile hormone, but this is due rather to the attractiveness of the idea than to any real evidence.

Under the influence of juvenile hormone, adult epidermal cells of several insects are capable of secreting juvenile cuticle, and a general correlation between degree of reversal of metamorphosis and wounding has been noted. The best case is that of adult implants into larval *Galleria* (PIEPHO, 1939; PIEPHO and MEYER, 1951), where the part of the epidermis that undergoes successive cell divisions and grows round to form a vesicle makes larval cuticle, while that region of the epidermis that descends more directly from the implant fails to do so and continues to make adult cuticle. Similarly, after grafting of adult patches onto larval *Oncopeltus*, the wounded edges of the graft make larval bristles before the centres do (LAWRENCE, 1966 b; 1967). Even in these cases the importance of cell divisions and/or DNA synthesis *per se* is far from being established, and wound effects independent of mitosis could still be the critical factor.

In spite of this continued lack of compelling evidence, the idea that insect cells are sensitive to juvenile hormone only when undergoing DNA synthesis has gathered its own momentum, and is now stated with increasing confidence in recent publications (METWALLY and SEHNAL, 1973; RIDDIFORD and AJAMI, 1973). In a paper on *Drosophila* abdomen by GUERRA et al. (1973) the argument has turned full circle: "the self-same abdominal epidermal cells secrete both a larval and a pupal cuticle. Apparently the DNA replication without cell division which occurs in the abdominal epidermal cells in the last larval instar permits the larval epidermal cells to be reprogrammed to secrete pupal cuticle (SCHNEIDERMAN, 1971)". Although there may yet prove to be something in the "quantal" mitosis hypothesis when applied to insect cellular metamorphosis, I believe such confidence to be inappropriate.

It may be that cell divisions are essential for maturation of imaginal disks of *Drosophila*. When 1st-stage and early 2nd-stage disks are implanted into metamorphosing hosts, they fail to differentiate and secrete cuticle (BODENSTEIN, 1939) whereas late 2nd-stage disks have acquired the competence to do so. MINDEK and NÖTHIGER (1973) have shown that the competence correlates with cell division; early 2nd-stage disks kept in an adult for a long time under conditions where they undergo no cell divisions will not aquire the competence to metamorphose. If they are kept under different conditions in an adult host so that they do divide, they become competent.

V. Pattern Formation and the Cell Cycle

Whether or not a cellular state is propagated to the daughter cells is clearly of importance to development. Let us consider two possible extreme situations to make this clear. At one extreme, each group of cells of the same type must always descend from one progenitor cell (a strictly clonal origin implying pattern-formation mechanisms that select individual cells, and thereafter a faithful propagation of the determined state). At the other extreme, the cells of one type have no more lineage in common than any other randomly selected group of adjacent cells (implying selection of cellular function purely on the position of the cells in question). As the following examples show, insect development fits neither of these extremes. SCHNEIDERMAN (1969 b) made an attempt to classify cell states as they relate to the cell cycle. He distinguished between 'firm biases' (propagated by cell heredity and equivalent to determination) and 'transitory biases' which have to be reestablished at each division. This is an educative distinction, as there are some cell states that can apparently be propagated indefinitely, for example 'legness' in imaginal disk cells (SCHUBIGER, 1968), while other cell states are more temporary; for instance, in *Galleria* the local position within an abdominal segment normally determines the type of cuticle produced. Grafting experiments have proved that the cells can change and, in association with cell division, make a different kind of cuticle (MARCUS, 1962; LAWRENCE, 1970). Grafting between different levels in the gradient of the cockroach femor can actually stimulate cell division and result in the generation of new tissue–the amount depending on the original gradient difference at the graft-host junction (BOHN, 1967).

Other grafting experiments on the stability of polarity have shown that polarity changes may be associated with cell division. After 90–degree rotation of square pieces of larval epidermis in *Rhodnius*, the polarity of the rotated cells (as expressed by the orientation of cuticular folds) progressively turns back toward normal. If the rotations were made through 90 degrees clockwise, the polarity rotates back anticlockwise and *vice versa* (LOCKE, 1959; LAWRENCE et al., 1972). If the adult *Rhodnius* is made to molt again artificially by the injection of molting hormone, the continued rotation is correlated with mitoses. Larger doses of molting hormone cause secretion of a new cuticle without cell divisions and in these cases the supernumerary cuticle bears an identical pattern of cuticular folds to the previous one (LAWRENCE et al., 1972). Recent evidence from three different insects has shown that this rotation is at least partly due to the bodily movement of the grafted cells (BOHN, LAWRENCE, NÜBLER-JUNG; unpubl. observations). Why this process should be correlated with cell divisions remains unclear.

The prospective fate of a cell may be much broader in experimental conditions than in its normal situation. In order to understand development, we need to know much more about the constraints that operate *in situ* to control the growth rate and distribution of cells. Some exciting experiments by GARCIA-BELLIDO and colleagues on *Drosophila* wing disc have just begun to describe these constraints (GARCIA-BELLIDO et al., 1973). X-ray-induced somatic crossing-over was used to generate marked clones (homozygous for the wild-type allele of a *Minute*; M^+/M^+) which had a considerably enhanced growth rate as compared to the remaining unmarked cells (M^+/M). It was found that the progeny of the marked cells were unable to spread as their growth rate should have allowed, but were instead confined to rigidly defined areas on the wing, with the borders of the clones consistently defining the edges of these "compartments". Irradiation at different times showed that progressively during wing development the wing disk is serially subdivided into groups of cells whose prospective fate thereafter is confined to smaller regions. This process is completed when the wing blade is subdivided into 8 compartments. Provided the clone is generated after the time of final subdivision, it will occupy only one compartment. Differently marked clones with the same growth rate as the background were found to be very variable in both shape and size, showing that within a compartment the lineage is not determined. Whether a cell forms, for example, a marginal bristle, a piece of vein, or a piece of wing surface, could perhaps depend on position and not on its ancestry. These experiments provide a new way of looking at determination *in vivo*. Characteristics that distinguish compartments are probably propagated by cell heredity, while cellular differences within a compartment may depend on cell position at the time of differentiation.

VI. Conclusion

While there are certain special divisions (asymmetric divisions) that are essential to generate diversity, the role of symmetric divisions is unclear. There is as yet no convincing evidence in insects that symmetric mitoses are *the* essential component in the subsequent differentiation of the daugther cells; no convincing evidence, that is, for "quantal mitoses" in HOLTZER's sense (HOLTZER et al., 1972).

Acknowledgements

I thank Drs. JOHN GURDON, GRAEME MITCHISON, and MICHAEL WILCOX for their constructive criticism of the manuscript.

References

ANDERSON, S. O.: Comparison between the sclerotisation of adult and larval cuticle in *Schistocerca gregaria*. J. Insect Physiol. **19**, 1603–1614 (1973).

BERNARD, F.: Recherches sur la morphogénèse des yeux composés d'arthropods. Bull. Biol. Fr. Belg. Supl. **23**, 1–162 (1937).

BODENSTEIN, D.: Investigations on the problem of metamorphosis. V. Some factors determining the facet number in the *Drosophila* mutant *Bar*. Genetics **24**, 494–508 (1939).

BOHN, H.: Transplantations Experimente mit interkalarer Regeneration zum Nachweis eines sich segmental wiederholenden Gradienten im Bein von *Leucophaea* (Blattaria) Zool. Anz. (Suppl.) **30**, 499–508 (1967).

CARLSON, J. G.: Microdissection studies of the dividing neuroblasts of the grasshopper *Chortophaga viridifasciata* (De Gees). Chromosoma **5**, 199–220 (1952).

CAVENEY, S.: Juvenile hormone and wound modelling of *Tenebrio* cuticle architecture. J. Insect. Physiol. **16**, 1087–1107 (1970).

CAVENEY, S.: Stability of polarity in the epidermis of a beetle, *Tenebrio molitor* L. Develop. Biol. **30**, 321–335 (1973).

CLEVER, U.: Untersuchungen zur Zelldifferenzierung und Musterbildung der Sinnesorgane und des Nervensystems im Wachsmottenflügel. Z. Morph. Ökol. Tiere **47**, 201–248 (1958).

CLEVER, U.: Der Einfluß der Sinneszellen auf die Borstenentwicklung bei *Galleria mellonella* L. Wilhelm Roux' Arch. **152**, 137–159 (1960).

DE WILDE, J., STAAL, G. B., DE KORT, C. A. D., DE LOOF, A., BAARD, G.: Juvenile hormone titer in the hemolymph as a function of photoperiodic treatment in the adult colorado beetle (*Leptinotarsa decemlineata* Say) Proc. Roy. Neth. Acad. Sci. **77**, 321–326 (1968).

GARCIA-BELLIDO, A., MORATA, G., RIPOLL, P.: Developmental compartmentalisation of the wing disk of *Drosophila*. Nature New Biol. **245**, 251–253 (1973).

GUERRA, M., POSTLETHWAIT, J. H., SCHNEIDERMAN, H. A.: The development of the imaginal abdomen of *Drosophila melanogaster*. Develop. Biol. **32**, 361–372 (1973).

HANSON, T. E., READY, D. F., BENZER, S.: Use of mosaics in the analysis of pattern formation in the retina of *Drosophila*. A. R. CalTech, p. 40 (1973).

HENKE, K.: Über die verschiedenen Zellteilungsvorgänge in der Entwicklung des beschuppten Flügelepithels der Mehlmotte *Ephestia kühniella*. Biol. Zbl. **65**, 120–136 (1946).

HINTON, H. E.: Metamorphosis of the epidermis and hormone mimetic substances. Science Progress **51**, 306–322 (1963).

HOLTZER, H., WEINTRAUB, H., MAYNE, R., MOCHAN, B.: The cell cycle, cell lineages and cell differentiation. Curr. Top. Dev. Biol. **17**, 229–256 (1972).

KRISHNAKUMARAN, A., BERRY, S. J., OBERLANDER, H., SCHNEIDERMAN, H. A.: Nucleic acid synthesis during insect development. II. Control of DNA synthesis in the *Cecropia* silk-worm and other saturniid moths. J. Insect. Physiol. **13**, 1–57 (1967).

KÜHN, A.: Vorlesungen über Entwicklungsphysiologie, Berlin-Heidelberg-New York: Springer 1965.

LAWRENCE, P. A.: Development and determination of hairs and bristles in the milkweed bug *Oncopeltus fasciatus* (Lygaeidae, Hemiptera). J. Cell Sci. **1**, 475–498 (1966 a).

LAWRENCE, P. A.: The hormonal control of the development of hairs and bristles in the milkweed bug *Oncopeltus fasciatus* Dall. J. Exp. Biol. **44**, 507–522 (1966 b).

LAWRENCE, P. A.: The insect epidermal cell—'a simple model of the embryo'. In: J. W. L. BEAMENT, J. E. TREHERNE (Eds.): Insects and Physiology, pp. 53–68. Edinburgh-London: Oliver and Boyd 1967.

LAWRENCE, P. A.: Mitosis and the cell cycle in the metamorphic moult of the milkweed bug *Oncopeltus fasciatus*. A radioautographic study. J. Cell Sci. **3**, 391–404 (1968).

LAWRENCE, P. A.: Cellular differentiation and pattern formation during metamorphosis of the milkweed bug *Oncopeltus*. Develop. Biol. **19**, 12–40 (1969).

LAWRENCE, P. A.: Polarity and pattern in the postembryonic development of insects. Advan. Insect Physiol. **7**, 197–266 (1970).

LAWRENCE, P. A.: The development of spatial patterns in the integument of insects. In: S. J. COUNCE, C. H. WADDINGTON (Eds.): Developmental Systems: Insects Vol. II, pp. 157–209. New York-London: Academic Press 1973.

LAWRENCE, P. A., CRICK, F. H. C., MUNRO, M.: A gradient of positional information in an insect, *Rhodnius*. J. Cell Sci. **11**, 815–853 (1972).

METWALLY, M. M., SEHNAL, F.: Effects of juvenile hormone analogues on the metamorphosis of the beetles, *Trogoderma granarium* (Dermestidae) and *Caryedon gonagra* (Bruchidae). Biol. Bull. **144**, 368–382, (1973).

MINDEK, G., NÖTHIGER, R.: Parameters influencing the acquisition of competence for metamorphosis in imaginal discs of *Drosophila*. J. Insect Physiol. **19**, 1711–1720 (1973).

OKAZAKI, K., HOLTZER, H.: Myogenesis: fusion, myosin synthesis, and the mitotic cycle. Proc. Natl. Acad. Sci. **56**, 1484–1490 (1966).

PETERS, W.: Die Sinnesorgane an den Labellen von *Calliphora erythrocephala* Mg. (Diptera). Z. Morph. Ökol. Tiere **55**, 259–320 (1965).

PIEPHO, H.: Raupenhäutungen bereits verpuppter Hautstücke bei der Wachsmotte *Galleria mellonella* L. Naturwissenschaften **27**, 301–302 (1939).

PIEPHO, H., MEYER, H.: Reaktionen der Schmetterlingshaut auf Häutungshormone. Biol. Zbl. **70**, 252–260 (1951).

RIDDIFORD, L., AJAMI, A. M.: Juvenile hormone: its assay and effects on pupae of *Manduca sexta*. J. Insect Physiol. **19**, 749–762 (1973).

SCHNEIDERMAN, H. A.: Endocrinological and genetic strategies in insect control. Symposium on potentials in crop protection. Proc. N. Y. Agric. Exp. Stat. (Geneva, N. Y.) 14–25 (1969 a).

SCHNEIDERMAN, H. A.: Control systems in insect development. In: DEVONS, S. (Ed.): Biology and the Physical Sciences, pp. 186–208, 1969 b.

SCHNEIDERMAN, H. A.: The strategy of controlling insect pests with growth regulators. Mitt. Schweiz. Entomol. Ges. **44**, 141–149 (1971).

SCHNEIDERMAN, H. A.: Insect hormones and insect control. In: MENN, J. J., BEROZA, M. (Eds.): Insect Juvenile Hormones: Chemistry and Action, pp. 3–27. New York-London: Academic Press 1972.

SCHNEIDERMAN, H. A., GILBERT, L. I.: Substances with juvenile hormone activity in Crustaceans and other invertebrates. Biol. Bull. (Woods Hole) **115**, 530–535 (1958).

SCHNEIDERMAN, H. A., KRISHNAKUMARAN, A., KULKARN, V. G., FRIEDMAN, L.: Juvenile hormone activity of structurally unrelated compounds. J. Insect. Physiol. **11**, 1641–1650 1965.

SCHUBIGER, G.: Anlageplan, Determinationszustand und Transdeterminationsleistungen der männlichen Vorderbeinscheibe von *Drosophila melanogaster*. Wilhelm Roux' Arch. **160**, 9–40 (1968).

SEHNAL, F.: Action of the juvenile hormone on the metamorphosis of *Galleria mellonella* epidermal cells. Gen. Comp. Endocr. **13**, Abstract 130 (1969).

SEHNAL, F.: Endocrines of arthropods. Chem. Zool. **6**, 307–345 (1971).

SEHNAL, F., NOVAK, V. J. A.: Morphogenesis of the pupal integument in the waxmoth (*Galleria mellonella*) and its analysis by means of juvenile hormone. Acta Entomol. Bohem. **66**, 137–145 (1969).

SHELTON, P., LAWRENCE, P. A.: J. embryol. exp. Morph. In press.

SMOLKA, H.: Untersuchungen an Kleinorganen im Integument der Mehlmotte. Biol. Zbl. **77**, 437–478 (1958).

STOSSBERG, M.: Die Zellvorgänge bei der Entwicklung der Flügelschuppen von *Ephestia kühniella* Z. Z. Morph. Ökol. Tiere **34**, 173–206 (1938).

WHITTEN, J. M.: Kinetosomes in insect epidermal cells and their orientation with respect to cell symmetry and intercellular patterning. Science **181**, 1066–1067 (1973).

WIGGLESWORTH, V. B.: The function of the corpus allatum in the growth and reproduction of *Rhodnius prolixus* (Hemiptera). Quart. J. Microsc. Sci. **79**, 91–121 (1936).

WIGGLESWORTH, V. B.: The determination of characters at metamorphosis in *Rhodnius prolixus* (Hemiptera). J. Exp. Biol. **17**, 201–222 (1940).

WIGGLESWORTH, V. B.: The structure and deposition of the cuticle in the adult mealworm, *Tenebrio molitor*. Quart. J. Microsc. Sci. **89**, 197–218 (1948).

WIGGLESWORTH, V. B.: The action of moulting hormone and juvenile hormone at the cellular level in *Rhodnius prolixus*. J. Exp. Biol. **40**, 231–245 (1963).

WILLIS, J. H., LAWRENCE, P. A.: Deferred action of juvenile hormone. Nature **225**, 81–83 (1970).

Nuclear Transplantation and the Cyclic Reprogramming of Gene Expression

J.B. GURDON

Medical Research Council, Laboratory of Molecular Biology, Cambridge, England

I. Introduction

The idea that some event in the cell cycle may be important in making available to a daughter cell genetic information that is not readily available in the parent cell has been interpreted in various more specific ways according to the cell system being studied. On the basis of nuclear transplantation results we have suggested (review by GURDON and WOODLAND, 1970) that nuclear swelling and chromosome condensation may be the important cell cycle events, rather than DNA synthesis or any other aspect of mitosis. In this article, we (A) outline the way in which the genetic reprogramming elicited by nuclear transplantation can in principle contribute to our understanding of the relationship between the cell cycle and the control of gene expression; (B) summarize the relevant results of nuclear transfer experiments so far performed, and (C) offer an interpretation of these and some other experiments, emphasizing the potential importance of cycles of nuclear swelling and chromosome condensation.

A. The Reprogramming of Gene Expression Elicited by Nuclear Transplantation

The proposal that a relationship exists between the cell cycle and cell differentiation may be phrased more precisely in the question: Is there some event in the cell cycle that must take place before a gene (or genes) inactive in a parent cell can be expressed in a daughter cell? The ideal condition in which to examine such a hypothesis is one in which the synthesis of a kind of RNA or protein can be demonstrated in a daughter cell but not in its parent cell. Such a condition is not commonly encountered under circumstances that are convenient for experimental analysis. This is partly because of the difficulty in recognising newly synthesized molecules in single cells, and partly because it is very hard to obtain a fully synchronized population of cells in which a clear change of gene expression takes place. From this last point of view nuclear transplantation is a potentially useful technique, since it provides an opportunity to study changes in gene expression in a single nucleus. Furthermore, the extent of the changes in gene activity elicited by nuclear transplantation is much greater than the changes that occur spontaneously in unmanipulated cells. The principal facts about nuclear transplantation, relevant to the present problem, are illustrated by the results of nuclear transfers from frog skin cells.

Cells growing out in culture from a small piece of skin taken from the foot-web of *Xenopus* spontaneously fill up with birefringent keratinous material about 8–10 days after outgrowth has commenced (REEVES and LASKEY, unpubl.). Recently, REEVES (unpubl.) has shown that over 99% of the cells that have grown out from skin tissue within 5 days contain cytoplasmic material which binds fluorescent anti-keratin antibody. Since the cells used as a source of nuclei for transplantation had been grown out from explanted skin for 6–7 days, we now know that virtually all such donor cells were differentiated in that they contained keratin. 8% of all nuclei transplanted from skin cells yield tadpoles with functional muscle, nerve, blood, and other cell types (LASKEY and GURDON, 1970). Therefore, in these cases it appears that a cell containing a nucleus engaged in the synthesis of keratin-mRNA was "reprogrammed" so that some of its mitotic progeny or daughter nuclei were involved in totally different kinds of activity, such as haemoglobin or myosin synthesis. There is no known way of converting a skin cell into other kinds of cells. Only very rarely does one type of specialized vertebrate cell give rise to daughter cells that spontaneously differentiate in a totally unrelated way (e.g. Wolffian regeneration); changes of this kind do not occur often or consistently enough to be very useful for analyzing the relationship between the cell cycle and changes in gene expression. Nuclear transplantation is one of very few experimental systems in which a major change from one type of specialized gene activity to another, quite different type can be experimentally induced in a consistent way.

The means by which nuclear transplantation can, in principle, be used to test the relationship between the cell cycle and cell differentiation involves the transplantation of skin cell nuclei to unfertilized eggs (in order to yield nuclear-transplant embryos, as just indicated) as well as to growing oocytes. When nuclei are transplanted to oocytes, as has been done for embryonic and adult–brain nuclei, they start to synthesize RNA if not already doing so, but never divide or undergo mitosis (GURDON, 1968). Such injected nuclei remain apparently normal for up to 3 days. In contrast, nuclei injected into eggs divide frequently, and have become part of a young tadpole within 3 days (even if a serial nuclear transfer step is included). Therefore a comparison of the synthetic activity of skin cell nuclei 3 days after transfer to eggs, on the one hand, and oocytes, on the other, could provide a clear test of the importance of cell division for eliciting a substantial reprogramming of gene expression.

In fact the experiment outlined has not yet been carried out, due to the difficulty that exists in recognising the continued synthesis of keratin mRNA in injected cells. This kind of mRNA might be recognised by use of complementary DNA, made from keratin mRNA, if enough keratin mRNA was synthesized in eggs. Alternatively, very small amounts of mRNA synthesized could be recognised if it was translated into keratin, which itself could be identified by immunological methods. As a result of direct message-injection experiments (LANE et al., 1971; GURDON et al., 1971), we now have reason to believe that all vertebrate messages are translatable in oocytes. It should therefore be possible to test for keratin synthesis in oocytes and eggs that received skin cell nuclei. If mitosis is a prerequisite for a major change in gene activity, such an experiment should result in keratin genes being switched off in eggs, in which transplanted nuclei undergo frequent mitoses, but not in oocytes, where they do not. In fact, as indicated in the next section, there are indirect reasons for thinking that gene activity is changed in both oocytes and eggs.

Various kinds of information have been obtained about the changes that immediately follow the transplantation of nuclei to eggs and oocytes. These are of some interest, since they are certainly connected temporarily, and may be connected causally, with induced changes in gene expression.

B. Changes Resulting from the Transplantation of Nuclei to Eggs and Oocytes

These have been reviewed in detail by GURDON and WOODLAND (1970). The relevant facts are these: Within half an hour of transplantation to eggs, nuclei cease all detectable RNA synthesis, and commence DNA synthesis. A few hours later when nuclear-transplant embryos have reached the blastula stage, the rate and frequency of DNA synthesis declines, and the sequential activation of the different classes of RNA synthesis characteristic of normal development takes place (Fig. 1). Therefore a few hours after injection into an egg, a transplanted nucleus and its daughter nuclei are indistinguishable from a zygote nucleus and its mitotic progeny. In contrast, nuclei injected into mature *oocytes* never synthesize DNA, but continue to synthesize RNA for up to 3 days. Blastula nuclei are of special interest because at this stage they are totally inactive in RNA synthesis. When injected into oocytes, such nuclei cease DNA synthesis and mitosis, and start synthesizing RNA. Within a few hours, nuclei injected into oocytes acquire prominent nucleoli (Fig. 2), or in the case of adult nuclei which already possess nucleoli, their nucleoli enlarge. The close correlation between the presence of nucleoli and the synthesis of rRNA in *Xenopus* and other amphibia suggests that these nuclei are synthesizing rRNA. If so, this means that blastula nuclei can be induced to start rRNA synthesis without any intervening mitoses or periods of DNA synthesis. In normal development, mid-blastula cells would pass through at least four cell divisions before reaching the stage when rRNA synthesis is first detectable (early gastrula stage).

In addition to changes in activity, transplanted nuclei also undergo changes in morphology and structure. Starting immediately after injection, nuclei transplanted to either eggs or oocytes undergo a very substantial enlargement of up to 50 times, and a dispersion of their chromatin, just as is observed in egg and sperm nuclei after fertilization. Correlated with the increase in volume and the degree of chromatin dispersion, there is a movement of cytoplasmic proteins into transplanted nuclei (GURDON and WOODLAND, 1969; ECKER and SMITH, 1971). Changes in synthetic activity and the ingress of cytoplasmic proteins are not observed in those few nuclei which fail (perhaps because they have been damaged) to undergo enlargement and chromatin dispersion. The transplantation of nuclei containing labelled proteins failed to show any substantial movement of these

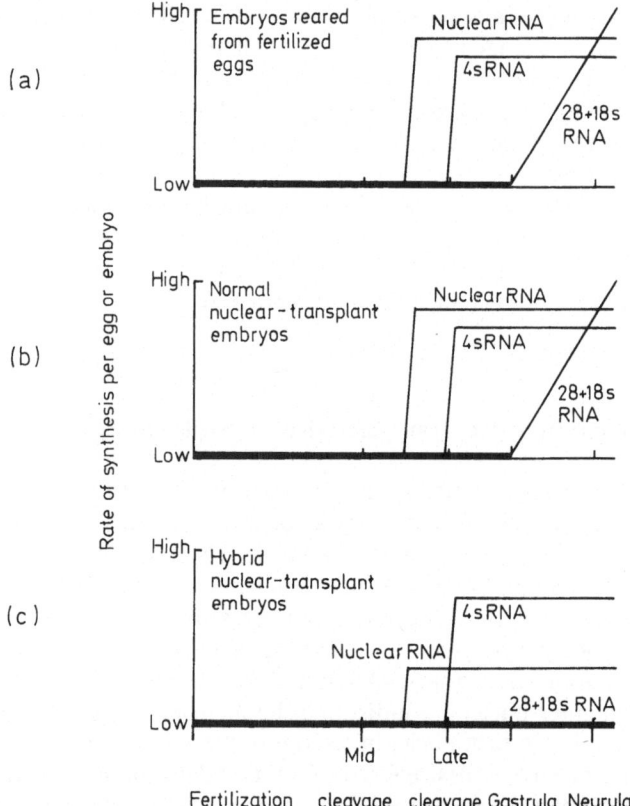

Fig. 1. (a) Sequential activation of RNA synthesis in *Xenopus* development. These diagrams show that three major classes of RNA synthesis are initiated independently in normal development. (b) Shows that a neurula nucleus, active in all classes of RNA synthesis, is reverted to an egg or embryo pattern of RNA synthesis, when transplanted to enucleated eggs; it and its daughter nuclei pass through the same sequence of events, even though the neurula nucleus has just completed this series of changes in activity in its development from a fertilized egg. (c) Shows that the factors which determine that sequential activations are truly independent, since some are normal and others abnormal when the nucleus of a foreign species is transplanted to *Xenopus* eggs. (From GURDON, 1974)

proteins from the nucleus into the surrounding egg cytoplasm at the time when the other events outlined above are taking place (GURDON, 1970).

The main conclusion from this type of study is that major changes in the structure, and in those synthetic activities which can be measured, take place in nuclei very soon after transplantation to eggs. In all respects these changes are the same as those which take place in egg, sperm, and cleavage nuclei normally present in fertilized eggs and embryos. The possibility clearly exists, though at present it is no more than a possibility, that these events are causally connected with the changes in gene activity known to take place at some time between nuclear transfer (or fertilization) and differentiation into an early embryo.

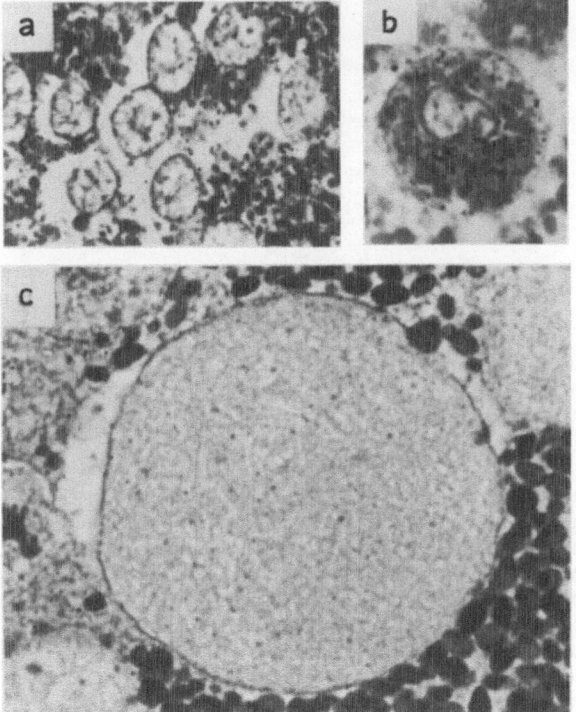

Fig. 2a–c. Changes in synthetic activity of a blastula nucleus transplanted to a growing oocyte, in Xenopus. A mid-blastula nucleus is inactive in RNA synthesis, but undergoes rapid replications about every 30 min. (a) Shows a blastula cell with enclosed nucleus. (b) Shows several blastula nuclei 2 hrs after injection into oocyte cytoplasm. (c) Shows one of such blastula nuclei 48 hrs after injection into an oocyte. During these two days it has failed to divide or synthesize DNA, but has undergone an enormous enlargement. Autoradiography shows it to have been active in RNA synthesis. Nucleoli are often seen in nuclei transplanted to oocytes. (Figure and further details from GURDON, 1968)

C. An Interpretation of Nuclear Transfer Results, Emphasizing the Role of Nuclear Enlargement and Chromatin Dispersion

We have seen that changes in RNA synthesis can apparently be induced in transplanted nuclei independently of mitosis and DNA synthesis. In view of the possibility that this may also be true of other (protein coding) genes, we now discuss, hypothetically, a type of regulation which would account for nuclear transfer results as well as for certain characteristics of gene expression in unmanipulated cells. Cytoplasmically induced changes in the activity of transplanted nuclei are presumed to be brought about by an exchange of chromosomal proteins that is itself promoted by cycles of nuclear enlargement and chromosome dispersion. These same events are also believed to take place during each mitosis of normal cells. The reasons for this hypothesis are outlined here.

Nuclear enlargement and chromosome dispersion take place at the end of each mitosis, as interphase nuclei are formed. The specific reason for regarding

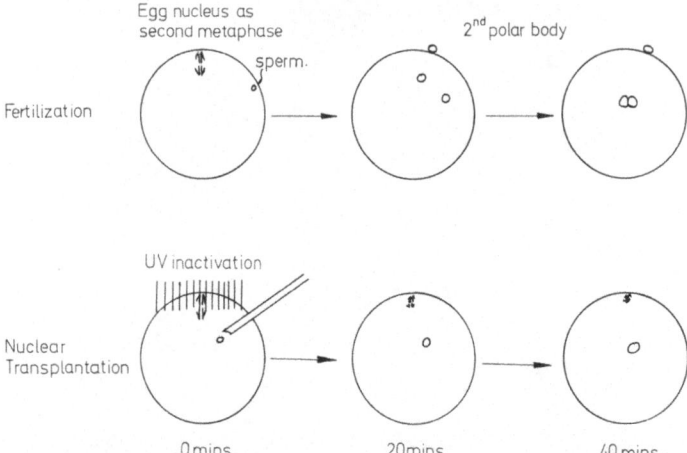

Fig. 3. Functional equivalence of post-mitotic nuclear reconstitution and nuclear swelling. Eggs of amphibia and of many other animal groups are laid in second meiotic metaphase. After fertilization (upper figure), the egg nucleus completes this division (which is comparable to mitosis), the second polar body is released, and the haploid egg pronucleus is reconstituted. The sperm nucleus, on the other hand, enters the egg as an interphase nucleus, as does a transplanted nucleus (lower figure). In all these cases, nuclei enlarge by up to fifty times in volume, and change their previous synthetic activity; they commence DNA synthesis simultaneously, and cease RNA synthesis if they had been active in this respect. From this time onward, the chromosomes of egg, sperm, or transplanted nuclei behave in exactly the same way. This correlation of chromosomal events suggests that the reconstitution of mitotic chromosomes may be functionally equivalent to the substantial swelling of sperm or transplanted nuclei

the enlargement of transplanted nuclei as functionally equivalent to the post-mitotic reconstitution of an interphase nucleus comes from a comparison of egg and sperm pronuclei after fertilization (Fig. 3). In unfertilized, unactivated eggs the egg nucleus is present as a metaphase spindle with associated chromosomes. Fertilization, or any other activation stimulus, causes the completion of this mitosis (sometimes called the second meiosis), so that the haploid egg-pronucleus is reconstituted, with decondensation of chromosomes and nuclear swelling taking place during the next 15–20 min. At the same time, the sperm nucleus, which enters the egg at fertilization as an interphase nucleus, also undergoes decondensation and enlargement. The behaviour of transplanted nuclei is like that of the sperm nucleus in that they enter the egg in interphase. Sperm and transplanted nuclei undergo the same morphological changes after egg activation, and are functionally equivalent thereafter. This is the reason for suggesting that the reprogramming believed to follow nuclear transplantation may also take place at each mitosis.

Reprogramming is visualized in the following way (Fig. 4): The activity of genes is assumed to depend on the association of regulatory proteins with particular regions of chromosomes. It is suggested that these proteins are dissociated from chromosomes as they become condensed for mitosis, and are released into the cell cytoplasm, where they become mixed with any other regulatory proteins that

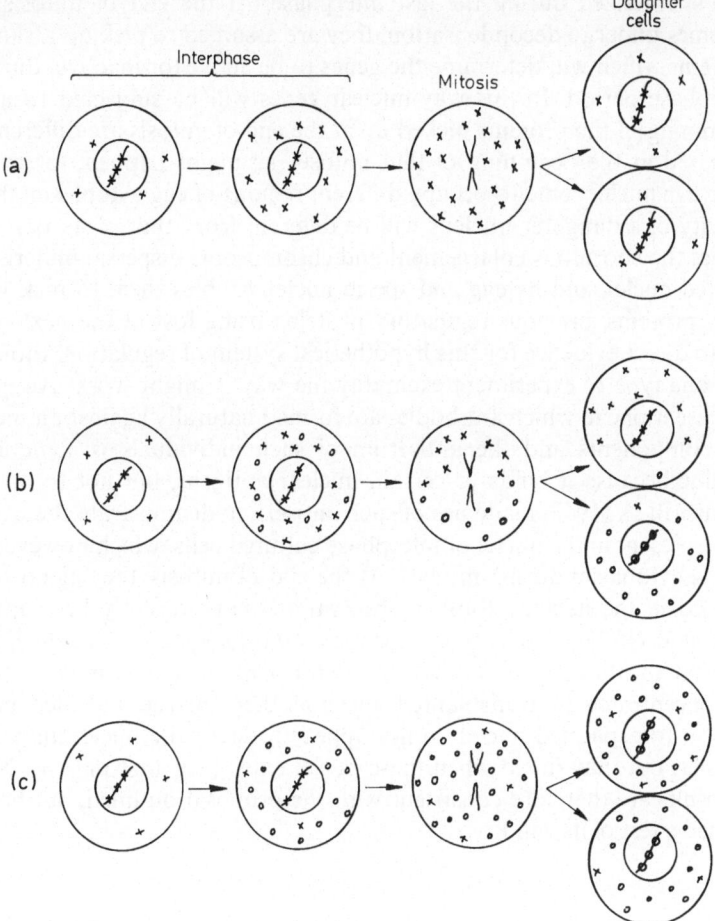

Fig. 4a–c. Hypothetical model of gene reprogramming and its association with chromosome condensation and with the nuclear enlargement that takes place after mitosis. It is suggested that gene activity is switched on or off by proteins which are synthesized in the cytoplasm (like all proteins), but which cannot pass through the nuclear membrane nor gain access to the genes which they regulate except during mitosis. (a) Represents a stably determined or differentiated state which is propagated through several divisions (e.g. HOLTZER and ABBOTT, 1968); in this case proteins which determine the activity of genes are synthesized throughout interphase, and reprogram the chromosomes of the daughter cells in the same way. (b) Represents an unequal mitosis, as happens in many plant tissues and in the meiotic divisions of eggs (where one daughter cell is the egg and the other a polar body); in this case a new gene becomes active during interphase, and its products are unequally distributed at mitosis, so that the two daughter cells have their chromosomes reprogrammed in different ways. (c) Shows a similar situation to that shown in (b), except that a large number of products of the newly active gene are synthesized, but are equally distributed at mitosis; as a result, both daughter cells are programmed for an activity different from that of the parent cell. It is assumed that when nuclei become distributed throughout an animal egg during cleavage, they will necessarily find themselves in different kinds of cytoplasmic environments; if some of the cytoplasmic materials that are unequally distributed in the egg influence gene activity, a situation comparable to that shown in (b) would exist. (See GURDON and WOODLAND, 1970, for further discussion)

had been synthesized during the last interphase. At the end of mitosis, when chromosomes undergo decondensation, they are assumed to pick up a sample of these proteins which will determine the genes to be active (or inactive) during the next interphase period. In this way nuclear genes will be subjected to a cyclic reprogramming; if the proteins picked up at the end of mitosis are different from those released at the beginning of that mitosis, as might happen, for example, when cleavage nuclei come to occupy different regions of egg cytoplasm, then the gene activity of a daughter nucleus will be different from that of its parent. We assume that the enormous enlargement and chromosome dispersal undergone by transplanted nuclei, and by egg and sperm nuclei, enables them to pick up new regulatory proteins, previous regulatory proteins being lost at the next mitosis. There is no direct evidence for this hypothetical system of regulation, though the results of one type of experiment exemplify the way it might work. An autoimmune disease is one in which antibodies are formed naturally against an individuals' own components and the antiserum of such individuals is sometimes of special value because it binds to certain nuclear antigens, but not to other cell components. BECK (1962) made use of such antisera to demonstrate the existence of antigens present in the nuclei of interphase cultured cells, which however move out into the cytoplasm during mitosis. At the end of mitosis, they again become concentrated in the nucleus. Similar observations have recently been made by RINGERTZ et al. (1971). With respect to nuclear transfer experiments, the labelling tests referred to above have shown that cytoplasmic proteins move into, and become concentrated in, transplanted nuclei as they enlarge. Labelled proteins contained in transplanted nuclei do not appear to leave the nuclei immediately after transfer, but may do so when these nuclei enter their first mitosis. Nuclear transfer results are therefore consistent with the proposal outlined, but have not yet provided a test of its validity.

II. Conclusions

Nuclear transplantation offers in principle a technique for examining the relationship between cell division and the control of gene activity. There is reason to believe that the activation and repression of ribosomal RNA genes can take place in the absence of cell division, and it is suggested that cycles of chromosome condensation and nuclear swelling may be important in "reprogramming" the activity of genes. This hypothesis has not yet been tested with respect to protein-coding genes.

References

BECK, J. S.: The behaviour of certain nuclear antigens in mitosis. Exp. Cell. Res. **28**, 406–418 (1962).
ECKER, R. E., SMITH, L. D.: The nature and fate of *Rana pipiens* proteins synthesized during maturation and early cleavage. Develop. Biol. **24**, 559–576 (1971).

GURDON, J. B.: Changes in somatic cell nuclei inserted into growing and maturing amphibian oocytes. J. Embryol. Exp. Morphol. **20**, 401–414 (1968).

GURDON, J. B.: Nuclear transplantation and the control of gene activity in animal development. Proc. Roy. Soc. B **176**, 303–314 (1970).

GURDON, J. B.: The Control of Gene Expression in Animal Development. Oxford: Clarendon Press 1974.

GURDON, J. B., LANE, C. D., WOODLAND, H. R., MARBAIX, G.: Use of frog eggs and oocytes for the study of messenger RNA and its translation in living cells. Nature **233**, 177–182 (1971).

GURDON, J. B., WOODLAND, H. R.: The influence of the cytoplasm on the nucleus during cell differentiation with special reference to RNA synthesis during amphibian cleavage. Proc. Roy. Soc. B **173**, 99–111 (1969).

GURDON, J. B., WOODLAND, H. R.: On the long-term control of nuclear activity during cell differentiation. Curr. Top. Develop. Biol. **5**, 39–70 (1970).

HOLTZER, H., ABBOTT, J.: Oscillations of the chondrogenic phenotype *in vitro*. In: URSPRUNG, H. (Ed.): Results and Problems in Cell Differentiation, Vol. 1, The Stability of the Differentiated State, pp. 1–16. Berlin-Heidelberg-New York: Springer 1968.

LANE, C. D., MARBAIX, G., GURDON, J. B.: Rabbit haemoglobin synthesis in frog cells: the translation of reticulocyte 9 s RNA in frog oocytes. J. Molu. Biol. **61**, 73–91 (1971).

LASKEY, R. A., GURDON, J. B.: Genetic content of adult somatic cells tested by nuclear transplantation from cultured cells. Nature **228**, 1332–1334 (1970).

RINGERTZ, N. R., CARLSSON, S.-A., EGE, T., BOLUND, L.: Detection of human and chick nuclear antigens in nuclei of chick erythrocytes during reactivation in heterokaryons with HeLa cells. Proc. Natl. Acad. Sci. **68**, 3228–3232 (1971).

Morphogenesis during the Cell Cycle of the Prokaryote, Caulobacter crescentus

Nancy B. Wood and Lucille Shapiro

Albert Einstein College of Medicine, Bronx, NY, USA

I. Introduction

The molecular machinery for the ordered unfolding of genetic potential, within the cell cycle, exists even in the simplest cells we know, the prokaryotes. Although all bacteria regulate their growth rate and assembly of internal structures, these events are difficult to study because they represent an organizational level too small to be resolved by traditional cytological methods and, until recently, too complex to dissect by purely biochemical methods. An example of such a multimolecular process is the replication of DNA in *E. coli*, which is now known to involve several protein components (Gefter et al., 1971; Scheckman et al., 1972; Wickner et al., 1972), probably occurs attached to the inside surface of the membrane (Ryter, 1968) and is expressed as a defined time function of the cell cycle (Donachie et al., 1973). Clearly, it would be advantageous to investigate a unicellular bacterium that undergoes gross morphological and biosynthetic changes during the cell cycle, so that events taking place within the cell can be conveniently monitored.

The dimorphic eubacterium, *Caulobacter crescentus*, which undergoes readily observable structural changes between cell divisions, is currently being studied in several laboratories. The unique morphogenesis exhibited by this bacterium, unlike the process of endospore-formation in other prokaryotes, occurs during each cell cycle. The normal cell cycle is diagrammed in Fig. 1. One of the daughter cells resulting from binary fission of *C. crescentus* is motile by means of a single polar flagellum. After 30 min in nutrient medium, the swarmer cell loses motility and then its flagellum; a flexible stalk is synthesized at the site of flagella attachment (Poindexter, 1964; Shapiro and Agabian-Keshishian, 1970). The stalked cell elongates and synthesizes a flagellum, several pili, and a phage receptor site at the pole distal to the stalk (Shapiro et al., 1971; Shapiro and Maizel, 1973). Division then produces a stalked and a swarmer cell. The essential characteristic of the *Caulobacter* cell cycle which makes it an interesting example of cell differentiation is that cell division yields daughter cells of differing phenotype. In addition to differing surface appendages, the DNA replication cycle is different in the daughter cells. The stalked cell immediately begins the replication of its DNA, whereas the swarmer cell delays DNA synthesis

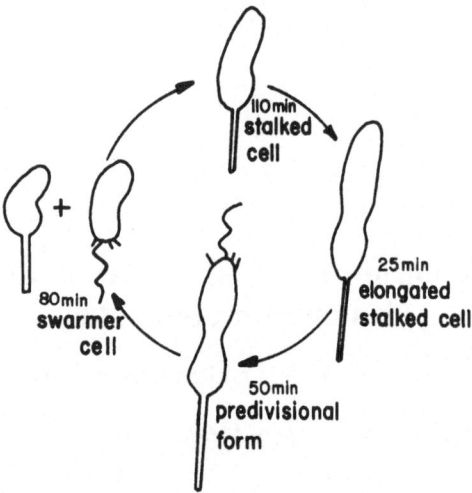

Fig. 1. Life cycle of *Caulobacter crescentus*

until later in the cell cycle (DEGNAN and NEWTON, 1972a). The *act* of cell division in *Caulobacter* is not solely responsible for the production of dissimilar daughter cells. The predivisional cell has already expressed the polarity which, upon binary fission, will yield two phenotypically different cells. How information is sequestered prior to cell division which is then expressed differentially in the progeny cells is a question basic to the understanding of differentiation in both prokaryotes and eukaryotes.

It is hoped that studies of this unique prokaryote will ultimately provide answers to questions dealing with the sequence and timing of morphogenic events, assembly of cell structures in proper spatial orientation, and the maintenance of cell polarity. Investigations of the *Caulobacter* cell cycle can be grouped into the following areas: (a) biochemical and structural definition of the cellular alterations exhibited during each cell cycle, (b) analysis of several classes of cell cycle mutants, and (c) studies of regulatory enzymes and effector molecules possibly related to cellular morphogenesis. In this article we review the structural and biochemical anatomy of the *Caulobacter* cell and then describe the experimental approaches being employed to define the regulatory mechanisms involved in *Caulobacter* morphogenesis.

II. The Caulobacter Cell Cycle

Structural and biochemical changes inherent in the *Caulobacter* cell cycle are diagrammed in Fig. 2. Since these data have been obtained from cultures growing in both nutrient broth and in defined minimal medium, and hence at both fast and slow rates, the time of appearance of each event is presented as a fraction of one division. The *Caulobacter* clock is constant in the sense that morphogenic events

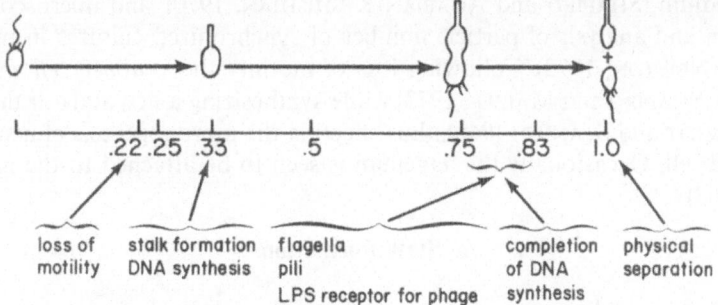

loss of   stalk formation   flagella   completion   physical
motility   DNA synthesis   pili   of DNA   separation
                                 LPS receptor for phage   synthesis

Fig. 2. Scheme of *Caulobacter crescentus* cell cycle events. 1.0 division units is the time between two successive separations of swarmer cells. (Data has been compiled from SHAPIRO and AGABIAN-KESHISHIAN, 1970 and NEWTON, 1972)

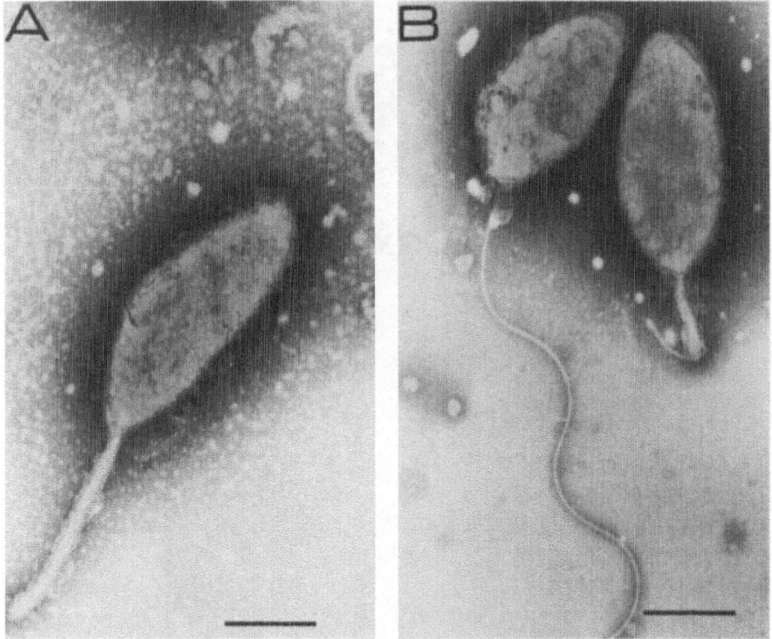

Fig. 3 A and B. Electron micrographs of *C. crescentus* cells grown in glucose minimal medium. (A) Stalked cell; (B) swarmer cell (left) and newly differentiated stalked cell (right) showing the flagellum still attached to the nascent stalk. Bar indicates 1 μm

appear to occur in the same order and in the same fraction of a life-time regardless of growth rate.

Beginning with the swarmer cell, the first event is a loss of motility at 0.22 division units. Evidence for this observation is provided by microscopic examination of clones in microculture (STOVE and STANIER, 1962), viable count and microscopic analysis of synchronized cultures of swarmer and stalked cells in nutrient medium (Shapiro and Agabian-Keshishian, 1970), and microscopic examination and analysis of particle number of synchronized cultures in minimal medium (Newton, 1972). Following loss of motility the *swarmer cell* sheds its flagellum (Shapiro and Maizel, 1973) while synthesizing a new stalk at the same site (Schmidt and Stanier, 1966), thus effecting the morphogenesis of a *swarmer* to *stalked* cell. Occasionally the flagellum is seen to be attached to the growing stalk (Fig. 3).

A. Stalk Formation

The stalk boundaries are a continuation of the cells' cytoplasmic membrane and cell wall (Poindexter, 1964) (Fig. 4). It has been demonstrated by autoradiographic analysis that new stalk-wall synthesis is restricted to the base of the nascent appendage (Schmidt and Stanier, 1966). This site-restricted wall syn-

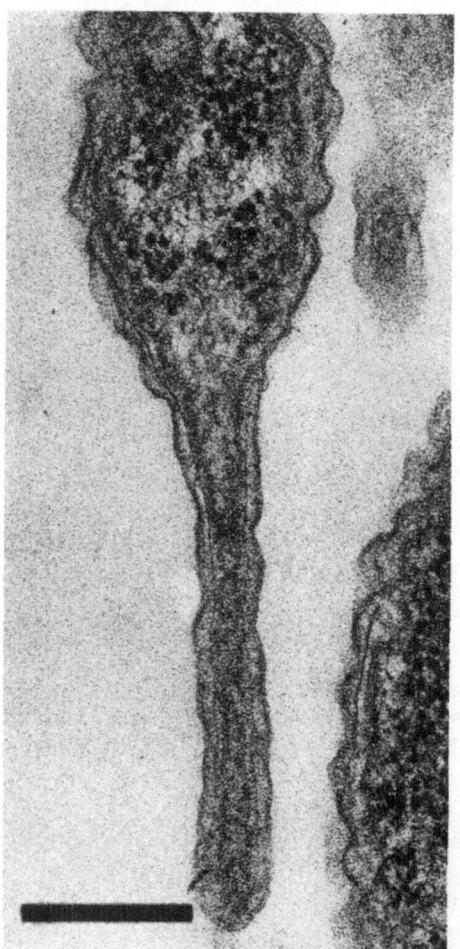

Fig. 4. Electron micrograph of a thin section of an osmium-fixed *C. crescentus* cell ($\times$ 81000; bar is 0.2 μm). (Courtesy of Dr. P. G. Model)

thesis, leading to the formation of a stalk, and occurring at a defined time during each cell cycle, represents a prokaryotic–morphogenic event unique to the *Caulobacter*, *Asticcacaulis* (PATE and ORDAL, 1965), and the *Hyphomycrobia* (MOORE and HIRSCH, 1973).

Fine structure analysis of *Caulobacter* stalks has revealed crossbands occurring at intervals perpendicular to the long axis of the stalk (POINDEXTER and COHEN-BAZIRE, 1964). The bands appear to be composed of murein, since they disappear upon treatment with lysozyme (SCHMIDT, 1973). Electron micrographs of the stalks indicate that the concentric rings of cross-band material form a barrier across the width of the stalk (JONES and SCHMIDT, 1973). Although it has been suggested that these bands constitute a support for the membranes that continue down the length of the stalk, the function of these structures is unknown. It has recently been suggested that the number of cross-bands reflects the age of the stalked cell (STALEY and JORDAN, 1973).

The stalk is a complex structure, yet very little is known about its function. Since stalked cells exhibit viscous drag when centrifuged, it has been suggested that stalks act as organs of flotation, maintaining cells near the air-water interface in their natural aquatic environment (POINDEXTER, 1964). Another suggestion is that the growth of a stalk provides an increased membrane surface area facilitating respiration and nutrient uptake (PATE and ORDAL, 1965). The surface area of *Caulobacter* cells and stalks was measured using electron micrographs of a population of *C. crescentus* CB 13, grown in nutrient medium (WOOD, unpubl.). Surface areas were calculated with the assumption that the cell is a prolate spheroid and the stalk is a right circular cylinder. Both the size of the cell and the length of the stalk varied in a mixed population. In the extreme case of the smallest cell (666 mm^2) with the longest stalk (285 mm^2) the stalk represents an increase in total surface area of about 40%. In the case of the largest cell (1153 mm^2) with the shortest stalk (66 mm^2), the increase is approximately 6%. Intracellular extension of the plasma membrane surface by the formation of structures such as mesosomes is known to occur in many prokaryotes, including the *Caulobacter*. In other groups of bacteria, special biochemical functions, such as photosynthesis, take place on intracellular lamellar or vessicular membrane structures. The elaboration of a complex extracellular structure, such as the *Caulobacter* stalk, may not simply be for amplification of existing membrane functions, but may provide some as yet unknown special function for the *Caulobacter* cell.

B. DNA Replication

Following the swarmer-to-stalked cell transition, an abrupt increase in DNA synthesis occurs, which is reduced prior to cell division. Experiments performed by DEGNAN and NEWTON (1972a) and in our laboratory (SUN and SHAPIRO, unpubl.) showed that incorporation of radioactive precursors (deoxyadenosine, deoxyguanósine, and adenine) into DNA was at a barely detectable level during the swarmer stage and increased coincident with the appearance of stalked cells. Whether the low level of incorporation by swarmer cell populations represents some degree of asynchrony, DNA repair, or a small amount of swarmer cell DNA synthesis has not been determined. It appears, however, that the *Caulobacter* genome is replicated during a limited time interval in the cell cycle and that the temporal control of DNA replication is linked to the division cycle (DEGNAN and NEWTON, 1972a). The absence of significant DNA synthesis during the swarmer stage of the cell cycle is, so far, unique to *Caulobacter;* uptake of labeled precursors has been observed during the entire life cycle of *Hyphomicrobium sp.,* a prosthecate organism that exhibits a pattern of morphological change similar to that of *Caulobacter* (MOORE and HIRSCH, 1973).

Rifampin, a drug known to inhibit bacterial RNA polymerase, will block *Caulobacter* DNA synthesis when added at the time of onset of replication (0.33 division units), or at any prior time during the swarmer stage (NEWTON, 1972). This finding implies that RNA synthesis may be coupled to or required for replication; but it is not known if the required RNA is used directly, such as in Okazaki fragments (SUGINO and OKAZAKI, 1973), or if it is the message for one or more proteins involved in the replication process.

C. Polar Morphogenic Events

The next event in the cell cycle is the simultaneous appearance of three surface structures at the pole opposite the stalk. These structures are the flagellum, a receptor site for the DNA phage ØCbK, and pili. The time of appearance of these structures on the elongated cell was measured by a variety of techniques.

1. Flagella

C. cresentus flagella are purified from the culture fluid since intact flagella are released into the growth medium during the normal cell cycle (Fig. 5). The *C. cresentus* filament terminates in a hook-like body. Protruding from the proximal end of the hook is a rod whose diameter differs from that of the filament. The structure of the intact *C. crescentus* flagella is analogous to the structure of the *E. coli* flagella complex, with the exception of the basal body (SHAPIRO and MAIZEL, 1973). It has been shown by DE PAMPHILIS and ADLER (1971) that the basal bodies of *E. coli* and *B. subtilis* flagella are made up of rings that attach to the cytoplasmic membrane, peptidoglycan, and lipopolysaccharide in the case of *E. coli,* and to the cytoplasmic membrane and peptidoglycan in the case of *B. subtilis.* Since basal bodies (rings) were never observed attached to the hooks and rods of *C. crescentus* flagella concentrated from the culture fluid, it is possible that the basal body remains on the cell during and after the development of the stalk, or that the rings composing the basal body are released into the medium. Attempts to prepare basal body-flagella complexes from spheroplasts of *C. crescentus* were unsuccessful.

The subunit protein of *C. crescentus* flagella (flagellin) comigrated with a marker protein of 25000 daltons on an SDS-polyacrylamide gel. The apparent molecular weight obtained in this way falls into a spectrum of molecular weights estimated for flagellin by other workers ranging from 14000 to 40000 daltons. Using purified flagellin as a marker, it was observed in synchronized cultures that

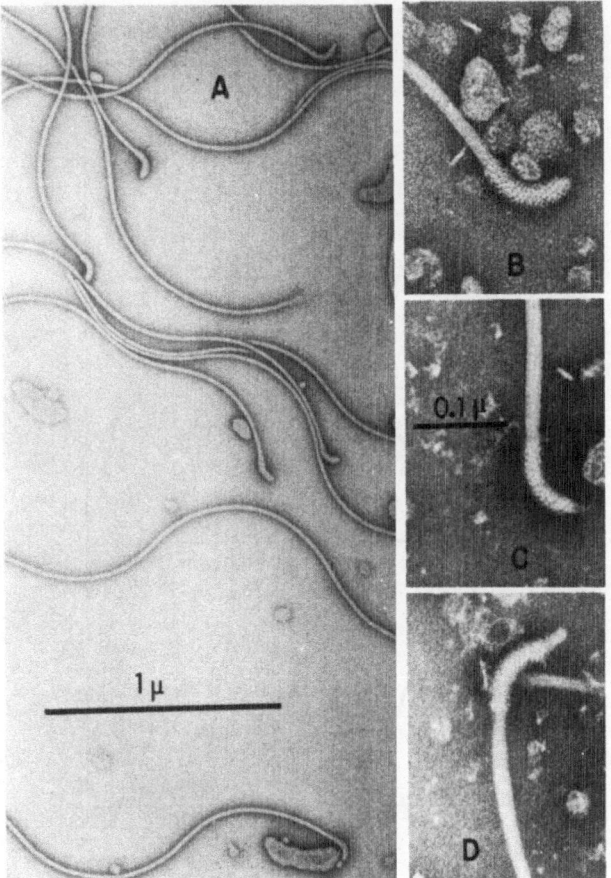

Fig. 5 A–D. Electron micrographs of purified *C. crescentus* flagella filaments and hooks. Flagella were prepared from culture supernatant fluid of *C. crescentus* grown to late exponential phase in glucose minimal medium. (A) Intact flagella; bar indicates 1.0 μm. (B, C, and D) Flagella hooks; bar indicates 0.1 μm. (From SHAPIRO and MAIZEL, 1973)

flagellin is synthesized between 0.67 and 0.73 division units (Fig. 6). The precise relationship between the time of synthesis of flagellin and its assembly into visible flagella has not yet been defined.

Since the flagella protein products are assembled in a precise location on the *Caulobacter* cell, at the cell pole, as well as at a precise point in the cell cycle, the type of controls which govern temporal transcription, as well as the spacial restrictions of the ultimate gene products, have become amenable to study. It was proposed many years ago that in prokaryotes a "localized" ribosome system which synthesizes flagellin might exist at the base of the flagellum, thereby controling site specificity (IINO and LEDERBERG, 1954). The factors that mediate site specificity in unicellular organisms, however, remain a mystery. Studies with *Caulobacter* provide a means for investigating the control of flagellin synthesis and assembly of flagella, as well as an approach to the problems of polarity in a unicellular prokaryote.

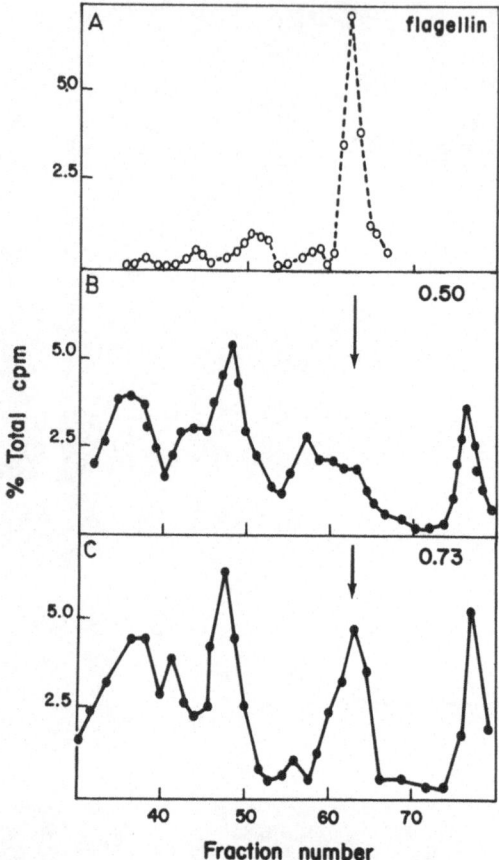

Fig. 6 A–C. Polyacrylamide gel electrophoresis pattern of ^{14}C proteins in synchronized cultures of *C. crescentus*. Synchrony was obtained by differential centrifugation and initiated with a population of stalked cells growing in nutrient medium. At 10 min intervals 2 ml samples were withdrawn and added to 100 μCi of ^{14}C-reconstituted protein hydrolysate. At the end of each 10 min pulse, 2 ml of 10% trichloroacetic acid was mixed with the culture and the samples were fractionated on 13% acrylamide gels for 18 hrs at 50 volts total. Direction of migration is to the right. (A) Purified flagellin. (B) Proteins from cells taken at 0.5 division units. (C) Cells taken at 0.73 division cycles. (Data adapted from SHAPIRO and MAIZEL, 1973)

2. Cell Envelope

It has been demonstrated that the *Caulobacter*-specific DNA bacteriophage ØCbK (AGABIAN-KESHISHIAN and SHAPIRO, 1970) selectively injects its DNA into the predivisional and swarmer cell (AGABIAN-KESHISHIAN and SHAPIRO, 1971). Electron micrographs of infected cells showed the phage selectively adsorbed to the flagella pole of the predivisional and swarmer cells. By measuring the time of injection of labeled phage DNA into synchronized host cells, it was determined that the phage receptor site was either synthesized or uncovered at 0.67 to 0.73 division units. That DNA phage infection occurs independently of the

presence of either flagella or pili was demonstrated by the successful infection of a
C. crescentus mutant lacking these structures (KURN et. al., 1974).

The lipopolysaccharide (LPS) from the cell wall of wild-type *C. crescentus* has
been analyzed by D. BUTTON and R. BEVILL. The overall composition of the LPS
for the swarmer cells did not differ significantly from that of the LPS from the
stalked cells. The neutral sugar composition determined by gas-liquid chromatography included D-glycero-D-mannoheptose, L-glycero-D-mannoheptose, D-glu-
cose, D-galactose, D-mannose and an unidentified polyhydroxy compound. The
LPS also contained 2-keto-3-deoxyoctulosonate (KDO), phosphate and glucos-
amine, although *o*-phosphoryl-ethanolamine and glucosamine-PO_4, components
of LPS in *E. coli* and *Salmonella*, were absent. The presence of myristic, β-hydrox-
ymyristic and lauric acids, usually found in the lipid moiety, could not be demonstrated. We concluded from these studies that although *C. crescentus* is a gram-
negative organism, it lacks the normal components of "lipid A" in its lipopolysaccharide.

3. Pili

Another event which occurs at 0.67 to 0.73 division units is the appearance of
pili, which can be observed at the flagella pole of the cell. Since the *Caulobacter*
RNA phage ØCb5 selectively adsorbs to pili (SCHMIDT, 1966), it was possible to
measure the time of pili appearance by mixing synchronized cultures with labeled
phage (Fig. 7). The adsorption of labeled RNA phage to synchronous cell popula-

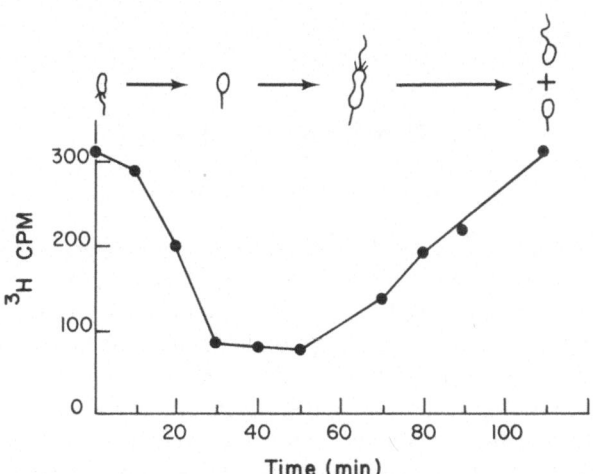

Fig. 7. Selective adsorption of RNA bacteriophage during synchronous growth of *C. crescentus*. A pure swarmer cell population was prepared, diluted with fresh nutrient medium and
incubated in a rotary shaker at 30° C. Aliquots (6×10^9 cells) were withdrawn at the times
indicated and mixed with ^{3}H-labeled bacteriophage ØCb5 (4×10^{10} phage; 5000 cpm total).
Adsorption was carried out 5 min at room temperature and the complex was collected,
washed, and counted on a millipore filter. ●—●—●, [^{3}H] phage adsorbed. [From SHA-
PIRO and BENDIS. In: RNA Phage (Eds. NORTON and ZINDER); Cold Spring Harbor (in Press)
1974]

tions revealed that pili adsorption sites were only available to the phage at the predivisional and swarmer cell stages (AGABIAN-KESHISHIAN and SHAPIRO, 1970). Thus, three different surface structures—flagellum, cell envelope phage receptor, and pili—appear at approximately the same time in the cell cycle and are assembled at the pole opposite the stalk of the parent cell.

D. Cell Division

The next event in the cell cycle is the completion of DNA replication which occurs by 0.83 division units (DEGNAN and NEWTON, 1972a). As is the case in *E.coli* (HELMSTETTER and PIERUCCI, 1968), it appears that the completion of the *Caulobacter* genome is required for cell division (DEGNAN and NEWTON, 1972b). Furthermore, it has been reported that in *C.crescentus* the parental and newly replicated genomes are randomly distributed to the two dissimilar sibling cells (OSLEY and NEWTON, 1973). Presumably the time between the completion of the genome and the physical separation of the two cells is occupied first by the segregation of the genomes, and second by the synthesis of a septum between the cells. After separation, which yields dissimilar daughter cells, the stalked cell presumably begins replication of DNA and repeats all of the events of the cycle from 0.33 division units on. It should be stressed that a given population of *Caulobacter* cells contains two types of stalked cells: those arising from the swarmer cell and those resulting from the asymmetric cell division. The pattern of DNA synthesis may vary in these two morphologically similar cells.

III. Regulation of Cell Cycle

A. Regulatory Mechanisms

Control mechanisms involved in prokaryotic morphogenesis must regulate both the timing of each cell cycle event and the structural basis of the observed polarity. Studies of bacterial cell regulation, accumulated over the past two decades, have shown that message transcription is the most likely target for a variety of control mechanisms. In attempting to define the mechanisms that organize the expression of a series of events within a given cell cycle, it cannot tacitly be assumed that transcription of the message for a given gene product at the proper time in the cell cycle automatically leads to the assembly of an intracellular or extracellular organelle. Further, the manner in which the cell dictates structure formation at one site rather than another may be several steps removed from the act of transcription. However, initial studies on the regulation of morphogenesis in *Caulobacter* questioned whether selective transcription was operative at some level in controlling the cell cycle. In this section we will consider experiments which relate transcriptional control to the temporal expression of morphogenic events.

1. In vivo RNA Synthesis

It has been demonstrated that continued RNA synthesis is required throughout the *Caulobacter* cell cycle (NEWTON, 1972). The drug rifampin, which inhibits the initiation of RNA synthesis, was added to synchronized cell cultures and several functions of the cell cycle were measured. RNA synthesis was apparently necessary for the loss of swarmer cell motility during the swarmer-stalked cell transition, initiation of chromosome replication, and cell division (NEWTON, 1972). Whether these effects are immediately attributable to the inhibition of specific enzymes or structural proteins, each relating to a given cell cycle function, or if they are a pleiotropic effect of the absence of some regulatory protein, awaits further investigation. An instance in which a specific protein is synthesized at a particular time in the cell cycle is that of flagellin synthesis (SHAPIRO and MAIZEL, 1973). The appearance of flagellin suggests specific transcription control, but does not establish it because the possibility of a stable messenger being formed at some earlier time has not been eliminated.

2. RNA Polymerase

It has been suggested that the subunit components of RNA polymerase change during sporulation in *Bacillus subtilis* (GREENLEAF et al., 1973). It is assumed that changes in template specificity of the purified enzyme reflect *in vivo* alterations in transcription patterns. The number of regulatory steps between alteration of transcription specificity and the vegetative-to-spore cell structural transition may, in fact, be multiple.

In order to determine if the structure and template specificity of the *Caulobacter* RNA polymerase varied during the cell cycle the polymerase was purified to homogeneity from swarmer, stalked, and predivisional cells (BENDIS and SHAPIRO, 1973). On SDS-polyacrylamide gels each of the enzyme preparations showed only four bands of molecular weight 165000, 155000, 101000, and 44000, respectively, corresponding to the β, β', α, and σ bands described for the *E. coli* polymerase. The *C. crescentus* enzyme appeared similar to the *Pseudomonas aeruginosa* enzyme and unlike the *E. coli* enzyme with respect to subunit molecular weights and failure to separate into core and sigma components upon phospho-cellulose chromatography. No other bands were visible even at high protein concentrations. The subunits of polymerase preparations from swarmer, stalked and predivisional cells co-migrated and there appeared to be no variation in the molecular weight of the polymerase subunits in the various *Caulobacter* cell types. Only small differences in template specificity, using dAT copolymer and phage ØCbK DNA, were observed. Studies of RNA polymerase from various *Caulobacter* cell forms have shown, therefore, that morphological and biochemical changes associated with development cannot be accounted for by major structural changes in DNA-dependent RNA polymerase.

B. Coordinate Control

A basic question relevant to the *Caulobacter* cell cycle is: to what extent are morphogenic events coupled—that is, how much does the appearance of an event depend on the previous or simultaneous appearance of other events? The components built at the cell pole, flagella, pili and LPS receptor site are all surface structures whose synthesis must be mediated by the cell's inner and outer membranes. A membranous structure has been observed at the base of the flagellum and ultimately the stalk (COHEN-BAZIRE et al., 1966), but the relation of this structure to the polar surface components is not known. Essential to defining the regulation of these morphogenic events is determining to what extent they are coordinately controlled.

1. Effect of Carbon Source on Cell Cycle

The first indication of possible coordinate control was the observation that flagella, pili, and receptor site not only appeared at a defined pole of the cell, but were coordinately expressed at a defined time in the cell cycle (SHAPIRO et al., 1971; SHAPIRO and MAIZEL, 1973). Additional evidence was obtained by interfering with the cell cycle physiologically. The normal pattern of morphogenesis is altered upon (a) maturation of the culture to stationary phase, or (b) changing the carbon source in minimal medium from glucose to lactose or galactose. In both cases, development in the culture is coordinately arrested at the non-motile elongated *stalked cell* stage, just prior to the expression of the various polar structures. It has been demonstrated that the normal development cycle of *C. crescentus* can be arrested by shifting cultures grown in minimal medium plus glucose to minimal medium containing lactose or galactose as the sole carbon source (SHAPIRO et al., 1972). Shifting mid-log phase cultures to conditions in which they must induce a series of enzymes in order to use a carbon source resulted in the accumulation of the cells prior to the predivisional stage; the cells stopped growth and cell division and did not display any of the surface differentiation events, including flagella, pili, and the phage receptor site. This block was eventually overcome spontaneously after 20–30 hrs (Fig. 8). Alternatively, growth, followed by the appearance of all three polar organelles, can be induced rapidly at any time during the block by adding 0.5% glucose or $3 \times 10^{-3} M$ dibutyrylcyclic AMP. During this diauxic lag there was no loss in viability. The effect of dibutyrylcyclic AMP on metabolically blocked cultures appears to be related to accelerated induction of the β-galactosidase system since: (a) using [^{3}H] lactose it was observed that lactose was not taken into the cells or utilized during tenure of the block (KURN and SHAPIRO, unpubl.), and (b) β-galactosidase activity was shown to be induced in the culture following addition of dibutyrylcyclic AMP (SHAPIRO et al., 1972).

Macromolecular synthesis during the diauxic lag in lactose and subsequent rescue by glucose was followed by incorporation of labeled precursors into RNA, protein, and DNA (FUKUDA and SHAPIRO, unpubl.). It was found that the rate of synthesis of DNA, RNA and protein decreased during the lag. Upon addition of glucose or dibutyrylcyclic AMP, the optical density of the culture increased immediately (within 5 min). Also, within 5 min the rate of incorporation of labeled

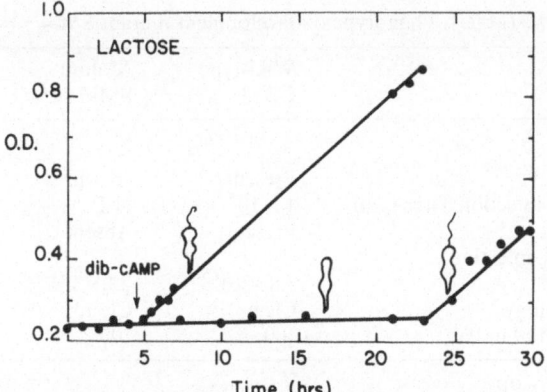

Fig. 8. The effect of dibutyrylcyclic AMP on β-galactosidase activity. *C. crescentus*, growing in minimal medium plus glucose (0.2%), was collected by centrifugation at 12000 × g for 10 min, washed and the pellet was gently resuspended in pre-warmed minimal medium containing 0.5% lactose. The absorbance at 660 nm was measured as a function of time. Dibutyrylcyclic AMP was added at 5 hrs to one part of the culture; the other part was maintained on lactose alone

precursors into RNA increased approximately 5-fold, followed by an increase in the rate of incorporation into protein, and lastly, by an increase in the rate of incorporation into DNA. This picture of macromolecular synthesis during a carbon source shift-down followed by a shift-up is esentially the same as that presented by MAALØE and KJELDGAARD (1966) for *Salmonella typhimurium*. What makes this physiological manipulation interesting in *Caulobacter*, however, is that the lag in growth during a nutritional shift-down is coincident with the arrest of morphological change at a particular stage in the life cycle; this is followed by a synchronous appearance of differentiated structures after the shift-up. The possibility exists that in the case of the glucose-to-lactose carbon source shift, the induction of the lactose operon is in some manner linked to the expression of the surface polar structures.

The DNA synthesis that takes place after the shift-up may represent the reinitiation of chromosome replication. It would be interesting to know if the completion of chromosome replication is coupled to the expression of the polar surface structures. The isolation of a pleiotropic mutant altered in surface structures, but with a normal cell cycle, suggests that the *assembly* of the polar structures is not required for normal cell growth and division (KURN et al., 1974). In addition, this pleiotropic mutant provides further evidence that the expression of the polar structures is, at some level, coordinately controlled.

2. Pleiotropic Morphogenic Mutant

Since mutants that lack motility and flagellin are known (SHAPIRO and MAIZEL, 1973), and mutants resistant to phage infection are likewise known (AGABIAN-KESHISHIAN and SHAPIRO, 1971), it appears that the various polar functions can occur independently of each other. A mutant exhibiting coordi-

Table 1. Phenotype of development mutant SM-4

	Wild-type CB13	Mutant SM-4 str^r	Revertant SM-4 str^r rev
Motility	+	–	+
Flagella	Present	Absent	Present
RNA phage ØCb5 infection Titer (pfu)	1×10^5	N.D.[b]	7.7×10^5
Pili[a]	Present	Absent	Present
DNA phage ØCbK infection			
Titer (pfu)	3.8×10^{10}	2.1×10^9	3.6×10^{10}
Plaque morphology	Clear	Cloudy	Clear
Generation time (min) in PYE	100	100	100

[a] By electron microscopy.
[b] Not detected.

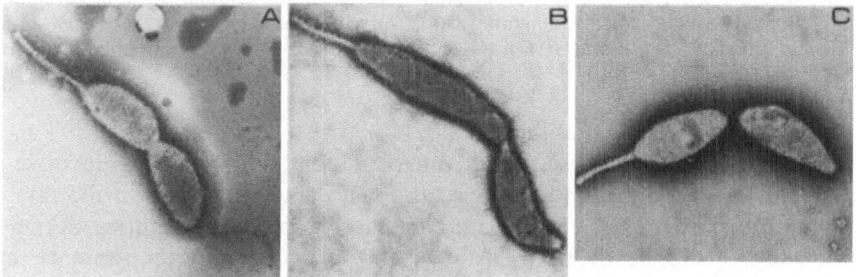

Fig. 9 A–C. Electron micrographs of wild-type *C. crescentus* CB13 (A), the polar coordinate mutant SM-4 (B), and the SM-4 revertant (C). Cultures were grown in nutrient medium to mid-log phase ($\times 10\,500$)

nate alteration of all surface polar functions, while maintaining normal growth and cell division, was isolated (Table 1). The SM-4 mutant did not support RNA phage infection, and infection by ØCbK yielded a 10-fold lower titer and cloudy plaques compared with the wild-type. Electron micrographs revealed that SM-4 lacked both flagella and pili (Fig. 9). An independent marker—streptomycin resistance—was introduced into the mutant strain and motile revertants of SM-4 str^{res} were selected on the basis of chemotaxis toward either glucose or lactose. Revertant strains were found simultaneously to regain normal polar structure and function.

The coordinate assembly of these polar structures is the only differentiation process found thus far that can be altered by a physiological change (nutritional shift-down). Thus, the response of the developmental mutant SM-4 to transfer from glucose to lactose-minimal medium and to external cyclic nucleotides should indicate the relative roles of growth and morphogenesis in this differentiation process. Mutant cultures supplemented with dibutyrylcyclic AMP did not regain motility or susceptibility to phage infection. As with wild-type cultures, the addition of dibutyrylcyclic AMP ($3 \times 10^{-3} M$) to mutant cultures blocked in lactose permitted normal growth and cell division, but did not rescue the expression of polar morphogenic events. These observations provided evidence that the differentiation of *Caulobacter crescentus* is regulated by at least two mechanisms. One of these, which responds to external addition of dibutyrylcyclic AMP, regulates cell growth and metabolism. Another mechanism regulates assembly of precursor molecules to form the polar structures. We suggest that the defect caused by the point mutation studied is not in gross cell wall structure, since it does not affect stalk formation, which is an extension of cell wall synthesis. Moreover, since stalk structure and site of formation are not affected, we suggest that the polarity of the cells is not impaired by this mutation.

In contrast to other non-motile mutants the coordinate mutation (SM-4) was found to synthesize the precursor molecule flagellin. At this stage we cannot identify other precursor molecules for the phage infection sites. The lack of pili in the mutant and regaining of pili in the motile revertant is supported both by susceptibility to RNA phage infection and electron microscopy. Since flagellin was present in mutant cells, we feel it is possible that the basic defect in this mutant is either in the mechanism involved in the assembly of the flagella or the mechanism regulating the assembly of the whole complex of these polar structures.

IV. Conclusions

A combination of biochemical studies and the analysis of mutants has shown that the *Caulobacter* cell cycle has two distinct categories of morphogenic alterations which are apparently controlled independently of one another. The synthesis of the surface polar structures, including flagella and pili, is under coordinate control and not obligatory to the life cycle, whereas the maintenance of cell polarity, exemplified by the proper orientation of stalk formation and temporal restriction of DNA synthesis, is essential.

Evidence for coordinate control of polar differentiation events includes: (a) flagella, phage receptor and pili are expressed coincidently at 0.67 to 0.73 division units, (b) a single mutant (SM-4) defective in all three polar structures has been isolated and, (c) in starved cultures, cells accumulate at a stage just prior to the formation of the three polar structures. That growth and cell division can be separated from surface morphogenesis during the cell cycle is suggested by the fact that surface morphogenesis is absent whereas growth and cell division are normal in the SM-4 mutant.

It appears at this point that although certain differentiation events can be deleted from the cell cycle, the order of events remains unchanged, and the polarity which leads to asymmetric cell division is preserved in all circumstances.

References

AGABIAN-KESHISHIAN, N., SHAPIRO, L.: Stalked bacteria: Properties of deoxyribonucleic acid bacteriophage ØCbK. J. Virol. **5**, 795–800 (1970).
AGABIAN-KESHISHIAN, N., SHAPIRO, L.: Bacterial differentiation and phage infection. Virology **44**, 46–53 (1971).

BENDIS, I. K., SHAPIRO, L.: Deoxyribonucleic acid-dependent ribonucleic acid polymerase of *Caulobacter crescentus*. J. Bacteriol. **115**, 848–857 (1973).

COHEN-BAZIRE, G., KUNISAWA, R., POINDEXTER, J. S.: The internal membranes of *Caulobacter crescentus*. J. Gen. Microbiol. **42**, 301–309 (1966).

DEGNAN, S. T., NEWTON, A.: Chromosome replication during development in *Caulobacter crescentus*. J. Molec. Biol. **64**, 671–680 (1972a).

DEGNAN, S. T., NEWTON, A.: Dependence of cell division on the completion of chromosome replication in *Caulobacter crescentus*. J. Bacteriol. **110**, 852–856 (1972b).

DE PAMPHILIS, M. L., ADLER, J.: Attachment of flagella basal bodies to the cell envelope: Specific attachment to the outer lipopolysaccharide membrane and the cytoplasmic membrane. J. Bacteriol. **105**, 396–407 (1971).

DONACHIE, W. D., JONES, N. C., TEATHER, R.: The bacterial cell cycle. In: Microbial Differentiation; Society for General Microbiology Symposia, Vol. 23. London: Cambridge University Press 1973.

GEFTER, M. L., HIROTA, Y., KORNBERG, T., WECHSLER, J. A., BARNOUX, C.: Analysis of DNA polymerases II and III in mutants of *Escherichia coli* thermosensitive for DNA synthesis. Proc. Natl. Acad. Sci. **68**, 3150–3153 (1971).

GREENLEAF, A. L., LINN, T. G., LOSICK, R.: Isolation of a new RNA polymerase-binding protein from sporulating *Bacillus subtilis*. Proc. Natl. Acad. Sci. **70**, 490–494 (1973).

HELMSTETTER, C. E., PIERUCCI, O.: Cell division during inhibition of deoxyribonucleic acid synthesis in *Escherichia coli*. J. Bacteriol. **95**, 1627–1633 (1968).

IINO, T., LEDERBERG, J.: Genetics of *Salmonella*. In: VON OYE, E. (Ed.): The World Problem of *Salmonellosis*, pp. 111–142. The Hague: Junk 1954.

JONES, H. C., SCHMIDT, J. M.: Ultrastructural study of cross-bands occurring in the stalks of *Caulobacter crescentus*. J. Bacteriol. **116**, 466–470 (1973).

KURN, N., AMMER, S., SHAPIRO, L.: A pleiotropic mutation affecting expression of polar developmental events in *Caulobacter crescentus*. Proc. Natl. Acad. Sci. **71**, 3157–3161 (1974).

MAALØE, O., KJELDGAARD, N. O.: Control of macromolecular synthesis. New York-Amsterdam: W. A. Benjamin 1966.

MOORE, R. L., HIRSCH, P.: First generation synchrony of isolated *Hyphomicrobium* swarmer populations. J. Bacteriol. **116**, 418–423 (1973).

NEWTON, A.: Role of transcription in the temporal control of development in *Caulobacter crescentus*. Proc. Natl. Acad. Sci. **69**, 447–451 (1972).

OSLEY, M. A., NEWTON, A.: Segregation of DNA during the development of *Caulobacter crescentus*. Federation Proc. **32**, 616 (1973).

PATE, J. L., ORDAL, E. J.: The fine structure of two unusual *Caulobacters*. J. Cell Biol. **27**, 133–150 (1965).

POINDEXTER, J. S.: Biological properties and classification of the *Caulobacter* Group. Bact. Rev. **28**, 231–295 (1964).

POINDEXTER, J. S., COHEN-BAZIRE, G.: The fine structure of stalked bacteria belonging to the family *Caulobacteraceae*. J. Cell Biol. **23**, 587–607 (1964).

RYTER, A.: Association of the nucleus and the membrane of bacteria: A morphological study. Bact. Rev. **32**, 39–54 (1968).

SCHMIDT, J. M.: Observations on the adsorptions of *Caulobacter* bacteriophages containing ribonucleic acid. J. Gen. Microbiol. **45**, 347–353 (1966).

SCHMIDT, J. M.: Effect of lysozyme on cross-bands in stalks of *Caulobacter crescentus*. Arch. Microbiol. **89**, 33–40 (1973).

SCHMIDT, J. M., STANIER, R. Y.: The development of cellular stalks in bacteria. J. Cell Biol. **28**, 423–436 (1966).

SHAPIRO, L., AGABIAN-KESHISHIAN, N.: Specific assay for differentiation in the stalked bacterium *Caulobacter crescentus*. Proc. Natl. Acad. Sci. **67**, 200–203 (1970).

SHAPIRO, L., AGABIAN-KESHISHIAN, N., BENDIS, I.: Bacterial differentiation. Science **173**, 884–892 (1971).

SHAPIRO, L., AGABIAN-KESHISHIAN, N., HIRSCH, A., ROSEN, O. M.: Effect of dibutyryladenosine 3′, 5′-cyclic monophosphate on growth and differentiation in *Caulobacter crescentus*. Proc. Natl. Acad. Sci. **69**, 1225–1229 (1972).

SHAPIRO, L., MAIZEL, JR., J. V.: Synthesis and structure of *Caulobacter crescentus* flagella. J. Bacteriol. **113**, 478–485 (1973).

SHEKMAN, R., WICKNER, W. T., WESTERGAARD, O., BRUTLAG, D., GEIDER, K., BERTCH, L. L., KORNBERG, A.: Initiation of DNA synthesis: Synthesis of ØX174 replicative form requires RNA synthesis resistant to rifampicin. Proc. Natl. Acad. Sci. **69**, 2691–2695 (1972).

STALEY, J. R., JORDAN, T. L.: Cross-bands of *Caulobacter crescentus* stalks serve as indicators of cell age. Nature **246**, 155–156 (1973).

STOVE, J., STANIER, R. Y.: Cellular differentiation in stalked bacteria. Nature **196**, 1189–1192 (1962).

SUGINO, A., OKAZAKI, R.: RNA-linked DNA fragments *in vitro*. Proc. Natl. Acad. Sci. **70**, 88–92 (1973).

WICKNER, R. B., WRIGHT, M., WICKNER, S., HURWITZ, J.: Conversion of ØX174 and fd single-stranded DNA to replicative forms in extracts of *Escherichia coli*. Proc. Natl. Acad. Sci. **69**, 3233–3237 (1972).

Cell Division and the Determination Phase of Cytodifferentiation in Plants[1]

FREDERICK MEINS, JR.

Department of Biology, Princeton University, Princeton, NJ, USA

I. Introduction

The development of higher plants involves progressive cytodifferentiation and precisely regulated cell division and cell enlargement. The process of cytodifferentiation can be divided into two phases (WEISS, 1950; HELSOP-HARRISON, 1967). First, cells become committed or determined for particular developmental fates, often without showing overt signs of specialization. Once established, determination can persist even in the absence of the factors that initiated the event, thus providing for the stable and progressive character of differentiation. Second, determined cells eventually express their differentiated properties in response to inductive influences. Unlike determination, this phase of cytodifferentiation is not stable. Thus, the expression phase is, in a sense, a physiological response which provides the basis for the regulation and integration of cellular activities already established by an earlier event in development (HOLTZER et al., 1972).

Differentiation of plant cells is often preceeded by an unequal cell division. Well-known examples of this include the formation of the stomatal complex in monocots (STEBBINS and SHAH, 1960), the differentiation of sieve tubes and companion cells in phloem (ESAU, 1969), unequal division of leaf cells of *Sphagnum cymbifolium* to yield hyaline and chlorenchymous cells (ZEPF, 1952) and, development of root hairs in angiosperms bearing a differentiated root epidermis (CORMACK, 1949). In some cases, there is a visible polarization of the cytoplasm prior to division suggesting that cell division, by physically isolating dissimilar regions of cytoplasm, initiates the cytodifferentiation process (BÜNNING, 1952, 1957; STEBBINS and JAIN, 1960). These observations raise a more general question: Is cell division, or an event in the cell cycle leading to division, required for cytodifferentiation? As pointed out by TORREY (1971), determination must depend, at least indirectly, upon division, since all cells ultimately arise by division from the zygote. Thus, the question of fundamental interest is whether or not a critical cell division is required as an immediate prelude to the initial event in cytodifferentiation, i.e. cell determination.

[1] Unpublished experimental work described here was supported by grants from the National Science Foundation, the National Institutes of Health, and the Merck Company Foundation.

This chapter deals with the role of cell division in the differentiation of plant cells. I will provide several lines of evidence that a critical division, probably in the presence of specific inducers, is required for determination. We will also see that determination, once established, can persist for many cell generations and thus represents an alteration in cellular heredity. Particular emphasis will be placed on the use of kinetic models to link division with cytodifferentiation and to provide a quantitative picture of changes in the differentiation of tissues. My intention is to illustrate concepts and experimental approaches with pertinent examples rather than to review the literature.

II. Differentiation of Tracheary Elements

There is convincing evidence that cell division is required for the formation of tracheary elements in higher plants. The reader is referred to excellent reviews on this topic by ROBERTS (1969) and TORREY et al. (1971). Tracheary element is a collective term for a group of cells with characteristic lignified secondary walls that are devoid of cytoplasm at maturity. Members of this group such as tracheids and vessel elements in xylem, wound vessel members produced in response to wounding, and tracheary elements found in cultured tissues, all arise in regions of meristematic activity, suggesting that cell division is required for their formation.

The formation of wound vessel members (WVM) in response to transverse incisions of *Coleus* stem segments is associated with cell division and an increased DNA content *per* cell (FOSKET, 1968). When stem segments are treated with fluorodeoxyuridine (FUdR), a potent inhibitor of DNA synthesis, cell division is inhibited and very few WVM, less than 10% of the control number, are formed. A similar inhibitory effect was observed with colchicine, which arrests cells at mitosis but does not affect DNA synthesis, indicating that mitosis or cell division rather than DNA synthesis is the critical event. FOSKET (1970) has studied the temporal relationship between cell division and xylogenesis by inducing WVM formation with the auxin, indoleacetic acid. After auxin treatment there is a burst of DNA synthesis which reaches a maximum after two days, followed by the beginning of WVM formation. If FUdR is administered during the burst, then DNA synthesis is blocked and WVM formation is inhibited by about 80%. However, if FUdR is administered after the burst but during the period of active xylogenesis, then WVM formation is not inhibited. These results show that cell division is required for the determination phase but not the expression phase of WVM differentiation.

Cell division in cultured root slices of pea occurs in two phases (TORREY and FOSKET, 1970). During the first 24 hrs in culture, division is confined to diploid cells of the pericycle. Later, beginning at 72 hrs, polyploid cells of the cortex enter division followed by the formation of tracheary elements. These observations were supported by pulse-labeling cells with ^{3}H-thymidine. Pulses given during the first 24 hrs of culture labeled only 4% of the cortical cells, whereas pulses given between 48 and 72 hrs labeled about 68% of the cortical cells. When cells were labeled in the later phase, 96% of the immature tracheary elements formed were labeled indicating that all tracheary elements arise from division of specific progenitors, in this case the cortical cells.

The question arises whether or not the initiation of tracheary element differentiation requires specific inducers in combination with cell division. A number of substances either stimulate or are necessary for tracheary element formation (TORREY et al., 1971). The difficulty is in distinguishing between effects of these substances directly on differentiation and indirect effects on cell division. For example, cytokinin in combination with auxin is required for proliferation of soybean callus and pea root cortical tissues in culture. In the absence of added cytokinin there is no cell division in these cultures, nor is there tracheary element formation (FOSKET and TORREY, 1969; PHILLIPS and TORREY, 1973). In the soybean callus system effects on cell division and on differentiation can be distinguished (FOSKET and TORREY, 1969). At low concentrations of the cytokinin, kinetin, the cells proliferate but do not form tracheary elements. Thus, with prolonged culturing it is possible to obtain a tissue line devoid of tracheary elements. Although the rate of cell division in these tissues increases gradually with increasing concentrations of kinetin, tracheary elements are only formed after a critical threshold concentration of kinetin has been reached. Thus, cytokinins appear to serve a dual role. Division of specific progenitor cells in an inductive environment is necessary for determination, but not for the subsequent expression of the tracheary element phenotype. Cytokinins provide both the specific stimulus for determination and necessary cell division.

III. Differentiation of Chlorophyll-Producing Cells

The elegant studies of tracheary element differentiation provide perhaps the best direct evidence for the involvement of cell division in the differentiation of plant cells. Another system that might profitably be studied in this regard is the differentiation of chlorophyll-producing cells, since chlorophyll is easily measured and its biosynthesis has been studied in detail. In culture, teratoma tissues of tobacco of single cell origin regularly develop highly abnormal leaves and shoots which green when exposed to light (BRAUN, 1959). When freshly cut explants of this tissue are incubated in liquid culture, cell division occurs primarily in bulbous outgrowths that arise at the cut surface. If explants from dark grown tissues are incubated in light for several days, the newly formed outgrowths appear as brilliant green spots at the surface of the explant, which remains white in color. Thus, chlorophyll production by teratoma cells is confined to regions of active cell division. Colchicine, after a short lag phase, inhibits chlorophyll production in these regions, indicating that cell division is required to initiate or to express the chlorophyll-phenotype (KIESEWETTER and MEINS, in preparation).

These observations were extended by pulse-labeling tissues with ^{3}H-thymidine to find out whether regions actively synthesizing DNA were also rich in chlorophyll. After a 12 hrs pre-incubation in the dark, tissues were grown for 15 days in light and then cut into numerous small pieces which were pooled at random to give samples with varying amounts of chlorophyll. There was a highly significant

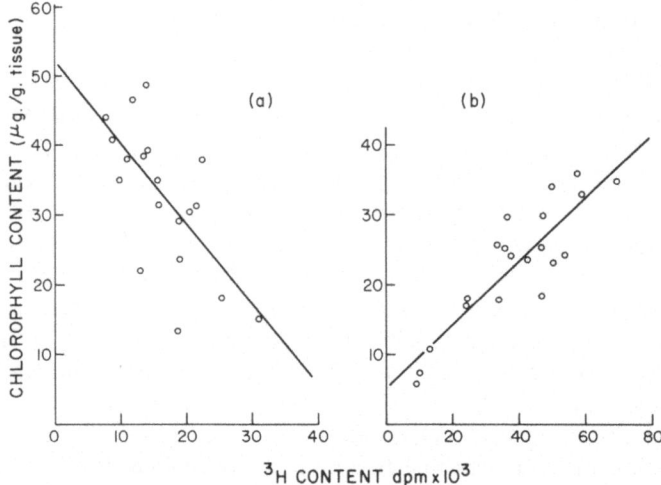

Fig. 1a and b. The correlation of DNA synthesis with chlorophyll production in teratoma tissues of tobacco. DNA synthesis is expressed as ³H-incorporated into DNA after pulse-labeling tissues with ³H-thymidine for 12 hrs. (a) Pulse administered during 12 hrs preincubation in the dark (coefficient of correlation = −0.681, n = 19, P < 0.01). (b) Pulse administered after 14.5 days in the light (coefficient of correlation = +0.875, n = 19, and P < 0.001). (From KIESEWETTER and MEINS, in preparation)

(P < 0.001) positive correlation between ³H-DNA content and chlorophyll content when tissues were pulse-labeled during the last 12 hrs in light, indicating that chlorophyll production is associated with regions of the tissue actively synthesizing DNA (Fig. 1b). In striking contrast, there is a significant (P < 0.01) negative correlation between DNA synthesis and chlorophyll production for tissues pulse-labeled during the preincubation period in the dark (Fig. 1a). The fact that different results are obtained with early and late labeling suggests that the teratoma tissues consist of at least two populations of cells. Cells that are labeled early have a low potential for subsequent production of chlorophyll and either predominate or preferentially synthsize DNA early in the incubation period. Cells that are labeled late are chlorophyll-producing cells or their immediate progenitors. This is consistent with the interpretation that the division of a specific group of progenitor cells is associated with chlorophyll production. A possible objection to the labeling experiments is that the correlation between DNA synthesis and chlorophyll content might result from replication of chloroplast DNA during the greening process rather than from nuclear-DNA synthesis in preparation for mitosis. This is unlikely. Although chloroplast DNA is synthesized in *Nicotiana* leaves (SHIPP et al., 1965; WOLLGIEHN and MOTHES, 1964), this DNA is a small fraction of the total cellular DNA. Thus, in exponentially growing cells, the bulk of the label should be incorporated into nuclear DNA. This was verified experimentally. Autoradiography of greening teratoma tissues labeled with ³H-thymidine shows that most of the ³H-label is confined to the nucleus.

The capacity to produce chlorophyll is closely linked to the differentiation of internal membranes in the plastid (KIRK and TILNEY-BASSETT, 1967; ROSINSKI and

ROSEN, 1972). Chlorophyll production by mature etiolated leaves in response to light occurs in several phases associated with the etioplast-to-chloroplast conversion and the enlargement and division of plastids (BOASSON and LAETSCH, 1969; BOASSON et al., 1972). Light-dependent chloroplast replication and chlorophyll production also occurs in mature green leaf tissues and is greatly enhanced by treatment with the cytokinin, kinetin (BOASSON et al., 1972). The important point is that the induction of these events occurs in the absence of discernible cell division or nuclear DNA synthesis (VERBEEK-BOASSON, 1969; WOLLGIEHN and MOTHES, 1964). The best interpretation of these findings appears to be that in the mature leaf, chlorophyll production reflects the expression of a differentiated state established sometime earlier in development. This process requires light, probably for membrane formation and plastid replication, but does not require cell division or nuclear DNA synthesis. In contrast, chlorophyll production in teratoma tissue actively engaged in organogenesis reflects determination of the chlorophyll-phenotype, which requires cell division, followed by expression of the newly established phenotype in response to light.

IV. Development of γ-Plantlets

FOARD (1970) has reviewed several cases in which organ formation and the appearance of specialized cells occur in the absence of cell division. The most dramatic of these is the development of γ-plantlets (HABER, 1968). Dry wheat grains irradiated with massive doses of γ-radiation are still able to germinate and develop into miniature plants (γ-plantlets) that never grow larger than about one-tenth the size of normal seedlings (FOARD and HABER, 1961). Development of these plantlets proceeds in the absence of cell division and, at least in the growing root tips, in the absence of DNA synthesis (HABER and LUIPPOLD, 1960; HABER et al., 1961). γ-plantlets carry out an impressive variety of functions associated with normal development, e.g.: correlative growth of leaves and roots, chlorophyll production, chlorophyll loss in leaf senescence, photosynthesis, growth expressed as increase in size, dry weight, and protein content, as well as the synthesis and nucleocytoplasmic transfer of RNA (HABER et al., 1961; FOARD and HABER, 1961; HABER, 1962; HABER and FOARD, 1964; FOARD and HABER, 1970). They differ from normal plants, however, in several significant ways: no new root or leaf primordia are initiated; no leaf trichomes are formed; and guard cells and their subsidiaries fail to develop, with the exception of morphologically distinct stomatal complexes that are formed in the apical region of the first foliage leaf (FOARD and HABER, 1961).

Superficially, these findings appear to argue against the involvement of cell division in cytodifferentiation. However, development from the seed, in fact, offers an extreme example of the separation of determination and the expression phase of cytodifferentiation. During embryogenesis, a period of active cell division, the basic bipolar structure of the plant, complete with root and shoot meristems is established (STEEVES and SUSSEX, 1972). Since γ-plantlets are generated by irradia-

tion after completion of embryogenesis, which ceases with onset of seed dormancy, it may be argued that development of the plantlets results from the expression
of differentiated characters which does not require cell division. This conclusion is
supported by the fact that development of organ primordia and formation of
specialized cells arising from unequal division, examples of differentiation believed to require cell division, are usually blocked in γ-plantlets.

V. Changes in Cytodifferentiation and the Potential
for Organogenesis in Culture

Plant tissues growing in culture undergo a bewildering variety of changes in
their pattern of differentiation and capacity for organogenesis (cf. GAUTHERET,
1959; REINERT, 1962; GAUTHERET, 1966; TORREY, 1966; HALPERIN, 1969; TOR-
REY, 1971). These changes fall into three broad catagories. Most commonly, the
cells placed in culture under conditions inducing rapid proliferation dedifferen-
tiate, i.e. the primary tissue explants lose their distinctive morphology and capacity for specialized function (GAUTHERET, 1966). In some cases, however, differentiated functions such as the production of pigments, alkaloids, and growth factors
are retained by the proliferating cells although regulation of these functions may
be altered (HALPERIN, 1969; KRIKORIAN and STEWARD, 1969). Finally, tissues
sometimes undergo successive changes in differentiation, first losing their characteristic functions and then acquiring new functions which are then inherited by
individual cells (ARYA et al., 1962; BLAKELY and STEWARD, 1964; SIEVERT and
HILDEBRANDT, 1965).

In addition to changes in overt differentiation, cultured materials may also
change in determination. For example, recently established cultures of asparagus
do not normally form organs. However, after the tissues had been allowed to
proliferate for 60 weeks on a medium containing naphthalene acetic acid, organ
formation could be induced with 2,4-dichlorophenoxyacetic acid even though this
treatment was ineffective prior to the long incubation period (STEWARD et al.,
1967). This suggests that cells committed to organ formation arise during prolonged periods of active division on the noninductive medium. This potential is
then expressed when tissues are subcultured on the appropriate inductive medium.

More commonly, the capacity rather than the potential, for organogenesis
undergoes change during prolonged culture. For example, teratoma tissues of
tobacco occasionally give rise to variants no longer able to form teratomatous
leaves and shoots. The unorganized phenotype is extremely stable; tissues have
been propagated for over 9 years in culture without reverting to the parental
phenotype. Nonetheless, such tissues retain the potential for organogenesis which
is only expressed when they are subcultured on an inductive medium (MEINS,
1969; BRAUN and MEINS, 1970). Similarly, lines of carrot root cells that have
apparently lost their capacity for embryogenesis after prolonged culture sometimes regain this capacity when the composition of the culture medium is changed
(REINERT, 1959; STEWARD et al., 1967). REINERT et al. (1971) have made quantita-

tive measurements of the loss in capacity and potential for embryogenesis in carrot root cell cultures. "Typical" cultures maintained on the inductive medium M_s begin embryogenesis after 4–6 weeks and reach a maximum at 14 weeks. Thereafter the incidence of embryogenesis gradually declines to zero by 36 weeks. No embryos are formed when cells are grown on the non-inductive medium M_w which supports a significantly lower growth rate than M_s. However, if cells growing for increasing intervals of time on M_w are subcultured on M_s, which induces embryo formation, the potential for embryogenesis is lost more slowly than cells grown continuously on M_s. "Atypical" cell lines have also been studied that lose their embryogenic potential more slowly than "typical" cultures on either M_s or M_w media. These experiments indicate that cells determined for embryo formation, i.e. "embryogenic" cells (HALPERIN, 1970), arise and are lost in a gradual process which depends upon the composition of the culture medium and line of cells used. The important point is that the rate at which embryogenic cells are lost corresponds roughly to the growth rate of the tissue. More recently REINERT (personal communication) has found a good correlation in various cell lines between the number of cell doublings and the time at which embryogenesis reaches a maximum. A less strict correlation was found in the declining phase of embryogenesis.

Two firm conclusions may be drawn from these types of experiments. First, the commitment and expression phases of cytodifferentiation can undergo independent variation in culture; and, second, once established, cell determination, e.g. the potential for organogenesis, may persist for long periods of time in a dividing cell population. Evidence that stable changes in determination depend on cell division is circumstantial. Tissues with a stable cell composition in the intact plant often either lose or gain differentiated properties when subjected to conditions that promote rapid cell division in culture. Moreover, the rate at which changes occur is roughly correlated to the growth rate of the tissue or cell culture. This suggests, at least, that changes in the developmental fate of cells are either hastened or initiated by cell division.

VI. Kinetic Models Relating Cell Proliferation and Cytodifferentiation [2]

The great difficulty in establishing a causal relationship between division and cytodifferentiation in culture is that events occurring within individual cells must be inferred from gross changes in the behavior of a tissue which may occur long after cells are committed to a new fate. Tissues explanted into culture consist of an heterogeneous population of cells. It is to be expected, therefore, that the balance of cell types growing at different rates will vary during prolonged culture. Thus,

[2] I should like to thank ART WINFREE, Purdue University; DAVID AXELROD, Douglas College–Rutgers University, and JOHN ENDLER, Princeton University for their criticism and stimulating conversations that have helped immeasurably in formulating the kinetic approach presented here.

changes in overt differentiation, as well as in developmental potential, could result from cell selection. To demonstrate that a change results from the conversion of individual cells and not from selection, it is necessary to study cell phenotypes readily detectable in the original tissue or to isolate cloned lines from tissues that have been cloned before the change is observed. Unfortunately, double-cloning methods have rarely been used (cf. SIEVERT and HILDEBRANDT, 1965). Even cloning is not a wholly satisfactory approach since delicate cell types might undergo further changes during the cloning procedure itself.

Ideally, an approach is needed in which basic features of the differentiation process can be investigated by measuring overt changes in the intact tissue. One wants to know to what extent tissue changes result from cytodifferentiation as opposed to cell selection; whether cytodifferentiation is linked to a critical cell division; what proportion of progenitor cells are involved; and, the relative abundance of differentiated and undifferentiated cells in a tissue at any given time. These problems can be approached by applying principles developed in studies of population growth and species interactions (cf. LOTKA, 1956). The basic idea is that tissues consist of sub-populations of cells, each with characteristic rates of proliferation and rates of conversion to other cell types. These relationships are used to specify a kinetic model that describes the abundance of cell types as a function of time. The validity of a particular model is then tested by comparing experimental data with values generated from the model.

In this section, I will describe a two-cell conversion (TCC) model which specifies that the conversion of progenitor cells to differentiated cells depends upon cell division. This model is then used to show that cell division is the rate limiting step in cytodifferentiation in three culture systems described earlier.

A. The Two-Cell Conversion Model

The TCC models specifies an idealized tissue cell suspension culture with the following properties:

1. It consists of two populations of cells, A and B.

2. The conversion of A cells to B cells requires cell division.

3. The rate of conversion of A to B is given by p, the probability of conversion per cell per division.

4. The proliferation of A and B cells obeys exponential growth laws.

5. The effect of A cells on the growth rate of B cells and *vice versa* is negligible.

An essential feature of this model is that it is only sensitive to the growth rate of the different cell populations. Thus, a tissue may be conveniently treated as if it consists of "marked" cells and "unmarked" cells without specifying the exact nature of the marker.

The proliferation rate of an exponentially dividing cell population is proportional to the number of cells present:

$$\frac{dn}{dt} = kn \qquad (1)$$

in which n is the number of cells at time t and k is the *cell doubling constant* with the dimensions time^{-1}. Integration of Eq. (1) gives the familiar exponential growth law:

$$n = n_o e^{kt} \tag{2}$$

where n_o is the initial number of cells. Equation (2) is a *state equation* which provides a description of the cell population at any time given the parameters k and n_o. When the cell number at several times is known, n_o and k may be evaluated by plotting the natural logarithm of n, ln n, versus time to give a straight line with slope k and y-intercept n_o. There is considerable evidence that increases in the fresh weight and cell number of cultured tissues obey an exponential growth law in which fresh weight is proportional to cell number (CAPLIN, 1947; STEWARD, 1958). Thus, even though growth in culture involves a combination of cell enlargement and cell division localized in meristematoid nests (TORREY, 1966), if fresh weight remains proportional to cell number, the doubling constant may still be measured by substituting fresh weight, W, for cell number in Eq. (2):

$$W = W_o e^{kt} \tag{3}$$

and

$$\ln W - \ln W_o = kt. \tag{4}$$

A tissue containing two independent populations of cells, A and B, is described by two rate equations:

$$\frac{da}{dt} = k_a a \tag{5}$$

and

$$\frac{db}{dt} = k_b b \tag{6}$$

in which, a and b are the numbers of A and B cells respectively, with doubling constants k_a and k_b. These rate equations can be modified to account for conversion of A cells to B cells. If p is the probability of A converting to B, then pa is the number of A cells that divide at rate k_a to become B cells. Thus, Eqs. (5) and (6) become:

$$\frac{da}{dt} = k_a a - k_a pa = k_a (1 - p)a \tag{7}$$

and

$$\frac{db}{dt} = k_b b + k_a pa. \tag{8}$$

It is convenient to combine these equations in a single expression which describes the proportion of cell types in a tissue as a function of time since this proportion is often more accessible from experimental data than is cell number. Defining the fraction of A cells in a tissue as $f = a/a + b$ and differentiating with respect to time gives:

$$\frac{df}{dt} = \frac{b\dfrac{da}{dt} - a\dfrac{db}{dt}}{(a+b)^2}$$

which, after substituting the expressions for da/dt and db/dt specified by Eqs. (7) and (8) becomes:

$$\frac{df}{dt} = f[k_a(1-p) - k_b + (k_b - k_a)f]. \tag{9}$$

Integration then provides two alternative state equations for the TCC-model in terms of f:

$$f = \frac{f_o[k_b - k_a(1-p)]}{f_o(k_b - k_a) + [k_b - k_a(1-p) + f_o(k_a - k_b)]e^{[k_b - k_a(1-p)]t}} \tag{10}$$

and

$$\ln\frac{f_o[k_b - k_a(1-p) + f(k_b - k_a)]}{f[k_b - k_a(1-p) + f_o(k_b - k_a)]} = [k_b - k_a(1-p)]t \tag{11}$$

where f_0 is the fraction of A cells in the tissue at $t = 0$.

Depending upon the values of k_a, k_b, and p, the model predicts that tissues should reach a histogenetic steady state in which the relative proportions of A and B cells no longer change with time. The value of f at steady state is obtained from Eq. (9) by setting the rate at which f changes equal to zero. The TCC model specifies two alternative steady states, one unstable and the other stable:

$$\text{State I (unstable): } f = 0 \text{ when } k_a(1-p) \neq k_b \tag{12}$$

and

$$\text{State II (stable): } f = \frac{k_a(1-p) - k_b}{k_a - k_b} \text{ when } k_a \neq k_b. \tag{13}$$

Since steady State II is always stable, tissues or cell suspensions containing any non-zero fraction of A cells will eventually return to the steady state value specifying a definite and fixed proportion of A and B cells. For a detailed discussion of stability in dynamical systems of biological interest see ROSEN (1970).

B. Application of the Model

1. Glutamine-Induced Changes in Teratoma Phenotype

Teratoma tissues of tobacco rapidly lose their capacity for organized growth and chlorophyll production when nitrate is replaced by glutamine in the culture medium (MEINS, 1971). This is illustrated in Fig. 2 which shows comparable branch-lets of teratoma tissue after two weeks growth on nitrate-medium, which pro-motes organoid development, and on glutamine-medium, which induces unorgan-

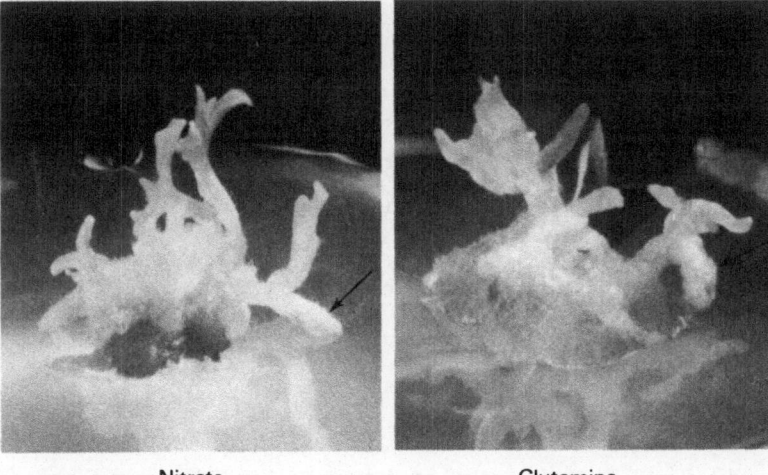

Nitrate   Glutamine

Fig. 2. Development of comparable branchlets of teratoma tissue placed on nitrate- and glutamine-media. The photographs show branchlets after two weeks growth on nitrate-medium (left panel) and on glutamine-medium (right panel). The arrows indicate shoots in contact with the agar substratum. On nitrate-medium the shoot has continued to grow in a polar fashion and produces chlorophyll. On glutamine-medium a comparable shoot as begun to lose its polar organization and no longer produces chlorophyll. (From MEINS, 1969)

ized growth. Early in the conversion process, unorganized growth appears at points where shoots are in contact with the agar substratum indicated by the arrow in Fig. 2. Later, these outgrowths appear elsewhere and eventually the highly organized branchlet is converted into a chaotic mass of dividing cells (MEINS, 1969). The gradual conversion of tissues to this unorganized form is accompanied by a parallel loss in the capacity of the tissue to produce chlorophyll in light (MEINS, 1971). The fact that changes in organization and the loss of chlorophyll are confined to the same regions on the tissue suggests that chlorophyll production can be used as a quantitative marker for the conversion process.

The TCC model can be applied to this system by assuming that teratoma tissue consists of cells with the capacity to form organs and produce chlorophyll (A cells) which are converted to unorganized cells that do not produce chlorophyll (B cells) (MEINS and WONG, in preparation). Since unorganized tissues subcultured for many cell generations on glutamine-medium grow at the same rate as organized tissues grow on nitrate-medium, it is reasonable to make the simplifying assumption that A and B cells proliferate at about the same rate, i.e. $k_a = k_b = k$. Equation (11) which describes the fraction of A cells as a function of time then becomes:

$$\ln f - \ln f_0 = -kpt . \tag{14}$$

To measure the fraction of A cells, we incubated tissues on glutamine medium for increasing intervals of time in the dark and then exposed them to light for a fixed period of time. Since only A cells can produce chlorophyll in light, the concentration of chlorophyll in the tissues will be proportional to the fraction of

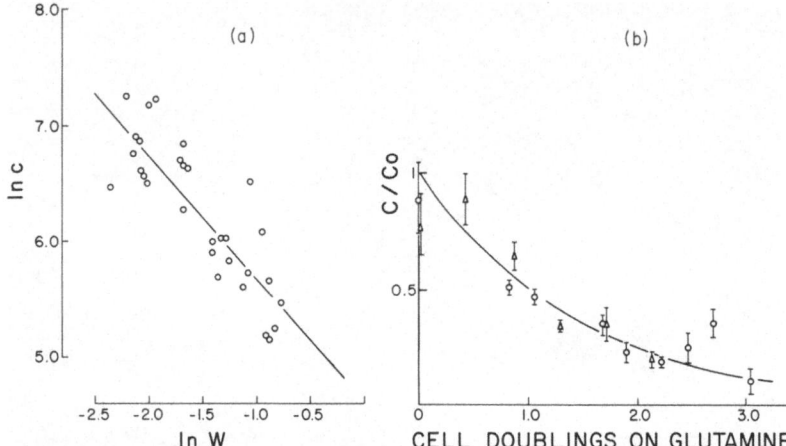

Fig. 3a and b. The TCC model applied to glutamine-induced changes in the capacity to produce chlorphyll. (a) Least squares fit of $\ln c$ to $\ln W$ [Eq. (16)]; probability of conversion = $1.06 \pm 0.125(30)$. (b) Reconstruction of the time course for chlorophyll loss using $p=1$ and parameters in Table 1 (solid line). Data points obtained from experiment I ($\circ$) and II ($\triangle$). (From MEINS and WONG, in preparation)

Table 1. Loss of chlorophyll-producing cells induced by glutamine treatment. Kinetic parameters obtained from two experiments (I and II)

Kinetic parameter	I	II
Doubling constant[a] (days^{-1})	$0.081 \pm 0.006(30)$[c]	$0.059 \pm 0.002(30)$
Initial chlorophyll content (μg/g fresh weight)	1373.	1170.
Conversion probability[b] (conversions/cell/doubling)	$1.022 \pm 0.162(30)$	$1.06 \pm 0.125(30)$

[a] Calculated from plots of $\ln W$ versus time.
[b] Calculated from graph in Fig. 3a.
[c] $\pm$ sample standard deviation of the regression coefficient (N).

A cells. Therefore Eq. (14) becomes:

$$\ln C - \ln C_0 = -kpt \tag{15}$$

where C is the concentration of chlorophyll in μg/g fresh weight of tissue. Replacing kt with $\ln(W/W_0)$ from Eq. (1) and combining constants provides a form of the TCC model which can be directly applied to the experimental data:

$$\ln C = -p \ln W + \text{constant}. \tag{16}$$

In a typical experiment, there was a highly significant ($P < 0.001$) linear correlation between $\ln C$ and $\ln W$ in agreement with the model (Fig. 3a). The probability of conversion obtained by least squares fit of the data in two experiments was within experimental error, equal to one (Table 1). The excellent agreement of the experimental data with values predicted from the model is illustrated in Fig. 3b in which a computer-generated graph of C/C_0 versus the number of cell doublings with $p = 1$ has been plotted along with the experimental points. The model predicts that changes in the growth rate of the tissue should affect the rate at which chlorophyll is lost but should not alter the probability of conversion. This was verified experimentally. Experiments in which the doubling constant differed by about 40% gave the same value for the conversion probability (Table 1).

These results provide strong kinetic evidence that cell division is the rate limiting step in the loss of the capacity of teratoma tissue to produce chlorophyll. Moreover, since $p = 1$, each cell that divides in the presence of glutamine yields either one cell unable to produce chlorophyll, or two daughter cells able to produce one half the amount of chlorophyll produced by the parent cell. As pointed out earlier, chlorophyll production is associated with the enlargement and division of plastids. Thus, a reasonable interpretation of our results is that glutamine treatment, in addition to preventing organogenesis, inhibits plastid replication. When the cells divide, preexistent plastids simply dilute out exponentially as described by Eq. (15).

The critical question is whether the TCC model provides a unique interpretation of the data. Although this cannot be answered unequivocally, it is possible to find out how well reasonable alternative models can account for the experimental data. In comparing the TCC model with models in which conversion does not depend on division or in which changes in f result from cell selection, we found that the TCC model fit the data best.

2. Embryogenesis in Carrot Root Cell Cultures

The potential for embryogenesis in carrot root cell cultures is gradually lost during prolonged culturing. REINERT et al. (1971) have pointed out that the rate at which embryogenic cells are lost is more rapid in fast-growing cultures than in slow-growing cultures, suggesting that cell division is involved. The TCC model may be applied to this system by assuming that embryogenic and non-embryogenic cells have roughly the same doubling constant. The fraction of embryogenic cells (A cells) as a function of time is then described by Eq. (14). The problem here is to obtain values of f from the incidence of embryogenesis in culture. If a population containing embryogenic cells is divided into a large number of samples containing approximately the same number of cells, the fraction of samples containing n embryogenic cells, $P(n)$, is given by the Poisson distribution:

$$P(n) = \frac{m^n e^{-m}}{n!} \tag{17}$$

where m is the average number of embryogenic cells in the samples. Data on the incidence of embryogenesis provides the fraction of cultures that do not form embryos, $P(0)$. Therefore:

$$P(0) = e^{-m}$$

and

$$\ln P(0) = -m. \tag{18}$$

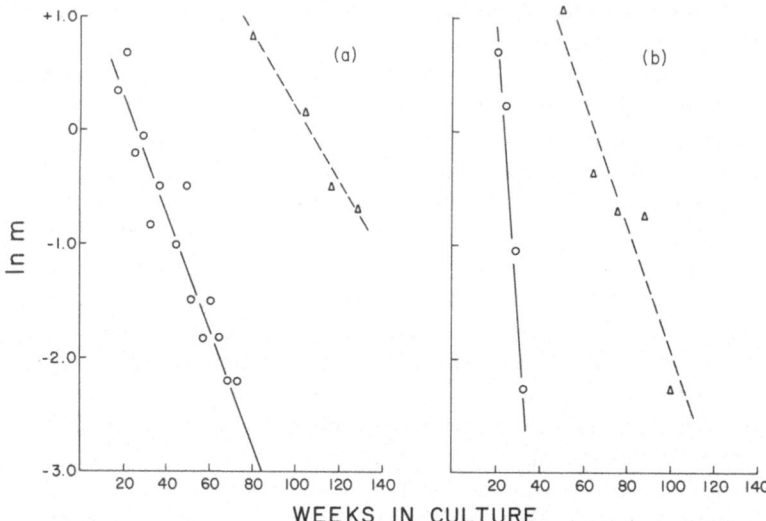

Fig. 4a and b. Time course for the loss in capacity for embryogenesis in carrot root cell cultures. Least squares fit of the experimental data using Eq. (19). (a) "Atypical" cultures: M_w medium, $pk = 0.033 \pm 0.005(4)$, ○—○; M_s medium, $pk = 0.051 \pm 0.004(15)$, △--△ (b) "Typical" cultures: M_w medium $pk = 0.059 \pm 0.013(5)$, ○—○; M_s medium, $pk = 0.256 \pm 0.038(4)$, △—△. pk values $\pm$ sample standard deviation of the regression coefficient (N). (Data from REINERT et al., 1971)

Assuming that the efficiency in detecting embryos and the cell number is approximately the same from culture to culture, then m will be proportional to f and Eq. (14) becomes:

$$\ln m = -pkt + \text{constant.} \tag{19}$$

Least squares fits of data obtained for "typical" and "atypical" cultures are shown in Fig. 4. Comparison of the slopes of the lines in Fig. 4 provides further support for the conclusions of REINERT et al. (1971), *viz.* (1) cell division is the rate limiting step in the loss of embryogenic cells; and, (2) the rate of loss is significantly less for "atypical" cultures than for "typical" cultures. Although further analysis of this system would require knowing the doubling constants of the cells, the important point is that analysis of embryogenesis in culture with kinetic models is feasible.

3. Tracheary Element Differentiation in Cortical Explants of Pea Root

An excellent opportunity for independent verification of the TCC model is afforded by the detailed studies of tracheary element differentiation in pea roots (cf. p. 152). The data obtained by PHILLIPS and TORREY (1973) from studies of cortical explants of pea root in culture are used in the analysis that follows. Since tracheary elements do not divide, the rate Eq. (17) for the conversion of

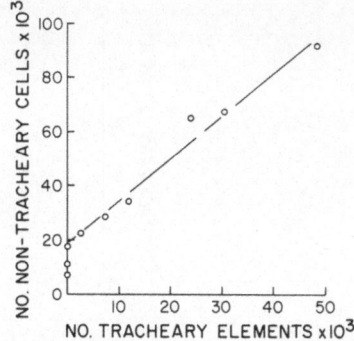

Fig. 5. Linear relationship between the number of non-tracheary cells and tracheary elements. Solid line is least squares fit of data from PHILLIPS and TORREY (1973). The value of p calculated using Eq. (22) is 0.385

nontracheary cells (A cells) to tracheary elements (B cells) takes the form:

$$\frac{db}{dt} = kpa. \tag{20}$$

Combining Eqs. (7) and (20) gives:

$$\frac{da}{db} = \frac{ka - kpa}{kpa} = \frac{1 - p}{p} \tag{21}$$

or

$$a - a_0 = \frac{(1 - p)(b - b_0)}{p} \tag{22}$$

where a_0 and b_0 are the initial numbers of A and B cells in the tissue. There is a highly significant ($P < 0.001$) linear correlation between the number of tracheary elements and the number of non-tracheary cells as predicted by the TCC model (Fig. 5). The probability of conversion of non-tracheary cells to tracheary elements calculated from the slope of the least squares fit line is 0.385. The doubling constant, k_a, for non-tracheary elements can then be obtained graphically using the integrated form of Eq. (7):

$$\ln a - \ln a_0 = k_a(1 - p)t . \tag{23}$$

A plot of $\ln a$ versus time provides a straight line with slope 0.122 ± 0.007 from which $k_a = 0.197$ days^{-1} corresponding to a doubling time of 3.5 days (Fig. 6). Non-tracheary cells divide in partial synchrony (PHILLIPS and TORREY, 1973). The doubling time of these cells estimated from the interval of time between peaks of mitotic activity is 3.4–3.5 days. This value is in very close agreement with the doubling time obtained independently with the TCC model.

The experimental values for the fraction of tracheary elements as a function of time and values obtained with the TCC model are shown in Fig. 7. With increas-

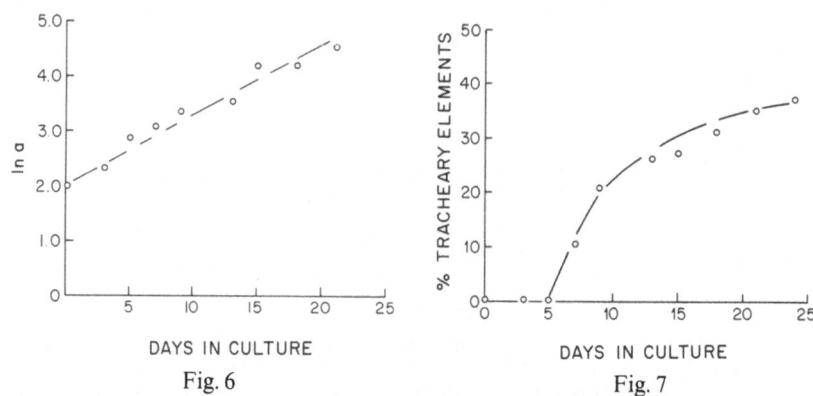

DAYS IN CULTURE   DAYS IN CULTURE

Fig. 6   Fig. 7

Fig. 6. Linear relationship between log of the number of nontracheary cells and time. Solid line is the least squares fit of data from PHILLIPS and TORREY (1973). The value of k_a calculated using Eq. (23) and p from Eq. (22) is 0.197 days^{-1}

Fig. 7. The fraction of cells that are tracheary elements as a function of time. Solid line represents values predicted by the TCC model reconstructed using Eq. (10) with the parameters $p=0.385$, $k_a=0.197$, and $f_0=1$. The model does not account for the lag phase preceeding tracheary element formation. Experimental values ($\circ$) are from PHILLIPS and TORREY (1973)

ing time, the proportion of tracheary elements approaches approximately 40%. This is also predicted by the model. Cortical tissues should reach a stable histogenetic steady state with 38.5% tracheary elements specified by Eq. (13).

For purposes of analysis, the simplifying assumption was made that non-tracheary cells are a homogeneous population. This is not the case. The cortical explants consist of cortical cells which differ in degree of ploidy (PHILLIPS and TORREY, 1973) as well as epidermal cells. The fact that Eqs. (22) and (23) still give straight-line relationships suggests two alternative interpretations of the parameter p. If most cells have the potential to become tracheary elements, then p is the probability that any one of these cells divides to yield a tracheary element. On the other hand, if a group of progenitor cells constituting a fixed proportion of the population were to convert with $p = 1$, then the observed p value specifies the fraction of progenitor cells in the population of non-tracheary cells. According to this interpretation, cells present in the tissue with an abundance less than p cannot be immediate progenitors of tracheary elements. Since diploid cells make up less than 10% of the population (PHILLIPS and TORREY, 1973), it is likely that the progenitor cells are polyploid, in agreement with the conclusions of TORREY and FOSKET (1970).

The two-cell model has been used here to make several predictions, subject to independent verification, based entirely on analysis of changes in the number of tracheary elements and non-tracheary cells with time. To summarize briefly: (1) The division of progenitor cells is the rate-limiting step in forming tracheary elements. (2) The doubling time for progenitor cells is 3.5 days (3) The tissues reach a histogenetic steady state in which the fraction of tracheary elements is 0.385. (4) The immediate progenitors of tracheary elements are polyploid cells.

VII. Epigenetic versus Genetic Mechanisms
for Cell Determination

One essential feature of determination is its stability. Newly acquired potentialities that arise in culture may persist for many cell generations. In principle, stability of this type could result either from alternative steady states in the balance of cell types in culture as described earlier, or, of more fundamental interest, from changes in the determination of individual cells. Cloning experiments have established that in some cases at least determination is inherited by individual cells (cf. HALPERIN, 1969). Two basic mechanisms could account for stability at the cell level: *genetic changes* resulting from rare, random point mutations and chromosomal rearrangements or deletions; and, *epigenetic modifications* resulting from self-perpetuated changes in the pattern of gene-expression without permanent alteration of the cell genome.

There are progressive changes in chromosomal constitution such as aneuploidy and polyploidy during rapid cell proliferation in culture (MITRA and STEWARD, 1961; MURASHIGE and NAKANO, 1967; SACRISTÁN and MELCHERS, 1969). These changes are often associated with a loss in differentiated function (BLAKELY and STEWARD, 1964) or in organogenic potential (MURASHIGE and NAKANO, 1967; TORREY, 1967). When organs do arise from aneuploid tissues, they differ from the parent tissue in chromosomal constitution, usually in the direction of the euploid state (TORREY, 1967; MURASHIGE and NAKANO, 1966). Nonetheless, complete, although abnormal, tobacco plants containing polyploid and aneuploid cells can be regenerated from callus cultures indicating that changes in chromosome complement do not preclude organogenesis (MURASHIGE and NAKANO, 1966; SACRISTÁN and MELCHERS, 1969). Chromosomal changes also accompany the acquisition of new differentiated functions by cultured cells; for example, the appearance of auxin-autotrophy in cultures of tobacco and *Crepis capillaris* (FOX, 1963; SACRISTÁN, 1967; SACRISTÁN and WENDT-GALLITELLI, 1971). Cells in plants regenerated from autotrophic cells of *Crepis capillaris* retain their characteristic chromosomal aberration but are no longer able to proliferate in the absence of added auxin. Experiments of this type clearly show that the association of a specific chromosomal change with a persistent alteration in cell phenotype is not sufficient to establish a causal relationship between the two events. The critical problem, therefore, is to devise tests that distinguish between genetic and epigenetic mechanisms in experimental systems where classical genetic analysis is not feasible.

There is compelling evidence that differentiated plant cells, with few exceptions, are totipotent (REINERT, 1968; BRAUN 1959; ZEPF, 1952; VASIL and HILDEBRANDT, 1965; STEWARD et al., 1966; TAKEBE et al., 1971; DUFFIELD et al., 1972, FREARSON et al., 1973). This provides the strongest possible support for the fundamental principle that cytodifferentiation results from epigenetic modifications and not from permanent genetic changes (DAVIDSON, 1968). Such modifications differ from genetic changes in several important ways: they are directed rather than random changes; they are potentially reversible; and, they leave the heritably altered cell totipotent (NANNEY, 1958; MEINS, 1972; MEINS, 1974). These

criteria have been applied in studies of *habituation*, a heritable change that occurs
in cultured tissues of various plant species (GAUTHERET, 1955).

Pith parenchyma tissues of tobacco normally require an exogenous supply of
auxin and a cell-division factor such as cytokinin for continued growth in culture
(JABLONSKI and SKOOG, 1954). These tissues sometimes lose their requirement for
either or both growth factors after a variable number of passages. Thereafter, such
habituated tissues can be propagated indefinitely without added growth factor,
which they are now able to produce (GAUTHERET, 1955; FOX, 1963; WOOD et al.,
1969; DYSON and HALL, 1972; EINSET and SKOOG, 1973). Moreover, cloning
experiments show that habituation is inherited by individual cells (LUTZ, 1971;
MELCHERS, 1971; BINNS and MEINS, 1973).

BINNS and MEINS (1973) have regenerated 62 complete tobacco plants from 19
different cytokinin-habituated lines of single cell origin. So far, plants from 18 of
the 19 clones have flowered and set fertile seed. Although plants regenerated from
both habituated and normal clones were aneuploid, indicating that the cultured
cells had undergone genetic changes, the important point is that complete, fertile
plants can be obtained from progeny of individual cytokinin-habituated cells.
Therefore, cells remain totipotent after the habituation process. Moreover, in
every case, tissues from regenerated plants had lost their cytokinin-habituation,
indicating that the habituation process is regularly reversible.

These results lead to the important conclusion that the heritable capacity to
produce specific growth factors, a differentiated function of cells in the intact
plant, results from epigenetic modifications and not from permanent genetic
changes. Although both mechanisms undoubtedly contribute to cellular variation
in culture, since differentiated cells in the intact plant remain totipotent, it is likely
that changes with an epigenetic basis most nearly reflect events in normal development.

So far, I have dealt with rather broad questions concerning the relationship of
cell division to determination, namely: Is division necessary for determination? Is
determination maintained when a cell divides and, if so, is this due to epigenetic
or genetic changes? A conceptual framework for posing more precise questions
dealing with the cellular basis for determination is provided by the "master
switch" metaphor. Complex processes such as organogenesis involve many separate events precisely coordinated in time and space. Nonetheless, the determination step in organogenesis often can be controlled in culture by a single chemical
substance suggesting that determination involves specific *master switches* that
regulate blocks of developmental events. KAUFMAN (1973) has been able to account for changes in the determination of imaginal disks in *Drosophila* using an
approach based, in part, on the assumption that distinct determined states result
from combinations of independent master switches that occupy only two possible
states, i.e. "on" or "off". According to this description, a cell with three switches
set in the combination on-off-on would express one determined state, while a cell
set in the combination on-on-off would express a different determined state.

Applying this metaphor to the problem of organ formation in culture leads to
the interpretation that changes in organogenic potential result from shifts of
switches between two alternative states. Since the switches occupy discrete positions, changes in the capacity for organ formation in response to an inducer

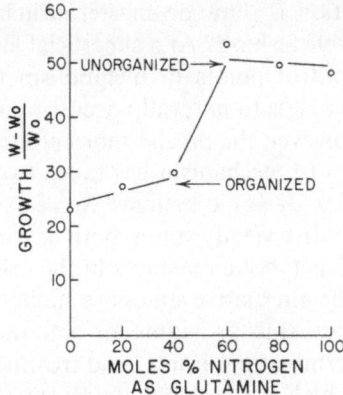

Fig. 8. Effect of the relative glutamine-and nitrate-nitrogen concentration on the growth rate of teratoma tissues. Note the discontinuous change in growth rate associated with the shift from organized to unorganized forms of the tissue. (Redrawn from MEINS, 1969)

should be discontinuous. Below a critical threshold concentration the inducer should be ineffective, while above this concentration the change in state should be complete. An example of this is provided by glutamine-induced changes in the phenotype of tobacco teratoma tissues (MEINS, 1969, 1971). Organized teratoma tissues were cultured on agar-media in which the total nitrogen content was kept constant and the relative proportions of glutamine and nitrate varied. At a concentration of about 50 moles-% nitrate-nitrogen tissues underwent an abrupt change from the organized form to the unorganized form. This was also reflected in the discontinuous shift in growth rate from the slowly growing organized form to the more rapidly growing unorganized form (Fig. 8).

If one accepts KAUFMAN'S (1973) assumption that switches are independent and confined to two states, then n switches may not specify more than 2^n different states. The number of distinct types of organogenesis encountered in carrot cultures is limited to the formation of callus, roots, and complete embryos. Similarly, organogenesis of tobacco pith induced by combinations of cytokinin and auxin involves four states: callus, roots, shoots and complete plantlets (SKOOG and MILLER, 1957). Thus the number of independent switches required for organogenesis is certainly greater than one and on the order two or three.

The question arises whether the initial event in organogenesis occurs in a single cell or involves the interaction of cells in different states. The studies of habituation described earlier (cf. p. 168) show that cells can lock into states in which they produce auxin, or cell division factor, or both growth regulators. This suggests that production of factors that control organogenesis in tobacco involves an "auxin" switch and a "cytokinin" switch that can exist in different states in different cells. In principle, therefore, the initial step in organogenesis could involve either changes in independent switches within the same cell or, alternatively, the interaction of cells with switches in different states. For a more detailed discussion of this problem within the context of cell interactions in carrot embryogenesis see REINERT (1968), BACKS-HÜSEMANN and REINERT (1970) and HALPERIN (1970).

The fundamental question is: how do master switches change the developmental fate of a cell in a stable fashion? At a superficial level, one can say that the switches correspond to control points in metabolism that regulate either the biosynthesis of or responsiveness to naturally occurring molecules that influence development in culture. However, the precise molecular mechanism for switching is unknown. Speculation as to mechanism has taken two directions. The first is based on studies of complex dynamic systems. Some systems of chemical reactions can exist in two alternative steady states, both of which are stable. Thus it is plausible that networks of metabolic reactions in the cell can function as biochemical switches. Because the alternative states are stable, cells could, in principle, divide without changing state, thus providing for both the heritability and potential reversibility of cell determination. For a lucid treatment of this approach, the reader is referred to papers by KACSER (1963) and ROSEN (1972).

The second approach, recently reviewed by FINCHAM (1973), is based upon the observation that certain regulatory gene loci in higher plants undergo directed and reversible changes in state. According to this view, determination results from reversible genetic changes. Transposable controlling elements, presumably regulatory genes, migrate to specific gene loci when changes in determination are induced. These elements are then inserted into the genome and regulate the transcription of adjacent structural genes. When the determined cell enters S phase, in preparation for division, the controlling elements simply replicate in their new positions and thus are distributed conservatively to the daughter cells.

VIII. Concluding Remarks

Contemporary studies of cytodifferentiation have focused, for the most part, on the molecular events accompanying overt differentiation. It may be argued that this approach has provided a great deal of information about the consequences of differentiation, but little understanding of the crucial event in which a cell decides between several alternative fates. We are, at present, just beginning to identify some of the factors and events associated with cell determination in plants. The determination and expression phases of cytodifferentiation can be separated experimentally and appear to be under independent control. The master switch metaphor provides a useful way of thinking about this problem. It is likely that determination involves a relatively small number of control points that probably regulate the production of specific low molecular weight compounds. Unlike the regulation encountered in bacterial systems or "housekeeping" functions of the cell, these control points probably exist in discrete alternative states.

From the few plant systems studied in detail, there is evidence that cell division, probably in combination with specific inducers, is required for determination. The fact that organogenic and embryogenic cells persist in growing cell populations suggests that determination can be inherited by individual cells. Studies of habituation provided direct evidence for this and show, moreover, that the inheritance of determination has an epigenetic basis. This system provides one of the very few opportunities to study the molecular basis for epigenetic modifications in cells whose developmental potentialities can be tested directly by regeneration of an entire organism.

The events in the cell cycle necessary for determination are not known. The polarization of cytoplasm preceeding unequal division suggests that cell division is necessary to create a physical barrier between dissimilar regions of cytoplasm, which then interact with the nucleus to evoke changes in gene activity. This is supported by a number of studies in both animals and plants showing that a change in the axis of nuclear division in a polarized cytoplasm may alter the developmental fate of the daughter nuclei (SAX, 1935; WILSON, 1896).

Cell division and mitosis are rather loosely linked in plant cells. Thus, either preceeding or accompanying cytodifferentiation, plant cells often undergo endomitosis without cell division to yield cells that are polyploid or polytene (D'AMATO, 1952, 1964; STANGE, 1965). At present, it is not certain whether the mitotic event is required for cytodifferentiation or a consequence of cytodifferentiation (STANGE, 1965).

Another possibility is that a critical round of DNA synthesis in the presence of specific inducers initiates determination. There is strong evidence for this from studies of animal systems (LOCKWOOD et al., 1967; TURKINGTON, 1968; HOLTZER, 1970; TSANEV and SENDOV, 1971; HOLTZER et al., 1972) and there is limited evidence that this is the case for some types of differentiation in plants as well (FOARD, 1970).

Several examples have been provided to illustrate how kinetic models can be used to get a quantitative picture of cytodifferentiation at the tissue level. This approach has several distinct advantages that could be profitably exploited. First, the rate at which differentiated cells arise can be measured by using cell products as markers when identification of individual differentiated cells is not experimentally feasible. Second, comparison of experimental data with values generated from models can establish whether or not cell division is the rate-limiting step in differentiation even when cell commitment occurs long before overt differentiation is detectable. Third, models can be used to predict conditions for histogenetic steady state. Fourth, models provide parameters such as the conversion probability which can be used to quantitate the effect of specific inducers. Finally, models are of heuristic value indicating how cell selection, changes in growth rate, etc. can affect the behavior of differentiating tissues.

References

ARYA, H. C., HILDEBRANDT, A. C., RIKER, A. J.: Growth in tissue culture of single-cell clones from grape stem and *Phylloxera* gall. Plant Physiol. **37**, 387–392 (1962).

BACKS-HÜSEMANN, D., REINERT, J.: Embryobildung durch isolierte Einzelzellen aus Gewebekulturen von *Daucus carota*. Protoplasma **70**, 49–60 (1970).

BINNS, A., MEINS, JR., F.: Evidence that habituation of tobacco pith cells for cell division-promoting factors is heritable and potentially reversible. Proc. Natl. Acad. Sci. **70**, 2660–2662 (1973).

BLAKELY, L. M., STEWARD, F. C.: Growth and organized development of cultured cells. VII. Cellular variation. Am. J. Botany **51**, 809–820 (1964).

BOASSON, R., BONNER, J. J., LAETSCH, W. M.: Induction and regulation of chloroplast replication in mature tobacco leaf tissue. Plant Physiol. **49**, 97–101 (1972).

BOASSON, R., LAETSCH, W. M.: Chloroplast replication and growth in tobacco. Science **166**, 749–751 (1969).

BOASSON, R., LAETSCH, W. M., PRICE, I.: The etioplast-chloroplast transformation in tobacco: correlation of ultrastructure, replication, and chlorophyll synthesis. Am. J. Botany **59**, 217–223 (1972).

BRAUN, A. C.: A demonstration of the recovery of the crown-gall tumor cell with the use of complex tumors of single-cell origin. Proc. Natl. Acad. Sci. **45**, 932–938 (1959).

BRAUN, A. C., MEINS, JR., F.: The regulation of the expression of cellular phenotypes in crown-gall teratoma tissue of tobacco. Symp. Int. Soc. Cell Biol. **9**, 193–204 (1970).

BÜNNING, E.: Morphogenesis in plants. Survey Biol. Progr. **2**, 105–140 (1952).

BÜNNING, E.: Polarität und inäquale Teilung des pflanzlichen Protoplasten. Protoplasmologia **8**, 1–86 (1957).

CAPLIN, S. M.: Growth and morphology of tobacco tissue cultures *in vitro*. Botan. Gaz. **108**, 379–393 (1947).

CORMACK, R. G. H.: The development of root hairs in angiosperms. Botan. Rev. **15**, 583–612 (1949).

D'AMATO, F.: Polyploidy in the differentiation and functioning of tissues and cells in plants. Caryologia **4**, 311–358 (1952).

D'AMATO, F.: Endopolyploidy as a factor in plant tissue development. Caryologia **17**, 41–52 (1964).

DAVIDSON, E. H.: Gene Activity in Early Development. New York: Academic Press 1968.

DUFFIELD, E. C. S., WAALAND, S. D., CLELAND, R.: Morphogenesis in the red alga, *Griffithsia pacifica*: regeneration from single cells. Planta **105**, 185–195 (1972).

DYSON, W. H., HALL, R. H.: N^6- (Δ^2-Isopentenyl) adenosine: its occurrence as a free nucleoside in an autonomous strain of tobacco tissue. Plant Physiol. **50**, 616–621 (1972).

EINSET, J. W., SKOOG, F.: Biosynthesis of cytokinin in cytokinin-autotrophic tobacco callus. Proc. Natl. Acad. Sci. **70**, 658–660 (1973).

ESAU, K.: The phloem. In: ZIMMERMANN, W., OZENDA, P., WULFF, H. D. (Eds.): Handbuch der Pflanzenanatomie Vol. V, part 2, p. 205. Berlin: Gebrüder Bornträger 1969.

FINCHAM, J. R. S.: Localized instabilities in plants — a review and some speculations. Genetics Suppl. **73**, 195–205 (1973).

FOARD, D. E.: Differentiation in plant cells. In: SCHJEIDE, O. A., DEVELLIS, J. (Eds.): Cell Differentiation pp. 575–602. New York: Van Nostrand 1970.

FOARD, D. E., HABER, A. H.: Anatomic studies of gamma-irradiated wheat growing without cell division. Am. J. Botany **48**, 438–446 (1961).

FOARD, D. E., HABER, A. H.: Physiologically normal senescence in seedlings grown without cell division after massive γ-irradiation of seeds. Radiation Res. **42**, 372–380 (1970).

FOSKET, D. E.: Cell division and the differentiation of wound-vessel members in cultured stem segments of *Coleus*. Proc. Natl. Acad. Sci. **59**, 1089–1096 (1968).

FOSKET, D. E.: The time course of xylem differentiation and its relation to DNA synthesis in cultured *Coleus* stem segments. Plant Physiol. **46**, 64–68 (1970).

FOSKET, D. E., TORREY, J. G.: Hormonal control of cell proliferation and xylem differentiation in cultured tissues of *Glycine max* var. Biloxi. Plant Physiol. **44**, 871–880 (1969).

FOX, J. E.: Growth factor requirements and chromosome number in tobacco tissue cultures. Physiol. Plant. **16**, 793–803 (1963).

FREARSON, E. M., POWER, J. B., COCKING, E. C.: The isolation, culture and regeneration of *Petunia* leaf protoplasts. Develop. Biol. **33**, 130–137 (1973).

GAUTHERET, R. J.: The nutrition of plant tissue cultures. Ann. Rev. Plant Physiol. **6**, 433–484 (1955).

GAUTHERET, R. J.: La culture des tissus végétaux. Paris: Masson and Cie. 1959.

GAUTHERET, R. J.: Factors affecting differentiation of plant tissues grown *in vitro*. In: Cell Differentiation and Morphogenesis, pp. 55–95. Amsterdam: North-Holland 1966.

HABER, A. H.: Ionizing radiations as research tools. Ann. Rev. Plant Physiol. **19**, 463–489 (1968).

HABER, A. H., CARRIER, W. L., FOARD, D. E.: Metabolic studies of gamma-irradiated wheat growing without cell division. Am. J. Botany **48**, 431–438 (1961).

HABER, A. H., FOARD, D. E.: Further studies of gamma-irradiated wheat and their relevance to the use of mitotic inhibition for developmental studies. Am. J. Botany **51**, 151–159 (1964).

HABER, A. H., LUIPPOLD, H. J.: Effects of gibberellin on gamma-irradiated wheat. Am. J. Botany **47**, 140–144 (1960).

HALPERIN, W.: Morphogenesis in cell cultures. Ann. Rev. Plant Physiol. **20**, 395–418 (1969).

HALPERIN, W.: Embryos from somatic plant cells. Symp. Int. Soc. Cell Biol. **9**, 169–191 (1970).

HESLOP-HARRISON, J.: Differentiation. Ann. Rev. Plant Physiol. **18**, 325–348 (1967).

HOLTZER, H.: Proliferative and quantal cell cycles in the differentiation of muscle, cartilage, and red blood cells. Symp. Int. Soc. Cell Biol. **9**, 69–88 (1970).

HOLTZER, H., WEINTRAUB, H., MAYNE, R., MOCHAN, B.: The cell cycle, cell lineage, and cell differentiation. Current Topics Develop. Biol. **7**, 229–256 (1972).

JABLONSKI, J. R., SKOOG, F.: Cell enlargement and cell division in excised tobacco pith tissue. Physiol. Plant. **7**, 16–24 (1954).

KACSER, H.: The kinetic structure of organisms. In: HARRIS, H. (Ed.): Biological Organization at the Cellular and Super-cellular Level, pp. 25–41. New York: Academic Press 1963.

KAUFMAN, S. A.: Control circuits for determination and transdetermination. Science **181**, 310–318 (1973).

KIESEWETTER, J., MEINS, JR., F.: The role of DNA synthesis and cell division in the differentiation of chlorophyll producing cells in crown-gall teratoma tissues of tobacco. (In preparation).

KIRK, J. T. O., TILNEY-BASSETT, R. A. E.: The plastids: their chemistry, structure, growth, and inheritance. London: Freeman 1967.

KRIKORIAN, A. D., STEWARD, F. C.: Biochemical differentiation: the biosynthetic potentialities of growing and quiescent tissues. In: STEWARD, F. C. (Ed.): Plant Physiology, a Treatise, Vol. VB, pp. 227–326. New York: Academic Press 1969.

LOCKWOOD, D. H., STOCKDALE, F. E., TOPPER, Y. J.: Hormone-dependent differentiation of mammary gland: sequence of action of hormones in relation to cell cycle. Science **156**, 945–946 (1967).

LOTKA, A. J.: Elements of mathematical biology. New York: Dover 1956.

LUTZ, A.: Aptitudes morphogénétiques des cultures de tissus d'origine unicellulaire. In: Les cultures de tissus de plantes. Colloq. Int. Cent. Nat. Rech. Scient. No. 193, pp. 163–168. Paris: Éditions du Centre Nat. Rech. Scient. 1971.

MEINS, JR., F.: Control of phenotypic expression in tobacco teratoma tissues. Thesis, Rockefeller University, New York 1969.

MEINS, JR., F.: Regulation of phenotypic expression in crown-gall teratoma tissues of tobacco. Develop. Biol. **24**, 287–300 (1971).

MEINS, JR., F.: Stability of the tumor phenotype in crown-gall tumors of tobacco. Progr. Exp. Tumor Res. **15**, 93–109 (1972).

MEINS, JR., F.: Mechanisms underlying tumor transformation and tumor reversal in crown-gall, a neoplastic disease of higher plants. In: KING, T. J. (Ed.): Developmental Aspects of Carcinogenesis and Immunity, pp. 23–39. New York: Academic Press 1974.

MEINS, JR., F., WONG, J.: Application of a quantitative model to glutamine-induced phenotypic changes of tobacco teratoma tissue in culture. (In preparation).

MELCHERS, G.: Transformation or habituation to autotrophy and tumor growth and recovery. In: Les cultures de tissus de plantes. Colloq. Int. Cent. Nat. Rech. Scient. No. 193, pp. 229–234. Paris: Éditions du Centre Nat. Rech. Scient. 1971.

MITRA, J., STEWARD, F. C.: Growth induction in cultures of *Happlopappus gracilis*. II. The behavior of the nucleus. Am. J. Botany **48**, 358–368 (1961).

MURASHIGE, T., NAKANO, R.: Tissue culture as a potential tool in obtaining polyploid plants. J. Heredity **57**, 115–118 (1966).

MURASHIGE, T., NAKANO, R.: Chromosome complement as a determinant of the morphogenetic potential of tobacco cells. Am. J. Botany **54**, 963–970 (1967).

NANNEY, D. F.: Epigenetic control systems. Proc. Natl. Acad. Sci. **44**, 712–717 (1958).

PHILLIPS, R., TORREY, J. G.: DNA synthesis, cell division, and specific cytodifferentiation in cultured pea root cortical explants. Develop. Biol. **31**, 336–347 (1973).

REINERT, J.: Über die Kontrolle der Morphogenese und die Induktion von Adventivembryonen an Gewebekulturen aus Karotten. Planta **53**, 318–333 (1959).

REINERT, J.: Morphogenesis in plant tissue cultures. Endeavour **21**, 85–90 (1962).

REINERT, J.: Morphogenese in Gewebe- und Zellkulturen. Naturwissenschaften **4**, 170–175 (1968).

REINERT, J., BACKS-HÜSEMANN, D., ZERBAN, H.: Determination of embryo and root formation in tissue cultures from *Daucus carota*. In: Les cultures de tissus de plantes. Colloq. Intern. Cent. Nat. Rech. Scient. No. 193, pp. 261–268. Paris: Éditions du Centre Nat. Rech. Scient. 1971.

ROBERTS, L. W.: The initiation of xylem differentiation. Botan. Rev. **35**, 201–250 (1969).

ROSEN, R.: Dynamical System Theory in Biology. New York: Wiley 1970.

ROSEN, R.: Mechanics of Epigenetic Control. In: ROSEN, R. (Ed.): Foundations of Mathematical Biology, Vol. 2, pp. 79–140. New York: Academic Press 1972.

ROSINSKI, J., ROSEN, W. G.: Chloroplast development: fine structure and chlorophyll synthesis. Quart. Rev. Biol. **47**, 160–191 (1972).

SACRISTÁN, M. D.: Auxin-Autotrophie und Chromosomenzahl. Molec. Gen. Genetics **99**, 311–321 (1967).

SACRISTÁN, M. D., MELCHERS, G.: The caryological analysis of plants regenerated from tumorous and other callus cultures of tobacco. Molec. Gen. Genetics **105**, 317–333 (1969).

SACRISTÁN, M. D., WENDT-GALLITELLI, M. F.: Transformation, auxin-autotrophy and its reversibility in a mutant line of *Crepis capillaris* callus culture. Molec. Gen. Genetics **110**, 355–360 (1971).

SAX, K.: The effect of temperature on nuclear differentiation in microspore development. J. Harvard University Arnold Arboretium **16**, 301–310 (1935).

SHIPP, W. S., KIERAS, F. J., HASELKORN, R.: DNA associated with tobacco chloroplasts. Proc. Natl. Acad. Sci. **54**, 207–213 (1965).

SIEVERT, R. C., HILDEBRANDT, A. C.: Variation within single cell clones of tobacco tissue cultures. Am. J. Botany **52**, 742–750 (1965).

SKOOG, F., MILLER, C. O.: Chemical regulation of growth and organ formation in plant tissues cultured *in vitro*. Soc. Exp. Biol. Symp. **11**, 118–131 (1957).

STANGE, L.: Plant cell differentiation. Ann. Rev. Plant Physiol. **16**, 119–140 (1965).

STEBBINS, G. L., JAIN, S. K.: Developmental studies of cell differentiation in the epidermis of monocotyledons. I. *Allium, Rhoeo* and *Commelina*. Develop. Biol. **2**, 409–426 (1960).

STEBBINS, G. L., SHAH, S. S.: Developmental studies of cell differentiation in the epidermis of monocotyledons. II. Cytological features of stomatal development in the *Gramineae*. Develop. Biol. **2**, 477–500 (1960).

STEEVES, T. A., SUSSEX, I. M.: Patterns in Plant Development. Englewood Cliffs, N. J. Prentice-Hall 1972.

STEWARD, F. C.: Growth and organized development of cultured cells. III. Interpretations of the growth from free cell to carrot plant. Am. J. Botany **45**, 709–713 (1958).

STEWARD, F. C., KENT, A. E., MAPES, M. O.: The culture of free plant cells and its significance for embryology and morphogenesis. Curr. Topics Develop. Biol. **1**, 113–154 (1966).

STEWARD, F. C., KENT, A. E., MAPES, M. O.: Growth and organization in cultured cells; sequential and synergistic effects of growth-regulating substances. Ann. N.Y. Acad. Sci. **144**, 326–334 (1967).

TAKEBE, I., LABIB, G., MELCHERS, G.: Regeneration of whole plants from isolated mesophyll protoplasts of tobacco. Naturwissenschaften **58**, 318–320 (1971).

TORREY, J. G.: The initiation of organized development in plants. Advan. Morphogenesis **5**, 39–92 (1966).

TORREY, J. G.: Morphogenesis in relation to chromosome constitution in long-term plant tissue cultures. Physiol. Plant. **20**, 265–275 (1967).

TORREY, J. G.: Cytodifferentiation in plant cell and tissue culture. In: Les cultures de tissus de plantes. Colloq. Int. Centre. Nat. Rech. Scient. No. 193, pp. 177–186. Paris: Éditions du Centre Nat. Rech. Scient. 1971.

TORREY, J. G., FOSKET, D. E.: Cell division in relation to cytodifferentiation in cultured pea root segments. Am. J. Botany **57**, 1072–1080 (1970).

TORREY, J. G., FOSKET, D. E., HEPLER, P. K.: Xylem formation: a paradigm of cytodifferentiation in higher plants. Am. Sci. **59**, 338–352 (1971).

TSANEV, R., SENDOV, B. L.: Possible mechanisms for cell differentiation in multicellular organisms. J. Theor. Biol. **30**, 337–393 (1971).

TURKINGTON, R. W.: Hormone dependent differentiation of mammary gland *in vitro*. Current Topics Develop. Biol. **3**, 199–218 (1968).

VASIL, V., HILDEBRANDT, A. C.: Differentiation of tobacco plants from single, isolated cells in microculture. Science **150**, 889–892 (1965).

VERBEEK-BOASSON, R.: Chloroplast replication and growth in tobacco. Groningen: V.R.B. Offsetdrukkerij 1969.

WEISS, P.: Perspectives in the field of morphogenesis. Quart. Rev. Biol. **25**, 177–198 (1950).

WILSON, E. B.: The Cell in Development and Inheritance. New York: Macmillian 1896.

WOLLGIEHN, R., MOTHES, K.: Über die Incorporation von H^3-thymidin in die Chloroplasten-DNS von *Nicotiana rustica*. Exp. Cell Res. **35**, 52–57 (1964).

WOOD, H. N., BRAUN, A. C., BRANDES, H., KENDE, H.: Studies on the distribution and properties of a new class of cell division-promoting substances from higher plant species. Proc. Natl. Acad. Sci. **62**, 349–356 (1969).

ZEPF, E.: Über die Differenzierung des Sphagnumblattes. Z. Botan. **40**, 87–118 (1952).

The Cell Cycle and Tumorigenesis in Plants[1]

ARMIN C. BRAUN

The Rockefeller University, New York, NY, USA

I. Introduction

The experimental oncologist is interested in persistently dividing cells and in discovering the reason why tumor cells divide continuously in their hosts while the growth of all normal cells is precisely regulated. The development by tumor cells of a capacity for essentially unrestrained or autonomous growth is the essential feature that characterizes the tumorous state for without it there would be no tumors. It is therefore clear that an understanding of those substances and mechanisms that regulate normal cell growth and division is important if insight is to be gained into how those regulatory mechanisms are affected when a cell is transformed to the neoplastic state. Other clinically important characteristics, such as, for example, the ability of certain tumor cells to invade and metastasize, may be acquired, sometimes relatively late, and can therefore be dismissed in a search for the earliest events that lead to the establishment and maintenance of the tumorous state.

In addition to the factors involved in the regulation of normal cell growth and division, specific events in the normal cell cycle appear to be importantly concerned in several other aspects of the plant tumor problem. These include, among others, the relationship that exists between a specific stage in the normal cell cycle and the ability of the tumor-inducing principle elaborated by a specific bacterium to transform normal plant cells into tumor cells in the crown gall disease of plants. Another quite distinct area in which a specific phase of the cell cycle appears to be involved is that concerned with the programming of cells for a form of terminal differentiation that results in the loss of neoplastic properties. It is with a consideration of these matters and their consequences as far as cellular metabolism is concerned that the present essay deals.

II. The Two-Phase Concept of Tumor Development

In any analysis of a complex series of events such as those that occur during tumor formation it is often convenient to subdivide, in so far as that is possible,

[1] Certain of these studies were supported in part by a research grant (PHS CA-13808) from the National Cancer Institute, U.S. Public Health Service.

the total event into a series of contributing events, each of which is essential for the consummation of the total process. In studying these events in the crown gall disease of plants two distinct phases have now been recognized (Braun and Laskaris, 1942).

A. The Inception Phase

1. Development of Cellular Competence for Transformation

a) Role of Irritation Accompanying a Wound

In the first or inception phase, normal plant cells are transformed into tumor cells which do not yet develop into a neoplastic growth. Two known requirements must be satisfied to complete the inception phase (Braun, 1952); these have been termed "conditioning" and "induction". By conditioning is meant that only those plant cells that have been rendered susceptible to transformation as a result of irritation accompanying a wound can be transformed into tumor cells. Of particular interest to this discussion is the finding that an excellent correlation exists between a specific stage in the normal wound-healing cycle in which normal cells are transformed into tumor cells and the size and rate of growth of the resulting tumors (Braun and Mandle, 1948; Lipetz, 1966). This is illustrated in Fig.1.

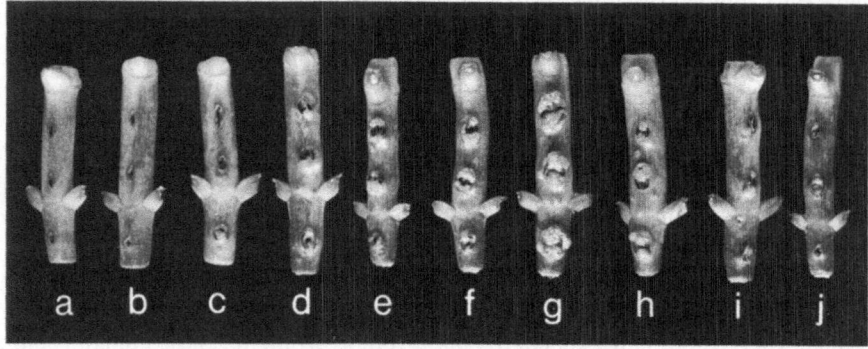

Fig. 1a–j. Picture illustrates the development and decline of competence for transformation in the crown gall disease following wounding. *Kalanchoe* stems were wounded with a sterile needle and the wounds were permitted to heal for (a) 0, (b) 3, (c) 6, (d) 12, (e) 18, (f) 24, (g) 48, (h) 72, (i) 96, (j) 120 hrs before being inoculated in the previously wounded areas with the transforming bacteria. The bacteria in all instances were allowed to act for only 24 hrs, after which the transformation process was stopped with a thermal treatment. Note that maximum response was obtained when cells in the wounded area were permitted to heal for 48 hrs prior to inoculation with the bacteria

b) Competence as a Transient Phenomenon

The conditioning effect was initially demonstrated by comparing the response of cells that had been allowed to heal for 48 hrs prior to inoculation with the inciting bacteria with the response obtained when the bacteria were introduced directly into previously unwounded tissue. In both instances the bacteria were allowed to act for only 24 hrs at 25° C, a period which is not in itself sufficient to permit the cellular transformation to occur when the bacteria are introduced into previously unwounded tissue. It was found in those studies that tissues wounded for 48 hrs prior to inoculation with the bacteria developed large, rapidly growing tumors, while those inoculated at the time of wounding showed no tumorous response. Conditioning was found, moreover, to take place gradually. The host cells reached maximum susceptibility to transformation between the second and third days after a wound was made, while competence for transformation declined again as wound healing progressed toward completion. If, therefore, the host cells are not adequately conditioned, as appears to be the case in the early and late stages of the normal wound-healing cycle as well as in cells not under the influence of a wound, the cellular transformation will not occur despite the presence of many virulent bacteria in intimate contact with the host cells.

c) Competence as a Temperature-Dependent Phenomenon

Conditioning is a temperature-dependent process and optimal competence of plant cells for transformation occurs earlier with increasing temperature and is of shorter duration (LIPETZ, 1965, 1966). It thus appears, as might reasonably be expected, to involve chemical reactions in which the reaction rate doubles for each ten-degree rise in temperature. Histological studies have shown, furthermore, that it is just before cytokinesis occurs that conditioned host cells are transformed into tumor cells of the most rapidly growing type (BRAUN, 1954; BRAUN and MANDLE, 1948; LIPETZ, 1966). Whether this optimal period for transformation occurs during the S phase of the cell cycle or during mitosis is, unfortunately, not yet known.

As a result of wounding, profound physiological and biochemical changes occur in the essentially resting cells present in the region of a wound. These changes include, among others, alterations in the permeability of the cell membranes, an increase in cyclosis, as well as respiratory modifications (BLOCH, 1941). The specific role, if any, that such changes play in rendering host cells susceptible to transformation is not known. It is possible, although this has not yet been demonstrated, that during specific phases of the cell cycle receptor sites are transiently exposed at the cell surface, as they are in certain animal cell types (BURGER and DOONAN, 1972), and that such newly exposed sites play a role in the conditioning process. Finally, as a result of irritation accompanying a wound, the responding cells switch their pattern of metabolism from one concerned with differentiated function and characteristic of normal resting cells to one concerned with cell growth and division. Wounding thus provides a mechanism for the reprogramming of genetic information present in responding cells, as evidenced by the fact that different genes or constellations of genes must be functional in the two alternative metastable states. This could mean that the pattern of metabolism concerned with cell growth and division must be established before cellular transformation can occur; that pattern is then fixed in a cell by the action of the tumor-inducing principle in the crown gall disease. Thus, instead of returning to the quiescent state following wound healing, as cells in the region of a wound normally do, the transformed cells continue to proliferate into a neoplastic growth. An alternative interpretation is that as a result of the transformation the pattern of metabolism concerned with cell growth and division is both established and maintained in a cell. That this second alternative offers a more likely explanation is suggested by the finding that in certain systems, e.g. decapitated plantlets of dwarf pea varieties, transformation and resulting tumor formation occur in the absence of cell divisions in the wound-healing cycle although changes in competence for transformation occur in such systems (KURKDJIAN et al., 1969). Although cell divisions do not occur at the wound site in the dwarf pea system, active synthesis of chromosomal DNA does occur (BEARDSLEY, 1972). It is unfortunately not yet known whether nuclear division also occurs in the absence of cytokinesis, as it does, for example, in excised tobacco-pith tissue treated with an auxin. This would be an important point to establish, for if mitosis did not occur it would rather specifically implicate the S phase of the cell cycle in the transformation process. The results obtained using an atypical system such as that of the dwarf pea nevertheless suggest that the transformation process itself establishes and maintains persistently the pattern of metabolism concerned with cell growth and division in conditioned host cells, rather than simply fixing a previously established pattern in a cell which is then perpetuated, leading to neoplastic growth.

Perhaps no event in the cell cycle is closer to being central to the initiation of the cell division process than is the duplication of chromosomal DNA. Once a cell starts to synthesize DNA it commonly (but not always as evidenced by the presence of specific blocks in the cell cycle to be described later) proceeds through cytokinesis without interruption until the next G_1 phase is reached in the daughter cells.

It is well known from results obtained in a number of different systems ranging from protozoa to mammalian cells—and it may be true for plant cells as well (SIMARD and ERHAN, 1972)—that chromosomal DNA synthesis is regulated in large part by a cytoplasmic initiator of that process. Since this cytoplasmic initiator is present only transiently during the S phase of the normal cell cycle, it could very well represent a fundamental point of departure between a normal cell and a tumor cell. While one would expect the cytoplasmic initiator to be present during the S phase in any dividing population of cells, whether normal or tumor, it may very well be that fundamental regulatory differences exist when the two cell types are present *in situ*. Since the replication of chromosomal DNA commonly represents the initial step that sets in motion the metabolic machinery required for cell growth and division, it might be appropriate to indicate what is known about DNA metabolism during the normal wound-healing cycle in plants. In 1961 KUPILA and STERN examined this question and found a significant increase in newly synthesized DNA by day one following a wound. The level of DNA remained constant between the first and second days and then dropped to zero day levels by day three postwounding. In an extension of this work, LIPETZ (1967) reported that two waves of DNA synthesis occur in the nuclei of cells in the region of a wound between the time of wounding and the first cell divisions. QUÉTIER et al. (1969) also observed two waves of DNA synthesis prior to cell division. They found a species of DNA rich in guanine and cytosine that appeared in nuclei as a result of wounding. They have termed this "stress" DNA and have found that its presence in a cell correlates with the conditioned state of a cell. These workers suggested further that this "stress" DNA may play an important role in both the conditioning process and transformation. Evidence obtained with the use of nucleic acid hybridization studies suggests that this "stress" DNA has nucleotide sequences homologous to those of DNA extracted from crown gall bacteria. GUILLÉ and QUÉTIER (1970) envisage three steps in the transformation process in the crown gall disease:(1)wounding results in the synthesis of a special type of satelite DNA—the "stress" DNA;(2)a transient complex is formed between the "stress" DNA and either bacterial or possibly phage DNA;(3)a part or all of this complex becomes integrated into the genome of the host cell where it replicates with the genome and transforms the cell. The interpretation presented above is based in large part on nucleic acid hybridization techniques. Recent studies (CHILTON et al., 1974; DRLICA and KADO, 1974) in this area raise very serious questions about the validity of those techniques as applied not only to this system but to the very extensive studies carried out in several laboratories (MILO and SRIVASTAVA, 1969; SCHILPEROORT et al., 1973) purporting to demonstrate that bacterial and/or phage DNA is integrated into the genome of the cell where it transforms the cell and replicates with the chromosomal DNA of the host cell. Thus, even though there is no doubt that wounding significantly increases nuclear DNA synthesis, the relationship between such synthesis and the competence of a cell for transformation appears not to have been established although such a relationship may well exist.

d) Development of Competence in Relation to the Normal Cell Cycle

The concept of conditioning as described above implies a state that is operationally unique to cells responding to the stimulus of a wound. The gradual acquisition of susceptibility to transformation by cells in the region of a wound and the gradual loss of that state as wound healing progresses toward completion are reminiscent of the phenomenon observed in developmental biology and for which the term "competence" has been used. The transformation of conditioned host cells into tumor cells by the tumor-inducing principle may also be conceived of as a developmental response comparable to induction leading to determination and cellular differentiation during the normal course of development. In both instances new patterns of metabolism are established which lead, in the normal situation, to the synthesis of substances required for differentiated function. In the abnormal situation, on the other hand, these new patterns lead to the persistent synthesis of the mitotic and enzymatic proteins, nucleic acids, and other substances specifically required for continued cell growth and division. From a developmental point of view the tumor problem may thus be considered one of anomalous differentiation. The basic cellular mechanisms underlying both normal and tumor growth doubtless have much in common, since both appear to depend on selective activation and repression of specific but, in part at least, different genes in the two types of cells. This, then, leads to the question as to how these new patterns of metabolism are established and maintained in a persistently dividing population of cells.

B. The Developmental Phase

1. The Regulation of the Normal Cell Cycle

The second or developmental phase of the crown gall disease is concerned with the continued abnormal and autonomous proliferation of the tumor cells once the cellular transformation has been accomplished. The basic problem here is in determining why the tumor cells grow in an unrestrained or autonomous manner while the growth of all normal cells is precisely regulated. This, in turn, requires an understanding of the factors that regulate normal cell growth and division. It is clear from an abundance of work using many different test systems that the normal cell cycle in both plants and animals proceeds through a continuously integrated series of closely related biochemical events beginning in early interphase and culminating in the division of one cell into two (MAZIA, 1961). The capacity of cells to regulate this sequence of events is apparent. It has, in fact, been suggested (MAZIA, 1961) that every cell that does not divide may be looked upon as one that is blocked at some point on the pathway to division. The implication of that statement, as far as the present discussion is concerned, is that in a persistently dividing population of cells such as those found in true tumors all of the naturally occurring blocks in the cell cycle have been removed, thus enabling such cells to divide continuously in an otherwise suitable environment. These blocks include: the G_1 phase block, at which point the normal cell cycle is most commonly arrested; the S phase block, which is determined in many systems at least partly by the absence of a cytoplasmic initiator of DNA synthesis; the G_2 phase block, from which the cells enter nuclear division directly without again replicating their chromosomal DNA; and finally, the block that occurs between mitosis and cytokinesis. The latter block is evidenced most clearly by the rather common occurrence in both animal and plant fields of binucleate or multinucleate cells in which mitosis has occurred without a corresponding cell division. All of the major blocks in the cell cycle appear to depend on the absence of some essential factor rather than on the presence of inhibitory substances. Yet naturally occurring tissue-specific substances inhibitory to cell division have been extracted from granulocytes (RYTÖMAA and KIVINIEMI, 1968a,b), liver and kidney tissue (SAETREN, 1963), and from mouse ear epidermal cells (BULLOUGH and LAURENCE, 1960). These inhibitory substances, collectively known as chalones, differ significantly from one another. The epidermal chalone is a basic glycoprotein having a molecular weight of about 35000 (BOLDINGH and LAURENCE, 1968), whereas the liver chalone is a polypeptide of low molecular weight (VERLY et al., 1971). The evidence presented thus far does not preclude the possibility that specific stimulators of chromosomal DNA replication, mitosis, and cytokinesis are synthesized by the cells following removal of the chalones. If this is true, as it appears to be, the concepts of specific stimulation *versus* specific inhibition of cell division are not mutually exclusive.

2. Removal of Natural Blocks in the Cell Cycle during Tumorigenesis

How, then, are these blocks that regulate events leading to division in the normal cell cycle persistently removed in plant tumor systems, thus permitting the tumor cells to grow and divide continuously in their hosts as well as under conditions of cell culture that do not encourage the continued growth of normal cells of the type from which the tumor cells were derived? Evidence now available indicates that during the transition from a normal cell to a fully autonomous, rapidly growing tumor cell a series of quite distinct but well-defined biosynthetic systems, which represent the entire area of metabolism concerned with cell growth and division, become progressively and persistently activated, and the degree of activation of those systems determines the rate at which a tumor cell grows (BRAUN, 1958).

a) Specific Substances Involved

The biosynthetic systems shown to be persistently activated may be divided operationally into two groups. The first of these groups includes the two growth-regulating substances, the auxins, which are concerned with cell enlargement and chromosomal DNA replication, on the one hand, and the cell division-promoting factors, which act synergistically with the auxins to promote growth accompanied by cell division, on the other. These are the two substances that not only remove all of the blocks that normally regulate the cell cycle in higher plant species, but in doing so also establish and maintain persistently the pattern of metabolism concerned with cell growth and division in a tumor cell. A second group of substances shown to be synthesized persistently by plant tumor cells includes, among others, such common nitrogenous metabolites as glutamine, asparagine, purines, and pyrimidines, as well as *myo*inositol. These metabolites are required to permit the pattern of metabolism established in the cells by the two growth-regulating substances to be expressed. Thus the plant tumor cells develop a capacity for essentially unrestrained or autonomous growth by virtue of the fact that such cells have acquired, as a result of their transformation, a capacity to synthesize persistently both the growth-regulating substances and the essential metabolites that their normal counterparts require, but cannot make, for cell growth and division.

It is clear that both the auxins and the cell division-promoting factors play a central role in the development by plant tumor cells of a capacity for autonomous growth. These substances have now been isolated from plant tumor tissues. The main auxin has been found to be β-indole-3-acetic acid. Two distinct classes of cell division-promoting factors have been isolated from tissues of higher plant species. The first of these are the 6-substituted adenylyl cytokinins of which kinetin (6-furfurylaminopurine) may be considered to be the prototype. Although this substance is a man-made artifact that has not been found to occur naturally, other 6-substituted purines possessing similar biological activity have been isolated not only from tissues of higher plant species but from bacterial and animal cells as well (HALL 1970). These compounds, which are found immediately adjacent to the anticodon in certain tRNAs, commonly contain an isopentenyl group at the 6 position of the adenine moiety. The isopentenyl derivative serves a specific transfer function in protein synthesis (GEFTER and RUSSELL, 1969). When these compounds are removed either chemically or enzymatically from the tRNA polymer, they possess cell division-promoting and other characteristic cytokinin activities. One 6-substituted purine, which has been given the trivial name zeatin (N^6-(trans-4-hydroxy-3-methyl-2-butenyl)adenine) and/or its riboside, has been found to occur as a free compound in certain plant tissues (Letham, 1963). The fact that a portion of $8\text{-}^{14}C(N^6\text{-}\varDelta^2$-isopentenyl)adenine can be hydroxylated by the fungus *Rhizopogon roseolus* to form zeatin suggests that zeatin may, in fact, also be derived from tRNAs (Miura and Miller, 1969). Whether a specific pathway for the synthesis of a compound such as zeatin exists in cells of higher plant species or whether such a compound is always derived from the degradation of certain tRNAs is not at present known. Miller (1974) has recently reported the isolation of ribosyl-trans-zeatin from crown gall tumor tissues.

A second, chemically quite distinct class of naturally occurring cell division-promoting factors has been isolated from both normal and tumor cells of a number of taxonomically very different plant species. Physical and chemical studies indicate that these compounds, now known as cytokinesins, are substituted hypoxanthines and are thus very different compounds from the 6-substituted adenine-containing cytokinins (Wood et al., 1974). Two cytokinesins, which have been given the trivial names cytokinesin I and cytokinesin II, have been isolated from crown gall tumor tissues.

b) Mode of Action of Those Substances

The question that arises here is how two different classes of compounds can both be effective in promoting cell division when used in association with an auxin. While tumor cells of *Vinca rosea* synthesize persistently both cytokinesin I and cytokinesin II, normal cells of that plant species do not commonly synthesize those compounds when planted in culture. It was found, however, that if the normal cells were forced into rapid growth with kinetin in an otherwise suitable culture medium, the normal cells synthesized a compound that was indistinguishable from cytokinesin I in its physical, chemical, and biological properties (Wood and Braun, 1967). This finding led to the suggestion that the 6-substituted cytokinins exert their cell division-promoting activities by activating the synthesis of cytokinesin I, and it is this substance, rather than the 6-substituted purine cytokinins, that is directly involved in promoting cell division. Thus the biosynthetic system responsible for the production of cytokinesin I may be temporarily activated by 6-substituted purine cytokinins or persistently activated as a result of the cellular transformation. The continued production of the cytokinesins by tumor cells could, therefore, result from the persistent synthesis by those cells of a 6-substituted purine cytokinin such as zeatin or zeatin riboside.

Since the cytokinesins appeared from physical and chemical evidence to be purinones and since the methylxanthines such as theophylline, theobromine, and caffeine are purinones and are known to be phosphodiesterase inhibitors, an attempt was made to learn whether the cytokinesins, like certain of the methylxanthines, also functioned as inhibitors of adenosine 3':5'-cyclic monophosphate [cAMP (3':5')] phosphodiesterases and, hence, possibly promoted cell division as regulators of cAMP (3':5'). The results of that study demonstrated that cytokinesin I is perhaps the most potent naturally occurring cAMP (3':5') phosphodiesterase inhibitor yet described, not only for plant but for the animal (bovine brain) enzymes as well (WOOD, LIN, and BRAUN, 1972). Zeatin riboside, by comparison, was a poor inhibitor and was not very much better in this respect than were adenine, adenosine, guanine, or guanosine, which are not known to promote cell division when used at physiological concentrations. These findings again suggest that the cytokinesins and the 6-substituted purine cytokinins act in different ways to promote cell division. If cAMP ($3':5'$) is, in fact, somehow concerned in the regulation of cell division in higher plant species and if, as postulated, the cytokinesins act specifically as inhibitors of the cAMP ($3':5'$) phosphodiesterases, then it should be possible to replace the growth-promoting effects of the cytokinesins with either cAMP ($3':5'$) or with a more stable, biologically active derivative of the compound. The results of those studies showed that the 8-bromo derivative of cAMP ($3':5'$), but not cAMP ($3':5'$) itself, promoted profuse growth accompanied by cell division when used in association with an auxin in excised tobacco-pith tissue and in Romaine lettuce parenchyma tissue, but not in soybean callus tissue (BASILE et al., 1973; WOOD and BRAUN, 1973). The 8-bromo-cAMP ($3':5'$) was thus able to replace completely both the cytokinesins and the 6-substituted purine cytokinins as cell division-promoting factors in two of the three experimental test systems studied. The 8-bromo derivative of cAMP ($3':5'$) was found to be far more resistant to degradation by the plant cAMP ($3':5'$) phosphodiesterases than was cAMP ($3':5'$) itself (WOOD and BRAUN, 1973). This property could account for differences in biological activity between the natural compound and its derivative. It has also been reported that 8-bromo-cAMP ($3':5'$) is at least as effective and can completely replace cAMP ($3':5'$) in the cAMP ($3':5'$)-dependent histone phosphorylation reaction (MUNEYAMA et al., 1971). The 8-bromo derivative of cAMP ($3':5'$) thus appears to be a stable, biologically active form of cAMP ($3':5'$). It could be exerting its biological action in promoting cell division either by functioning as cAMP ($3':5'$) in metabolism or, because it is a stable structural analogue of cAMP ($3':5'$), it could act as a competitive inhibitor of cAMP ($3':5'$) for binding sites on the phosphodiesterases and thus serve to protect the cAMP ($3':5'$) synthesized by the plant cells. The experimental evidence presently available favors the second possibility.

Doubt has been expressed in the recent past concerning the presence of the cyclic $3':5'$ nucleotides in tissues of higher plant species (NILES and MOUNT, 1974; AMRHEIN, 1974). There is, nevertheless, evidence available to suggest that the cyclic $3':5'$ nucleotides are not only present in tissues of higher plant species but that they play a regulatory role.

In an attempt to determine whether cAMP ($3':5'$) is present in tissues of higher plant species we have used the following criteria (LUNDEEN et al., 1973; and unpublished results). Plant tissue, free of contaminating microorganisms, was weighed, thoroughly ground in liquid nitrogen, and treated with 12% TCA in an ice bath. The TCA was removed by shaking with ethyl ether and the resulting material was passed successively through two columns. The first of these was either a Bio-Rad AG 50W-X8 resin or in later work a Dowex 1 X 8 anion exchange resin, while the second column contained neutral alumina. The resulting material, which was free of contaminating nucleotides, was then assayed with the use of the radioimmunoassay of STEINER et al. (1969). Tissue extracts that had been passed

successively through two columns and that had reacted positively in the radioimmunoassay were tested for kinase activity. The results of that study showed
that that material was not only effective in promoting the phosphorylation of
histone but that phosphorylation was a function of the concentration of the
unknown material present in the tissue extract (LUNDEEN et al., 1973). Since the
unknown material in the tissue extract not only reacted positively in the radioimmunoassay but that it phosphorylated histone in a concentration-dependent
manner strongly suggests that the unknown material present in the plant
extracts is cAMP (3':5') or possibly a derivative of that compound. In a
continuation of these studies it was found that intracellular levels of cAMP (3':5')
were determined by auxin during cell enlargement and cell division in the tobacco
pith system (LUNDEEN et al., 1973). A correlation existed between the levels of
cAMP (3':5') in a cell and chromosomal DNA replication. Whether this finding
represents a correlation or a causal relationship is not yet known.

3. Underlying Mechanism Epigenetic or Mutational

The question that arises next is whether we are dealing in plant tumor systems
with inductive mechanisms involving persistent changes in the pattern of gene
expression, as suggested earlier in this discussion, or whether it is necessary to
invoke changes in the integrity of the genetic information, such as, for example,
the mutation of a regulatory gene(s) to account for the continued abnormal and
autonomous proliferation of a tumor cell (PRESCOTT, 1972). In an attempt to
answer these questions it would appear necessary to demonstrate that all of the
essential properties that characterize the tumorous state are controlled by nuclear
gene function and are inherently present in both normal cells and tumor cells.
This could be accomplished most convincingly by demonstrating that the tumorous state is a completely reversible process, and thus to show that a change in the
integrity of the genetic information present in a cell is not a prerequisite for
establishment or maintenance of the tumorous state.

a) The Crown Gall Disease

Evidence now available indicates that three different and quite distinct plant
tumor systems are potentially reversible. In the crown gall disease this was demonstrated with cloned teratoma tissue (BRAUN, 1959). The tumor-inducing
principle in this disease appears to be a transmissible viral-like entity (AARON-DA
CUNHA, 1969; DE ROPP, 1947; MEINS, 1973). Crown gall teratomas result
when totipotential cells found in certain plant species are transformed to a moderate degree (BRAUN, 1953). Such tumor cells, when isolated and carried under
appropriate conditions of culture, may retain indefinitely a capacity to organize
abnormal buds and shoots (see Fig. 2, A). That these teratoma tissues are composed entirely of tumor cells and not of a mixture of normal and tumor cells was
demonstrated by cloning the tissues and showing that they behaved in every way
like the teratoma tissue from which they were derived. When such tumor shoots,
derived from tumor buds present in the teratoma tissues, were forced into rapid

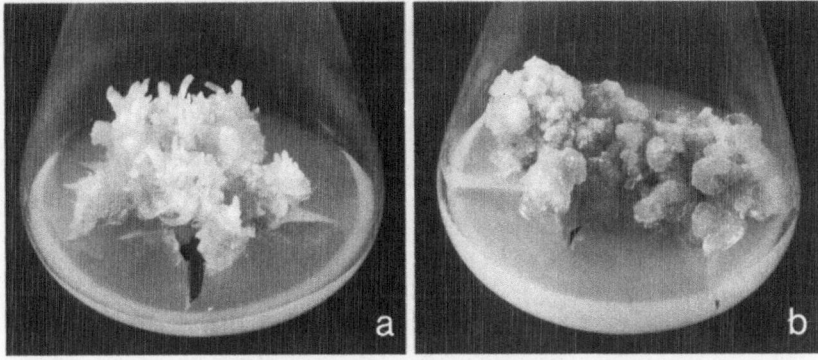

Fig. 2. (a) Crown gall teratoma tissue of tobacco. (b) Typical unorganized crown gall tumor tissue of tobacco. Both tissues were grown on a minimal medium containing agar

but organized growth as a result of graftings to healthy plants, some gradually recovered from the tumorous state and some of these ultimately flowered and set fertile seed. Figure 3 illustrates recovery. In this instance tumor shoots from cloned Havana 38 teratoma tissue were tip-grafted onto Turkish tobacco stocks, thus providing morphological markers to distinguish between the tumor tissue and the plant to which it was grafted.

Tumor shoots derived from a number of cloned tobacco teratoma-cell lines of both Turkish and Havana origin have now been analyzed in some detail. In the Havana-derived teratoma clone pictured in Fig. 3, A, 1163 tumor shoots were grafted to Turkish tobacco stock plants over an approximately three-year period. Of these often well-organized shoots somewhat more than 90% failed to establish suitable vascular connections with the host plants and, being tumorous, developed into typical teratomatous growths, the rapidly dividing cells of which often forced what remained of the implanted shoot away from the main axis of the stem (Fig. 4, A). Fifty-seven of the implanted shoots established vascular connections with the host and appeared to develop well until they reached a height of 4-8 cm (Fig, 4, B). Although the stems of these shoots appeared quite normal when they first developed, as they grew further such stems became quite abnormal and were clearly tumorous. The growth of these tumor cells often caused a severe disruption in the organization of the stem and of the vascular system, as evidenced by a wilting of the leaves and ultimate cessation of growth of the implanted shoot. The leaves found on many such teratoma shoots appeared normal and they often retained their characteristic morphology following a complete breakdown of the cellular organization of the stem (see Fig. 4, A). Histologically the well organized teratoma leaves contained all of the differentiated cell types found in normal Havana tobacco leaves. There was a well developed upper and lower epidermis containing both typical epidermal hairs composed of one to several cells as well as typical glandular hairs. The stomata found in the upper and lower epidermis appeared normal, showed a characteristic distribution on the leaf surfaces and appeared to be functional. There was often a well defined palisade layer the cells of which contained what appeared to be normal chloroplasts. Typical spongy parenchyma cells as well as histologically normal-appearing veins were found in

Fig. 3 A–D. Composite picture showing stages in the recovery of crown gall teratoma tissue of tobacco. (A) Pieces of cloned teratoma tissue of Havana tobacco were placed in a liquid medium on a reciprocal shaker to permit development of tumor shoots. (B) Shoot of Havana origin was grafted at cambial level into a tobacco plant of Turkish origin, thus providing a morphological marker to distinguish between the grafted tumor shoot and the stock plant. (C) Development of the grafted shoot. Note characteristic Havana leaf morphology present in the new growth. (D) An organized shoot, such as is shown in (C), was grafted to a Turkish tobacco stock plant. The shoot grew abnormally—note the development of spontaneous tumors just above the graft union; but as growth proceeded it became normal in appearance, ultimately flowered, and set fertile seed

the teratoma leaves. The cells of such leaves are clearly not exhibiting neoplastic properties and they appear normal by all generally accepted criteria. If such highly differentiated leaf cells are, in fact, normal, then it may be concluded that a selective and spontaneous self-healing of the tumor cells in a teratoma shoot may occur during the development of a teratoma leaf. It is, nevertheless, possible that although the differentiated teratoma leaf cells appear to be structurally and functionally normal, they may retain their neoplastic properties although those properties are clearly suppressed. It now appears well established in the animal

Fig. 4a–c. Picture showing how implanted teratoma shoots may develop. (a) Tumor cells at the base of the implanted shoot proliferated before a vascular connection could be established and the growing tumor cells forced what remained of the implanted shoot away from the main axis of the stem. (b) An example of a teratoma shoot that established a vascular connection with the host and appeared initially to grow quite normally. As the shoot developed, tumor cells in the stem proliferated actively and caused a serious disruption of the cellular organization, ultimately causing a cessation of growth in the length of the shoot. (c) An example of a teratoma shoot that was morphologically quite normal in appearance, had recovered from the tumorous state, but failed to flower. Note the development of a spontaneous tumor at the graft union

field that cloned lines of neoplastic cells transformed by either DNA or RNA oncogenic viruses may revert in significant numbers to the normal phenotype. This is true despite the fact that the viral DNA or, in the case of the RNA viruses, a DNA copy of the viral RNA is integrated into and replicates with the genome of the host cell. If, therefore, the tumor-inducing principle responsible for the establishment of the neoplastic state in the crown gall disease is present in the differentiated teratoma leaf cells but its influence on the cell is suppressed, then the fundamental problem involved here is concerned with the characterization of the substances and cellular mechanisms responsible for the suppression of the tumorous state in this instance. An attempt is therefore now being made to determine whether the differentiated teratoma leaf cells are, in fact, normal or whether they will revert to the neoplastic state if they are freed from the morphogenetic controls that govern the growth of all normal cells in an organism. This problem is now being approached with the use of cloned cell lines derived from several of the most highly differentiated cell types (glandular hairs, stomatal cells, etc.) present in the normal-appearing teratoma leaves. The leaves on many of the shoots appeared both morphologically and histologically normal and the stomata in many such leaves were functional.

An additional 42 teratoma shoots established vascular connections with the host, and although often abnormal at their bases, as evidenced by the appearance of spontaneous tumors at the graft union and/or on the lower part of the stem, all shoots recovered from the tumorous state as shown by an inability of tissues isolated from their upper parts to grow on a minimal culture medium (Fig. 4,C). Of the 42 teratoma scions that had recovered from the tumorous state, 11 flowered and set fertile seed (Fig. 3, D). Recovery in these latter instances was complete; these examples provide evidence that the nuclei of the normal and tumor cells are genetically equivalent. They suggest further that neoplastic growth in this instance arises from epigenetic changes involving a persistent change in the expression of the genetic potentialities present in the tumor cell nucleus rather than from a change involving the loss or permanent rearrangement of some of that information.

Recent studies suggest that a line of the typical unorganized crown gall tumor (see Fig. 2, B) may also be made to recover from the tumorous state. SACRISTÁN and MELCHERS (1969) used a typical unorganized tumor line that had been carried continuously in culture for more than 15 years. The authors of this study have, nevertheless, expressed reservations concerning an interpretation of the results obtained because of the possibility that the uncloned line used may have contained untransformed as well as tumor cells.

b) Kostoff Genetic Tumors

Recovery has now been demonstrated in a second neoplastic disease of plants, the Kostoff genetic tumors. When two plant species such as *Nicotiana glauca* ($2n=24$) and *Nicotiana langsdorffii* ($2n=18$) are crossed and the seed of the hybrid ($2n=21$) sown, the resulting plants commonly develop normally during the period of their active growth. Once the plants reach maturity a profusion of tumors develop spontaneously from all parts of the plant. When tumor tissues from such hybrid plants are isolated and planted on a simple, chemically defined culture medium, they grow profusely and indefinitely on that medium that does not support the continued growth of normal cells isolated from either parent. There is now some evidence to indicate that the hybrid tumor-forming ability is provided by individual parental contributions (HAGEN, 1969). A capacity for the synthesis of a cell division-promoting factor and thiamine appears to be provided by the *N. glauca* parent, while a capacity for *myo*inositol synthesis is derived from *N. langsdorffii*. The tumor cells in this instance, like the crown gall tumor cells described earlier, have acquired a capacity to synthesize all of the factors needed for their abnormal proliferation from a very simple, chemically defined culture medium consisting of mineral salts, sucrose, and vitamins.

In an elegant experiment CARLSON et al. (1972) isolated leaf mesophyll cells from the two parent species, made protoplasts, and fused the protoplasts in culture. Where successful fusion was achieved the hybrid cells grew, while the unfused or fused cells of either parent did not. The hybrid cells ($2n=42$) were thus behaving in culture like the tumor cells isolated from the interspecific hybrid plant in that they grew on a simple, chemically defined culture medium which served to selectively obtain the desired hybrids. When these were then cultured further as callus tissue, shoots were organized; when those shoots were grafted to appropriate stock plants they grew in an organized manner, ultimately flowered, and set fertile seed. Recovery was complete.

The genetic tumors of plants may be considered to provide a model for the so-called chromosomal imbalance theory of cancer, which postulates that tumors arise because of an upset in the balance between those genes that determine growth and those that are concerned with the regulation of growth. The findings reported above, which demonstrate that despite serious chromosomal imbalance the tumorous state is completely reversible, would appear significant and indicate that epigenetic mechanisms are also ultimately involved in those systems.

c) Habituation

A third example in plants in which a reversal of the neoplastic change can be achieved regularly is found in the phenomenon known as habituation. When plant cells are grown in culture they commonly acquire spontaneously capacities to synthesize growth-regulating substances, metabolites, or vitamins that the cells did not possess at the time of their isolation from the plant. The biosynthetic systems, shown to be newly activated, may become unblocked individually or sometimes more than one biosynthetic system may become persistently activated (Fox, 1963). When the two growth-regulating systems, the auxins and the cell division-promoting factors which, it will be recalled, establish the pattern of metabolism concerned with cell growth and division together with those metabolites discussed above required to permit that pattern to be expressed, are all persistently activated, habituated cells behave like typical plant tumor cells in that they grow profusely in culture on a minimal medium, and when implanted into a suitable host, develop into typical tumors (LIMASSET et GAUTHERET, 1950). Once these essential biosynthetic systems are activated in a cell the newly acquired capacities may be retained indefinitely in culture. This, then, is a heritable cellular change that occurs randomly and spontaneously and is very similar to the cellular transformation resulting in the crown gall disease and in the genetic tumors. Of particular interest is a preliminary finding of BINNS and MEINS (1973) that habituation for a cell division factor appears to occur with a frequency of between two and three orders of magnitude greater than is commonly found for spontaneously occurring random mutations.

That this phenomenon is completely reversible with the use of techniques described above has now been demonstrated using a significant number of cloned lines of habituated tissues (BINNS and MEINS, 1973; LUTZ, 1971). These studies demonstrate, then, that profound and heritable changes may occur spontaneously in the phenotype without corresponding changes in the integrity of the cellular genome. All of the genetic information required to account for the heritable cellular change that characterizes the neoplastic state in plants is present in the normal cell genome and needs only to be persistently activated. Any new genetic information added to a cell at the time of transformation (for example, in BLACK's wound tumor disease, which is caused by an RNA-containing virus, and possibly in the crown gall disease as well) might be considered to somehow persistently activate that segment of the host cell genome concerned with contin-

ued cell growth and division rather than with coding specifically for one or more
of the substances required for the autonomous proliferation of the plant tumor
cell.

It is clear from the three different plant tumor systems described above that
the tumorous state in plants is a potentially reversible process. The diversity of
tumors found in a wide spectrum of animals, ranging from the LUCKÉ adenocarci-
noma of the frog to neuroblastomas in man that have now also been found to be
reversible suggests in the strongest possible manner that the phenomenon of
reversibility has broad biological implications (BRAUN, 1969, 1974). In fact, this
spectrum of examples is now so broad it is tempting to assume that the neoplastic
state generally is potentially reversible. If this is true it opens, in principle at least,
entirely new avenues of approach to the tumor problem. If these avenues are now
to be exploited, it will be necessary to learn how to switch the pattern of synthesis
in a cell from one that makes it grow as a tumor cell to one that restores it to
normal or, in those instances in which extensive deletions or significant aneuploidy has occurred, at least to the nonmalignant state. This, in turn, will require
an ability to manipulate nuclear gene function at will.

4. Programming Cells for Terminal Differentiation under Defined Conditions. Relation to the Tumor Problem

It now appears well established that terminal differentiation resulting in the
loss of neoplastic properties may, in fact, occur regularly in certain tumors in both
the animal and plant fields (BASILE et al., 1973). That plant tumor cells may
undergo a form of terminal differentiation resulting in death has been observed
not only in tumors growing in their hosts but it has also been seen in certain
cloned tumor cell lines growing under defined conditions in culture. The problem
of control would thus appear to resolve itself into learning how to bring about the
controlled expression of those potentialities in all cells of the tumor. To accomplish this would require a complete understanding of the factors involved for each
tumor system studied. An attempt was therefore made to characterize more precisely the factors involved in the differentiation of parenchyma cells into tracheary
elements in order to gain insight into the cellular mechanisms that underlie at
least one example of this type of cytodifferentiation. The experimental test system
used in those studies was one developed by DALESSANDRO and ROBERTS (1971).
These workers found that Romaine lettuce (*Lactuca sativa* L. var. Romana) pith
parenchyma cells could be made to differentiate into tracheary elements in short
periods of time by exposing the excised parenchyma cells to an exogenous source
of auxin and kinetin in an otherwise suitable culture medium.

Since a substance such as kinetin is man-made, not found to occur naturally,
an attempt was made to replace the inductive effects of kinetin or other 6-substi-
tuted cytokinins with known but chemically quite distinct substances which, if
effective, might provide insight into the mechanisms that underlie this type of
terminal cellular differentiation. These studies demonstrated that both the cytokinesins [potent inhibitors of plant and animal cAMP (3':5') phosphodiesterases]
and the 8-bromo derivative of cAMP (3':5') [a stable, biologically active form of

cAMP (3':5')] are highly effective in encouraging the differentiation of parenchyma cells into tracheary elements with accompanying death (Basile et al.,
1973). Since cAMP (3':5') and theophylline (a known phosphodiesterase inhibitor)
were also found to be effective when used together in a culture medium, the results
obtained in these studies suggest that cAMP (3':5') is somehow involved in the
conversion of parenchyma cells into tracheary elements in this system. The important question that, of course, remains unanswered is precisely where in metabolism does the cyclic nucleotide exert the effects that accomplish a reprogramming of the genetic information in a cell required to achieve this unique type of
terminal cellular differentiation. Another significant aspect of this problem is to
pinpoint precisely the stage in the cell cycle that is critical for tracheary element
differentiation. Does induction occur before chromosomal DNA replication, is
such replication essential, or does the reprogramming of information occur only
during mitoses, and is, in addition, cytokinesis essential to fix the program that
leads to the differentiation of tracheary elements, as Fosket's studies (1968)
suggest?

To summarize, a number of interesting concepts have emerged from a study of
plant tumor systems. It is demonstrated that specific, but not yet completely
characterized, stages in the normal cell cycle are importantly involved in the
transformation of a normal cell into a tumor cell in the crown gall disease of
plants, as well as in the programming of cells for a form of terminal differentiation
that results in the loss of neoplastic properties. These studies also reveal how the
several naturally occurring blocks that normally regulate the cell cycle are persistently removed during the transition of a normal plant cell to a tumor cell, and
why, following the removal of those blocks, the tumor cell grows in an autonomous manner. These studies show further that the persistent removal of the
blocks does not require a change in the integrity of the genetic information
normally present in a cell but merely a persistent change in the expression of that
information. The plant tumor problem appears, therefore, to be one of anomalous
differentiation, since the basic cellular mechanisms underlying the tumor transformation and normal cellular differentiation are similar in that both depend for
their expression on the selective activation and repression of genes. There is little
in the animal literature that argues strongly against similar mechanisms being
involved in those systems and much in favor of such mechanisms.

References

Aaron-Da Cunha, M. I.: Sur la libération, par les rayons X, d'un principe tumorigène con-
tenu dans les tissus de crown gall de tabac. C. R. Acad. Sci. Paris, Ser. D, **268**, 318–321
(1969).
Amrhein, N.: Evidence against the occurrence of adenosine-3':5'-cyclic monophosphate in
higher plants. Planta (Berl.) **118**, 241–258 (1974).
Basile, D. V., Wood, H. N., Braun, A. C.: Programming of cells for death under defined experimental conditions: Relevance to the tumor problem. Proc. Natl. Acad. Sci. **70**, 3055–3059
(1973).
Beardsley, R. E.: The inception phase in the crown gall disease. Progr. Exp. Tumor Res. **15**,
1–75 (1972).

194 A. C. Braun

BINNS, A., MEINS, JR., F.: Habituation of tobacco pith cells for factors promoting cell division is heritable and potentially reversible. Proc. Natl. Acad. Sci. **70**, 2660–2662 (1973).

BLOCH, R.: Wound healing in higher plants. Botan. Rev. **7**, 110–146 (1941).

BOLDINGH, W. H., LAURENCE, E. B.: Extraction, purification and preliminary characterisation of the epidermal chalone: A tissue specific mitotic inhibitor obtained from vertebrate skin. European J. Biochem. **5**, 191–198 (1968).

BRAUN, A. C.: Conditioning of the host cell as a factor in the transformation process in crown gall. Growth **16**, 65–74 (1952).

BRAUN, A. C.: Bacterial and host factors concerned in determining tumor morphology in crown gall. Botan. Gaz. **114**, 363–371 (1953).

BRAUN, A. C.: Studies on the origin of the crown-gall tumor cell. In: Abnormal and Pathological Plant Growth. Brookhaven Symp. Biol. **6**, 115–127 (1954).

BRAUN, A. C.: A physiological basis for autonomous growth of the crown-gall tumor cell. Proc. Natl. Acad. Sci. **44**, 344–349 (1958).

BRAUN, A. C.: A demonstration of the recovery of the crown-gall tumor cell with the use of complex tumors of single-cell origin. Proc. Natl. Acad. Sci. **45**, 932–938 (1959).

BRAUN, A. C.: The cancer problem: A critical analysis and modern synthesis, pp. 209. New York-London: Columbia University Press 1969.

BRAUN, A. C.: The biology of cancer, pp. 169. Reading, Mass.: Addison-Wesley Publ. 1974.

BRAUN, A. C., LASKARIS, T.: Tumor formation by attenuated crown-gall bacteria in the presence of growth-promoting substances. Proc. Natl. Acad. Sci. **28**, 468–477 (1942).

BRAUN, A. C., MANDLE, R. J.: Studies on the inactivation of the tumor-inducing principle in crown gall. Growth **12**, 255–269 (1948).

BULLOUGH, W. S., LAURENCE, E. B.: The control of epidermal mitotic activity in the mouse. Proc. Roy. Soc. London Ser. B, **151**, 517–536 (1960).

BURGER, M. M., DOONAN, K. D.: Cell surface alterations in transformed cells as monitored by plant agglutinins. In: HARRIS, R., ALLIN, P., VIZA, D. (Eds.): Cell Differentiation. Proc. 1st International Conference on Cell Differentiation, pp. 182–185. Copenhagen: Munksgaard 1972.

CARLSON, P. S., SMITH, H. H., DEARING, R. D.: Parasexual interspecific plant hybridization. Proc. Natl. Acad. Sci. **69**, 2292–2294 (1972).

CHILTON, M.-D., CURRIER, T. C., FARRAND, S. K., BENDICH, A. J., GORDON, M. P., NESTER, E. W.: *Agrobacterium tumefaciens* DNA and PS8 bacteriophage DNA not detected in crown gall tumors. Proc. Natl. Acad. Sci. **71**, 3672–3676 (1974).

DALESSANDRO, G., ROBERTS, L. W.: Induction of xylogenesis in pith parenchyma explants of *Lactuca*. Am. J. Botan. **58**, 378–385 (1971).

DRLICA, K. A., KADO, C. I.: Quantitative estimation of *Agrobacterium tumefaciens* DNA in crown gall tumor cells. Proc. Natl. Acad. Sci. **71**, 3677–3681 (1974).

FOSKET, D. E.: Cell division and the differentiation of wound-vessel members in cultured stem segments of *Coleus*. Proc. Nat. Acad. Sci. **59**, 1089–1096 (1968).

FOX, J. E.: Growth factor requirements and chromosome number in tobacco tissue cultures. Physiol. Plant. **16**, 793–803 (1963).

GEFTER, M. L., RUSSELL, R. L.: Role of modifications in tyrosine transfer RNA: modified base affecting ribosome binding. J. Molec. Biol. **39**, 145–157 (1969).

GUILLÉ, E., QUÉTIER, F.: Le "crown gall": modèle expérimental pour l'étude du mécanisme de la transformation tumorale. C. R. Acad. Sci. Paris, Sér. D, **270**, 3307–3310 (1970).

HAGEN, G. L.: Tumor growth in hybrid tobacco: The parental contributions. 11th Internat. Bot. Congr. Seattle, Wash. Abstracts, p. 82 (1969).

HALL, R. H.: N^6-(Δ^2-isopentenyl)adenosine: Chemical reactions, biosynthesis, metabolism, and significance to the structure and function of tRNA. In: Progress in Nucleic Acid Research and Molecular Biology, Vol. 10, pp. 57–86. New York: Academic Press 1970.

KUPILA, S., STERN, H.: DNA content of broad bean (*Vicia faba*) internodes in connection with tumor induction by *Agrobacterium tumefaciens*. Plant Physiol. **36**, 216–219 (1961).

KURKDJIAN, A., MANIGAULT, P., BEARDSLEY, R.: Crown gall: Effect of temperature on tumorigenesis in pea seedlings. Canad. J. Botan. **47**, 803–808 (1969).

LETHAM, D. S.: Zeatin, a factor inducing cell division isolated from *Zea mays*. Life Sci. **2**, 569–573 (1963).

LIMASSET,P., GAUTHERET,R.: Sur le caractère tumoral des tissus de tabac ayant subi le phé-
nomène d'accoutumance aux hétéro-auxines. C. R. Acad. Sci. (Paris) 230, 2043–2045
(1950).

LIPETZ,J.: Crown-gall tumorigenesis: Effect of temperature on wound healing and conditioning. Science 149, 865–867 (1965).

LIPETZ,J.: Crown gall tumorigenesis. II. Relations between wound healing and the tumorigenic response. Cancer Res. 26, 1597–1605 (1966).

LIPETZ,J.: Wound healing in Kalanchoë. Ann. New York Acad. Sci. 144, 320–325 (1967).

LUNDEEN,C.V., WOOD,H.N., BRAUN,A.C.: Intracellular levels of cyclic nucleotides during
cell enlargement and cell division in excised tobacco pith tissues. Differentiation 1, 255–
260 (1973).

LUTZ,A.: Morphogenetic aptitudes of tissue cultures of unicellular origin. In: Les Cultures de
Tissus de Plantes; Colloq. internat. centre nat. recherche sci. (Paris), No. 193 (Proc. 2nd
internat. Conf. Plant Tissue Culture, Strasbourg, France), pp. 163–168. Paris: Centre National de la Recherche Scientifique 1971.

MAZIA,D.: Mitosis and the physiology of cell division. In: BRACHET,J., MIRSKY,A.E. (Eds.):
The Cell, Vol. 3, pp. 77–412. New York-London: Academic Press 1961.

MEINS,JR., F.: Evidence for the presence of a readily transmissible oncogenic principle in
crown gall teratoma cells of tobacco. Differentiation 1, 21–25 (1973).

MILLER,C.O.: Ribosyl-trans-zeatin, a major cytokinin produced by crown gall tumor tissue.
Proc. Natl. Acad. Sci. 71, 334–338 (1974).

MILO,G.E., SRIVASTAVA,B.I.S.: RNA-DNA hybridization studies with the crown gall bacteria and the tobacco tumor tissue. Biochem. Biophys. Res. Commun. 34, 196–199 (1969).

MIURA,G.A., MILLER,C.O.: 6-(γ,γ-dimethylallylamino)purine as a precursor of zeatin. Plant
Physiol. 44, 372–376 (1969).

MUNEYAMA,K., BAUER,R.J., SHUMAN,D.A., ROBINS,R.K., SIMON,L.N.: Chemical synthesis
and biological activity of 8-substituted adenosine 3′, 5′-cyclic monophosphate derivatives.
Biochemistry 10, 2390–2395 (1971).

NILES,R.M., MOUNT,M.S.: Failure to detect cyclic 3′,5′-adenosine monophosphate in healthy
and crown gall tumorous tissues of Vicia faba. Plant Physiol. 54, 372–373 (1974).

PRESCOTT,D.M.: Biology of cancer and the cancer cell: Normal and abnormal regulation of
cell reproduction. CA 22, 262–272 (1972).

QUÉTIER,F., HUGUET,T., GUILLÉ,E.: Induction of crown-gall: Partial homology between
tumor-cell DNA, bacterial DNA and the G+C-rich DNA of stressed normal cells.
Biochem. Biophys. Res. Comm. 34, 128–133 (1969).

DE ROPP,R.S.: The growth-promoting and tumefacient factors of bacteria-free crown-gall
tumor tissue. Am. J. Bot. 34, 248–261 (1947).

RYTÖMAA,T., KIVINIEMI,K.: Control of granulocyte production. I. Chalone and antichalone,
two specific humoral regulators. Cell Tissue Kinet. 1, 329–340 (1968a).

RYTÖMAA,T., KIVINIEMI,K.: Control of granulocyte production. II. Mode of action of chalone
and antichalone. Cell Tissue Kinet. 1, 341–350 (1968b).

SACRISTÁN,M.D., MELCHERS,G.: The caryological analysis of plants regenerated from tumorous and other callus cultures of tobacco. Molec. gen. Genet. 105, 317–333 (1969).

SAETREN,H.: The organ-specific growth inhibition of the tubule cells of the rat's kidney. Acta
Chem. Scand. 17, 889 (1963).

SCHILPEROORT,R.A., VAN SITTERT,N.J., SCHELL,J.: The presence of both phage PS8 and
Agrobacterium tumefaciens A₆ DNA base sequences in A₆-induced sterile crown-gall tissue cultured in vitro. European J. Biochem. 33, 1–7 (1973).

SIMARD,A., ERHAN,S.: Incorporation of tritiated thymidine in tobacco pith tissues induced
by a substance isolated from Ehrlich ascites fluid. Canad. J. Botan. 50, 719–722 (1972).

STEINER,A.L., KIPNIS,D.M., UTIGER,R., PARKER,C.: Radioimmunoassay for the measurement of adenosine 3′, 5′-cyclic phosphate. Proc. Natl. Acad. Sci. 64, 367–373 (1969).

VERLY,W.G., DESCHAMPS,Y., PUSHPATHADAM,J., DESROSIERS,M.: The hepatic chalone. I.
Assay method for the hormone and purification of the rabbit liver chalone. Canad. J.
Biochem. 49, 1376–1383 (1971).

WOOD, H. N., BRAUN, A. C.: The role of kinetin (6-furfurylaminopurine) in promoting division in cells of *Vinca rosea* L. Ann. N.Y. Acad. Sci. **144**, 244–250 (1967).

WOOD, H. N., BRAUN, A. C.: 8-Bromoadenosine 3′:5′-cyclic monophosphate as a promoter of cell division in excised tobacco pith parenchyma tissue. Proc. Nat. Acad. Sci. **70**, 447–450 (1973).

WOOD, H. N., LIN, M. C., BRAUN, A. C.: The inhibition of plant and animal adenosine 3′:5′-cyclic monophosphate phosphodiesterases by a cell-division-promoting substance from tissues of higher plant species. Proc. Nat. Acad. Sci. **69**, 403–406 (1972).

WOOD, H. N., RENNEKAMP, M. E., BOWEN, D. V., FIELD, F. H., BRAUN, A. C.: A comparative study of cytokinesins I and II and zeatin riboside: A reply to Carlos Miller. Proc. Natl. Acad. Sci. **71**, 4140–4143 (1974).

Cell Cycle and Liver Function

R. TSANEV

Institute of Biochemistry, Bulgarian Academy of Sciences, Sofia, Bulgaria

I. Introduction

The liver with its multiple functions is one of the most difficult cases to establish the relationship between proliferation and function. Nevertheless, a detailed analysis of the literature shows that some tentative conclusions can be made.

There is evidence that during embryonic development mitotic divisions play a critical role in determining the competence of the cells to synthesize tissue-specific proteins (HOLTZER et al., 1973). When considering a tissue already differentiated, such as the liver, the problem of the relationship between mitotic divisions and differentiated function acquires other aspects which can be summarized as follows: (1) What is the relationship between fetal and postnatal growth and the development of differentiated functions? (2) Are these functions preserved during the mitotic cycle? (3) Is the differentiated cellular type maintained upon continuous cellular proliferation? (4) How is the transition between the state of quiescence and mitotic cycle controlled? Although these questions cannot be answered satisfactorily at present, we shall try to discuss this problem from the point of view of the possible mechanisms controlling cellular reprogramming in eukaryotic cells.

II. Cellular Reprogramming

Some misunderstandings in the literature and some apparently controversial results regarding the relationship between cellular proliferation and differentiation arise from the tendency to classify all biological phenomena in which the protein pattern of the cell is altered without changes in DNA as "cytodifferentiation". Such a concept makes it difficult to elucidate the nature of cell differentiation and the role of such factors as embryonic induction and the mitotic cycle.

Many data show that the molecular mechanisms regulating protein synthesis in eukaryotic cells are much more complex than in prokaryotes and can operate at different levels (SCHERRER and MARCAUD, 1968; HARRIS, 1968; RECHCIGL, 1971 a). It seems rational to apply the general term of "cellular reprogramming" to all phenomena in which the protein pattern of a cell is altered without changes in

Table 1. Cellular reprogramming. (1) Altered protein pattern. (2) No changes in DNA (TSANEV, 1973)

I. Changes in the functional state of the cell	II. Changes in the cellular type
1. Reversible	1. Irreversible under normal anatomical and physiological conditions
2. Monophasic	2. Multiphasic
3. Short latent period	3. Long latent period
4. Inducer permanently needed	4. Inducer temporarily needed
5. Mitosis-independent	5. Mitosis-dependent

its DNA. Then at least two groups of biological phenomena can be distinguished: *changes in the functional state* of the cell and *changes in the cellular type* (TSANEV and SENDOV, 1971a; TSANEV, 1973).

As seen from the brief enumeration of their biological characteristics (Table 1), these two groups exhibit not only different but quite opposite features. It would be difficult, therefore, to assume that both types of transitions are controlled by the same molecular mechanism.

The first group is most probably a heterogeneous group including phenomena regulated at different levels. Generally speaking, all properties of this group can be explained by mechanisms based on the formation of molecular complexes easily dissociated by changing the concentration of regulatory substances. Thus, at the level of transcription, the DNA-repressor complex can be dissociated either by lowering the concentration of the repressor (if the dissociation constant of the complex is high), or by increasing the concentration of other molecules (effectors, derepressors, inducers) which increase the dissociation constant of the complex by binding to the repressor. At the level of translation, a similar mechanism can also operate through the production of molecules (translational repressors or derepressors) which stabilize or destabilize mRNAs (TOMKINS et al., 1970, 1972).

The second type of cellular reprogramming appeared later in evolution and permitted the development of multicellular organisms with different cellular types. There are reasons to believe that the regulation of this process occurs in the nucleus at the level of transcription (TSANEV and SENDOV, 1971a), although the opinion is maintained that a post-transcriptional control also exists (HARRIS, 1968; RUTTER et al., 1973). The expression of different opinions is partly due to the fact that no clear distinction is made between the two types of cellular reprogramming.

The mechanism controlling the emergence of different cellular types and especially the need of mitotic divisions, can hardly be explained by the above-mentioned formation of reversible complexes and it is reasonable to accept the existence of a different control mechanism. To distinguish the latter from the mechanisms of repression–derepression we have used the term blocking–deblocking mechanism (TSANEV and SENDOV, 1971a).

The main characteristics of the second type of cellular reprogramming can be explained if we assume that the mechanism of blocking–deblocking operates through an irreversible binding of macromolecules to DNA, i.e. the formation of complexes which can not be dissociated unless the cell goes through a mitotic cycle.

According to this view, the genome of prokaryotic cells is unblocked and can be repressed or derepressed, while in eukaryotic cells the genome is blocked, but during embryonic development a process of sequential deblocking of different genes takes place. Depending on the region of the genome which remains finally deblocked, different cellular types emerge. In the differentiated cellular type processes of repression and derepression take place within the limits of the deblocked regions of the genome, thus determining different functional states of the cells.

Analyzing such a model by computer simulation, we have shown that the final stage of cellular differentiation can only be reached through a chain of events in which both processes of repression–derepression (realized by the embryonic induction) and of blocking–deblocking (realized by the mitotic cycle) are successively involved (TSANEV and SENDOV, 1971a).

On the basis of recent data on the structure and properties of histones (WILHELM et al., 1971), we have assumed that the mechanism of blocking consists in the complexing of DNA with histones in such a way that specific histone arrangements are formed which mark different sites of the genome (TSANEV and SENDOV, 1971a). This makes possible the recognition of different genetic regions by nonhistone proteins, which seem to play an important role in genetic regulations (MCCLURE and HNILICA, 1972). In our model some of these proteins are supposed to bind irreversibly but specifically to different combinations of histones, thus deblocking the corresponding regions of the genome without eliminating the histones. Different cellular types, therefore, should have different nonhistone deblocking proteins. Alterations in the pattern of these proteins in the chromatin (i.e. changes of the cellular type) can be obtained only through the combined effect of two factors: (1) a mitotic division and (2) a factor (inducer) which can repress the synthesis or can inactivate some deblocking proteins (TSANEV and SENDOV, 1971a).

In this way the irreversible binding of some deblocking proteins to the chromatin can explain the requirement for mitotic divisions to obtain a cellular type, while the induction of different synthetic activities in a competent cellular type can be explained by processes of repression–derepression where mitotic divisions are not necessary.

Experimental evidence in favor of this model has been reviewed elsewhere (TSANEV and SENDOV, 1971a) and it has recently been supported by some new data. In isolated chromosomes of Chinese hamster fibroblasts we have demonstrated a specific distribution of nonhistone proteins corresponding to the localization of the newly synthesized RNA (DJONDJUROV et al., 1972) and showing a cell-cycle-dependent turn-over (TSANEV et al., 1974). This finding supports the idea that the requirement for mitotic divisions in cellular differentiation may be due to the need to change the pattern of these proteins in the chromosomes.

Another important consequence of this model is that the histone arrangements, which are the basis or recognition, should be conserved and reproduced during DNA replication. It seems difficult to explain the mechanism of this process when we consider the cytoplasmic synthesis of histones and the absence of specific recognition between histones and DNA (WILHELM et al., 1971). Recently we have studied the behavior of the histones during DNA replication in regenerating rat liver by separating the replicated DNA strand from the parental strand with 4 of the histone species still bound to DNA (TSANEV and RUSSEV, 1974). These experiments have shown that the old histones remained bound to the parental DNA strand, while the new histones joined the new DNA strand. These data strongly suggest the importance of maintaining the histone arrangements unchanged in the cell and that the histone arrangements along DNA are maintained during the process of replication by a process of specific interactions between DNA-bound (parental) and free (newly synthesized) histones.

This conservative distribution of histones during DNA replication, together with the striking evolutionary conservatism of their primary structure and the presence of long, nonbasic, amino acid sequences in all histone species (WILHELM et al., 1971), is consistent with the idea that specific interactions between histone arrangements and nonhistone proteins may determine the localization of the latter in the chromatin. According to our model, this will determine the deblocked region of the genome and therefore the cellular type.

Another point which is important in understanding the relationship between cell proliferation and differentiated functions is the possibility of functional interactions between different genes by means of regulatory molecules which influence the expression of gene activity. The problem of whether in eukaryotic cells there are genetic units similar to the operons of prokaryotic cells is still unsolved. The existence in the nucleus of giant, probably polycistronic, precursors of messenger RNA as a transcriptional unit favors a positive answer (GEORGIEV, 1969). In such a case the possibility exists that several genes may be under the same control and can act as a functional unit containing the information for both functional and regulatory proteins. Another possibility of functional interactions between genes is that single genes control the synthesis of regulatory proteins, which are activated as repressors or derepressors by metabolic products resulting from the activity of different structural genes.

In both cases the transcriptional control should be dependent on the functional interactions between different genetic units, which we shall arbitrarily call operons. These interactions form a complex functional genetic net which in eukaryotic cells may comprise three possible types of interaction: repression, derepression, and deblocking. The concept of genetic nets opens the possibility of constructing mathematical models of gene regulation and development (KAUFMAN, 1969; TSANEV and SENDOV, 1971a, b).

Since almost nothing is known about the functional links between different operons in eukaryotic cells, data on the interrelations between different functional states of a cell may throw some light on this problem.

For our discussion it is useful to classify the operons of eukaryotic cells into three major groups:

A. Operons for universal proteins which are common to all cellular types, being indispensible for the life of the cell.

B. Operons for mitotic proteins necessary for the mitotic cycle.

C. Operons for tissue-specific proteins which determine differentiated functions.

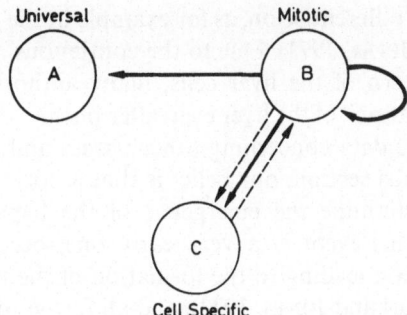

Fig. 1. Scheme of the possible functional interrelations between the three major groups of operons in eukaryotic cells; ⟶ deblocking; ---→ repression (TSANEV and SENDOV, 1971b)

In differentiated tissues all these regions of the genome should be deblocked, with the exception of some highly differentiated cells such as neurons, where the mitotic region seems to be blocked, or nucleated erythrocytes where the whole genome is blocked. The first two regions A and B, may be common to all tissues, but they may also be different, as indicated by the presence of isoenzymes. The region C should be at least partly different in different tissues.

On the other hand, the operons of group A should be continuously or at least periodically derepressed while those of group B and C could be repressed or derepressed depending on different intra- and extracellular factors.

Some general considerations regarding the interrelations between these regions of the genome can be easily suggested. In order to have a cellular population able to proliferate without changing its cellular type, it is obvious that it is necessary for the mitotic genome to contain the information for all deblocking proteins specific for a given cellular type (TSANEV and SENDOV, 1971a,b). To keep a balance between functional activity and cell proliferation, a mutual repression should exist between the mitotic operons and some of the tissue-specific operons (BULLOUGH, 1965; TSANEV and SENDOV, 1966) (Fig. 1).

III. Developmental Changes of Hepatocytes

Hepatocytes are an example of a tissue in which it seems difficult to fix an exact time for the appearance of a cellular type (CAMERON and JETER, 1971). In many developing systems in which the functional activity of the cells is more restricted and suitable markers of cytodifferentiation exist, it has been possible to establish the time of cell-cycle-dependent transitions (HOLTZER et al., 1973). This also makes it easier to distinguish between the two types of cellular reprogramming.

The growing embryonic and postnatal liver undergoes so many biochemical and morphological changes that some authors find it difficult even to define what the differentiated state of the hepatic cell is (CAMERON and JETER, 1971). Morphological criteria may be misleading, since biochemical differentiation may occur before morphological differentiation, as for example in the case of glycogen deposition (CAMERON and JETER, 1971). Due to the continuous developmental changes in the metabolic pattern of the liver cells, many authors prefer to speak of a "continuing differentiation" of the liver even after birth.

On the basis of the data concerning other tissues and for theoretical reasons outlined in the previous section, our belief is that a critical cell-cycle-dependent event should also determine the emergence of the hepatic cellular type. The problem is whether this event is a very early one, occurring during the first outburst of proliferation leading to the formation of the hepatic primordia (ZAVARZIN, 1967; JOHNSON and JONES, 1973), and all further developmental changes are changes in the functional state of an already established cellular type, or whether developmental changes are due to further alterations in the cellular type determined by the continuing cellular proliferation of the fetal and postnatal liver. Although there are no experiments designed to answer such a question, we shall review three main groups of developmental changes which may present suggestions: (1) changes in the enzyme pattern; (2) changes in other antigens, and (3) changes in the proliferative activity.

Enzyme Pattern. Many liver enzymes show typical developmental changes (RUTTER et al., 1963; GREENGARD, 1970). The rapid and large alterations in the enzyme pattern cannot be related to the slow and gradual changes in the proportion between hemopoietic and hepatic cells which also occur during liver development (GREENGARD, 1969; GREENGARD et al., 1972). A number of experiments indicate that most of these enzymic activities represent a *de novo* enzyme synthesis dependent on the synthesis of new RNA (GREENGARD, 1969, 1970; SCHIMKE and DOYLE, 1970).

All data suggest that the appearance of new enzymes in developing liver is determined by mechanisms which have the full characteristics of the first type of cellular reprogramming. Thus, the induction of tyrosine-aminotransferase (TAT) by corticosteroid hormones can be reproduced in explants of fetal liver where the response is maintained only in the presence of the steroid (WICK, 1968). In normal development this enzyme is induced by the combined action of at least three factors appearing sequentially: the cAMP system, glucagon and corticosteroids (GREENGARD, 1969, 1970). An important fact is the finding that a premature accumulation of TAT can be induced in embryonic liver by the combined action of these factors (GREENGARD, 1970).

Corticosteroids were also shown to stimulate the synthesis of cholesterol-7α-hydroxylase, and its appearance on the 20th day after birth was supposed to depend on the reestablishment at that time of the pituitary–adrenal axis (HOUBRECHTS and BARBASON, 1972). The hormonal dependence of glycogen synthesis in liver was also demonstrated (MONDER and COUFALIK, 1972). The increase in glucose-6-phosphatase around the 18th day of gestation was associated with the thyroid gland beginning to function, since injection of thyroxine enhanced the prenatal development of the enzyme (GREENGARD, 1969).

On analyzing the literature, GREENGARD (1970) came to the general conclusion that in rat liver the developmental appearance of different enzymes was clustered in three different stages of development which corresponded to the establishment of different hormonal systems in the organism. A first group of

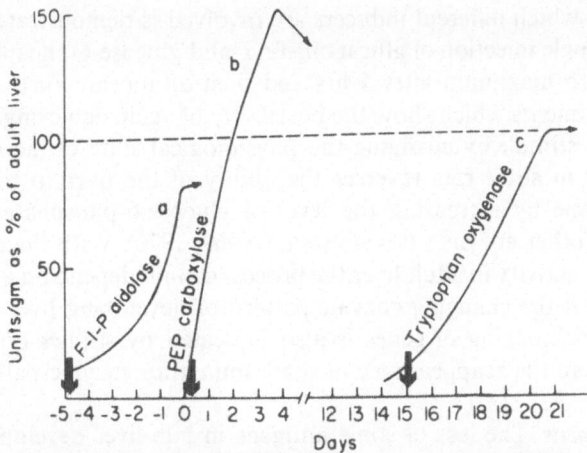

Fig. 2. Examples of the rate of enzyme accumulation at the late fetal, early postnatal and late suckling stage (GREENGARD, 1970). The arrows indicate the times of a sharp decrease in the proliferative activity of the hepatocytes according to data from: ZAVARZIN, 1967; GRISHAM, 1969; HOUBRECHTS and BARBASON, 1972; STÖCKER et al., 1972

enzymes appears during the late fetal period (16–20th day of gestation), a second during the first days after birth, and a third during the third week of postnatal life (Fig. 2). The development of the first group of enzymes coincides with the time when the pituitary, adrenal cortex, thyroid and β-cells of the pancreas start to function; the appearance of the second group is associated with the secretion of glucagon and epinephrine induced by neonatal hypoglycemia and possibly mediated through cAMP; the third group of enzymes seems to be correlated with the reestablishment of the pituitary–adrenal axis and the effect of changing nutritional conditions (GREENGARD, 1970).

Important metabolic changes were found also in the amphibian liver during metamorphosis (FRIEDEN and JUST, 1970; TATA, 1970). All data suggest that here again the changes have the characteristics of the first type of cellular reprogramming, being induced by thyroid hormones and other environmental conditions. That no cellular divisions play a role in this process is indicated by the fact that the low incorporation of P^{32} into liver DNA in this case is not increased by thyroid hormones (FRIEDEN and JUST, 1970).

The conclusion that the developmental changes of the enzyme pattern in liver are due to mechanisms of the reversible type may be questioned on the basis of some data which apparently disagree with the reversibility of these processes. Thus, it has been shown that the lack of the stimuli which evoked the synthesis of enzymes in developing liver does not eliminate these enzymes in adult liver. For example, the cessation of postnatal hypoglycemia only decreases the TAT concentration to the adult level; thyroidectomy causes only a slight decrease in the level of glucose-6-phosphatase in adult liver, etc. (GREENGARD, 1969, 1970). This can be explained by the fact that during development additional regulators come into play which replace the original stimuli (GREENGARD, 1969). For this reason the whole process may appear as irreversible. However, the reversibility of the indi-

vidual steps in which different inducers are involved is demonstrated experimentally. After a single injection of glucagon, TAT and glucose-6-phosphatase in fetal
liver increase to maximum after 5 hrs and then disappear. Further evidence is
given by experiments which show the possibility of again achieving a responsiveness to earlier stimuli by changing the physiological state of the animal. Thus,
thyroidectomy in adult rats reverses the ability of the liver to respond to the
thyroid hormone by increasing the level of glucose-6-phosphatase. The same
holds true for other enzymes (GREENGARD, 1969, 1970). With the extremely low
level of mitotic activity in adult liver the process cannot depend on a mitotic cycle.

The fact that the changing enzyme pattern of developing liver is not due to
blocking and deblocking of genes is also indicated by studies on regenerating
adult liver where the reappearance of some immature enzyme pattern has been
observed (see p. 221–222).

Other Antigens. The loss of some antigens in late liver development is illustrated by the so-called embryo-specific or "transitory" antigens (ABELEV et al.,
1963; URIEL, 1971). In rat liver these antigens are detectable in 12-day-old embryos and their synthesis continues for at least 14 days after birth. One of these
antigens—the αM-fetoprotein—can be induced in the adult liver by different
short-term hepatic lesions leading to inflammatory and proliferative processes
(URIEL, 1971). The α fetoprotein (α FP) was believed to reappear only in liver
cancer. However, by using more sensitive quantitative tests, it has been detected
at low concentrations in infectious hepatitis and in other non-malignant hepatic
lesions (ABELEV, 1971; DeNÉCHAUD and URIEL, 1971). These data show that in
this case also the genes for the embryo-specific antigens are not completely shut
off by a process of blocking but are only repressed.

Other developmental antigen changes in the hepatocytes have also been detected by using immunological techniques (RAFTELL and PERLMANN, 1968a, b).
More important for our discussion is the changing capacity of the developing
liver to produce serum albumin. The concentration of serum albumin increases
after the 16th day of gestation until one month after birth (DeNÉCHAUD and
URIEL, 1971) and an inverse relationship was established between the levels of
αFP and of serum albumin (Fig. 3). Such data support the concept of mutual
interactions between different genes.

Cell Proliferation. In embryonic and postnatal liver the proliferative activity is
very high and then, during a short period of time a sudden decrease occurs.
Summarizing the literature, this transition from a fast to slow mode of proliferation is characterized by the following features: (a) the fraction of cycling cells
dramatically falls; (b) a small non-growth fraction appears; (c) the length of
the cell cycle phases increases; (d) a circadian rhythm of the mitotic activity
appears; (e) the diploid hepatocytes begin to be transformed into polyploid cells.

a) In liver development the proliferative activity of the hepatocytes constantly
decreases. In mouse embryonic liver the highest labeling index is observed between the 10th and 12th day when the endodermal cells increase their proliferation rate to form a thickening in the ventral floor of the foregut. At this time the
labeling index of the hepatic primordia increases to 70% as compared with 54%
of the endodermal cells of the foregut (ZAVARZIN, 1967). From the 12th day on
this index gradually falls.

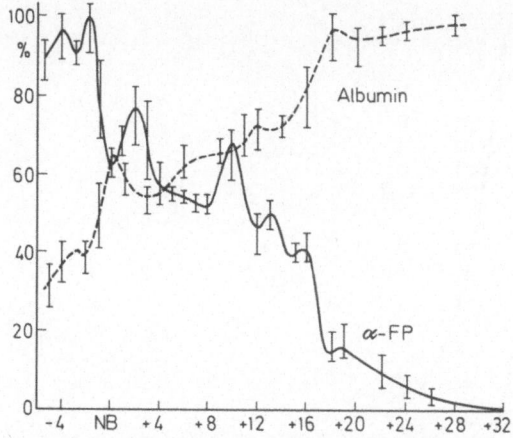

Fig. 3. Developmental changes in the concentration of albumin and of α-fetoprotein in the serum of rats (DeNéchaud and Uriel, 1971)

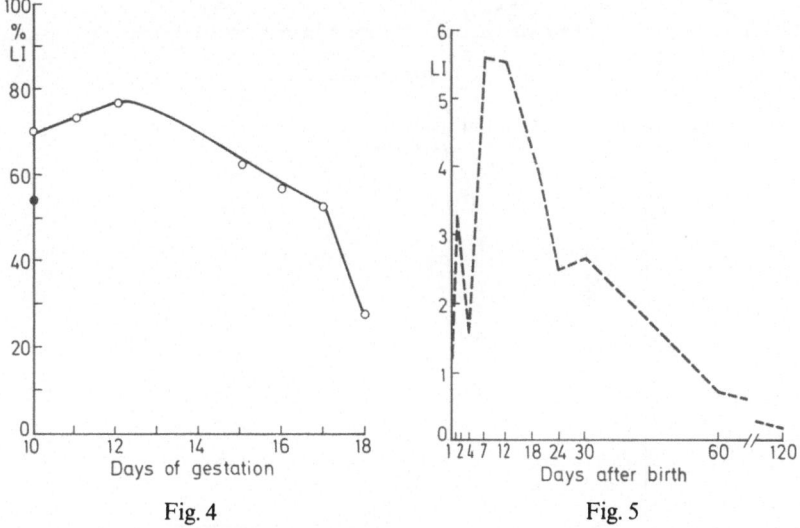

Fig. 4   Fig. 5

Fig. 4. Developmental changes in the labeling index of mouse embryonic hepatocytes. The closed circle at the 10th day refers to the endodermal cells of the foregut. (According to data from Zavarzin, 1967)

Fig. 5. Labeling index of rat liver parenchymal cells as a function of age (Stöcker et al., 1972)

The first sharp decrease in the mitotic activity of the liver is observed between the 17th and 18th day of embryonic development (Fig. 4). A second steep decrease occurs immediately after birth (Figs. 5 and 6) when the labeling index falls from 23% in the 20-day-old rat fetus to 1–5% during the first day after birth (Grisham, 1969a,b; Stöcker et al., 1972). A third sharp decrease takes place during the first month (Grisham, 1969a; Houbrechts and Barbason, 1972; Stöcker et al., 1972) (Figs. 5–7). The low level of mitotic activity characteristic of the adult liver is

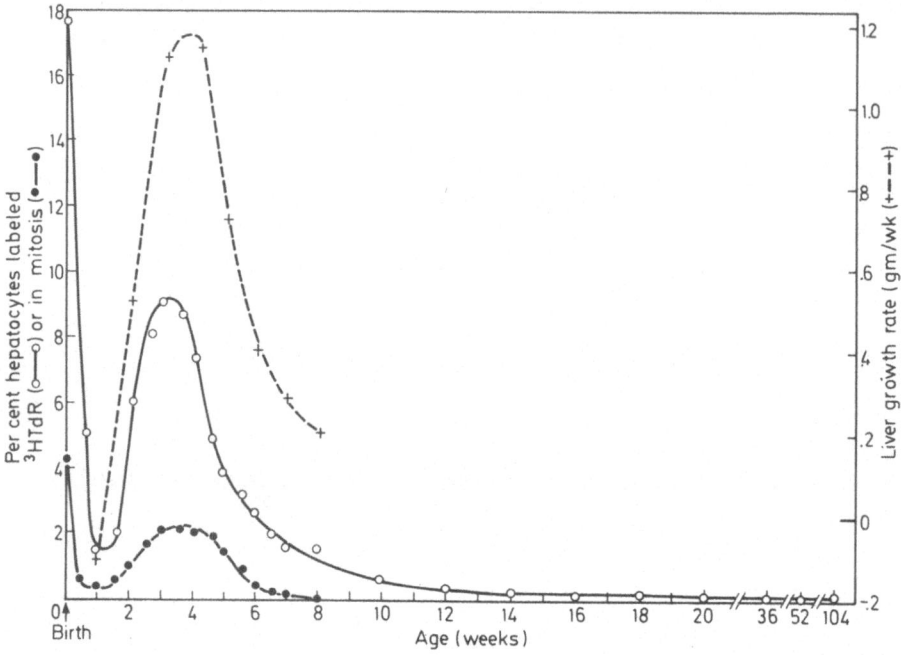

Fig. 6. Labeling index, mitotic index and growth rate of rat liver as a function of age (GRIS-
HAM, 1969)

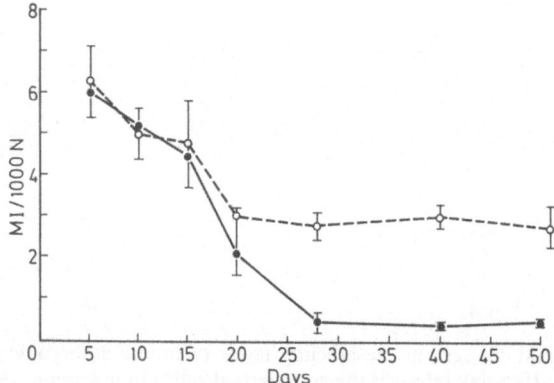

Fig. 7. Changes in the mitotic index of rat liver as a function of age and of the time of the day.
— o — points at 10 A.M., — ● — points at 8 P.M. (HOUBRECHTS and BARBASON, 1972)

then gradually established during the next 2 to 3 months. The decrease in the
mitotic activity after birth is connected with a gradual change in the spatial
distribution of proliferating cells within the liver lobule—from a random distribution of mitotic cells before birth, a marked decline in DNA-synthesizing cells at
the central zone takes place (GRISHAM, 1973).

A decrease in the labeling index can indicate either a decrease in the fraction of
cycling cells or changes in the cell cycle parameters—a decrease in the length of S

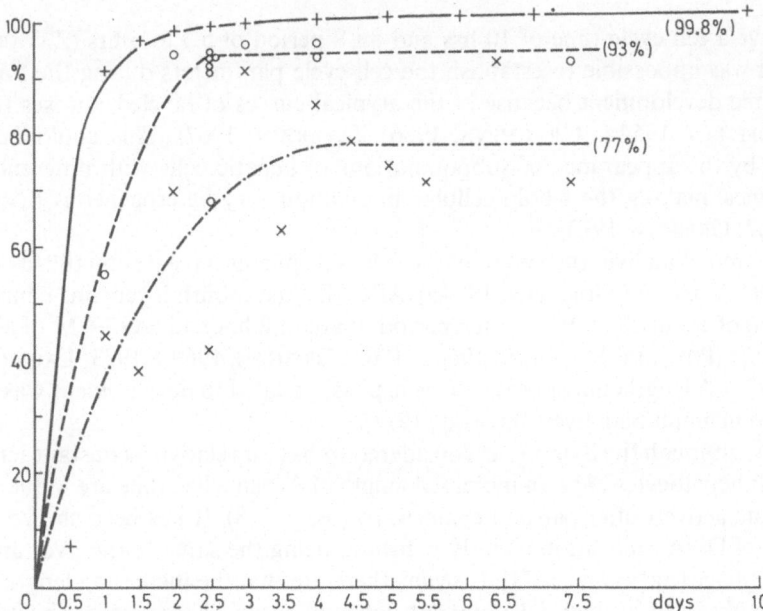

Fig. 8. Percentage of labeled hepatocytes as a function of time after $^2/_3$ partial hepatectomy. $-+-+-$ 3 to 4-week-old rats; $-o-o-$ 2 to 4-month-old rats; $-\overset{x}{\underset{x}{-}}$ senescent rats (STÖCKER et al., 1972)

or an increase in the cell cycle time. Some data concerning the cell cycle param-eters of young rats (POST, HUANG and HOFFMAN, 1963) were not confirmed by recent investigation (SCHULTZE and MAURER, 1973). From the few available data (see later) it can be concluded that in embryonic liver, the cell-cycle time and the length of S do not change significantly. The sharp decrease in the labeling index is due therefore, to the decrease of the proliferating fraction as shown also by the labeling indices after repeated injections of ^{3}HTdR (ZAVARZIN, 1967).

It can be concluded that the transition from a fast to slow mode of prolifera-tion at the end of the embryonic development and after birth, is mainly due to the appearance in the cell life-cycle of a period of quiescence which progressively becomes longer. This leads to a decrease in the proliferating fraction, which falls from about 36% on the first day of life to 4–6% in rats 6 to 8 weeks old (POST and HOFFMAN, 1964a) and to 0.5% in adult liver (GRISHAM, 1969a).

b) During the postnatal development of the liver a fraction of cells appears which seem unable to proliferate even after a strong proliferative stimulus such as partial hepatectomy. This "nongrowth" fraction can be detected by repeated labeling of the liver cells with ^{3}HTdR after partial hepatectomy (Fig. 8). From such data it has been calculated that the non-growth fraction in normal livers of 3 to 4-week-old, 2 to 4-month-old and 2.5-year-old rats is 0.6%, 21%, and 69%, respectively (STÖCKER and HEINE, 1971). It seems most probable that this is the result of the appearance of liver cells with a high level of ploidy which have a lower capacity to proliferate.

c) Some changes in the cell cycle parameters of the liver cells occur during development. The rapidly proliferating hepatocytes of the 11-day-old mouse embryo have a cell cycle time of 10 hrs and an S period of 5.5 to 6 hrs (ZAVARZIN, 1967). It was impossible to establish the cell cycle parameters during the further embryonic development because of the atypical curves of labeled mitoses (POST and HOFFMAN, 1964a; GRACHEVA, 1966; ZAVARZIN, 1967). This could be explained by the appearance of subpopulations of hepatic cells with different cell cycle times, making the whole cellular population very heterogeneous (ZAVARZIN, 1967; GRISHAM, 1973).

In mammalian liver the length of the cell cycle phases increases on the first day after birth (POST and HOFFMAN, 1964a). After the first month it remains constant, with a Gl of about 20 to 26 hrs, an S period of about 8 hrs and a G2 + M of about 3.5 to 4 hrs (POST and HOFFMAN, 1964a, 1965; GRISHAM, 1969a, 1973; LAGUCHEV et al., 1972). A lengthening of the cell cycle phases related to development was also observed in amphibian liver (BRUGAL, 1971).

Thus, although the S period is considered to have a relatively constant length, the adult hepatocytes have an increased length of S even when they are induced to proliferate actively after partial hepatectomy (see p. 218). It has been shown that the rate of DNA replication is fairly constant, being the same in the liver and in bacteriophage (BRESNICK, 1971). It seems therefore that the increasing length of S which is observed in liver development depends on an altered time schedule of different DNA replication sites.

d) The decrease in the proliferating fraction in the 3rd week coincides with the appearance of diurnal variations in the mitotic activity of rat liver (Fig. 7). This may be due to the appearance of circadian variations in the level of corticosteroids related to the maturation of the pituitary–adrenal axis at this time (HOUBRECHTS and BARBASON, 1972).

e) An important event taking place in the development of postnatal liver is the appearance of binucleated and polyploid cells which usually appear during the first three weeks of life (DOLJANSKI, 1960; CARRIERE, 1969; RYABININA and BENYUSH, 1973) when the proliferating fraction gradually decreases. In mammalian liver the proportion of polyploid cells increases with age to reach figures where the predominant cells are tetraploid with a small proportion of octaploid cells (Fig. 9).

There is strong evidence indicating that polyploidization depends on mitotic activity of the liver cells. Binucleated cells originate from mitotic divisions with a failure of cytoplasmic cleavage and seem to be the first step to the next level of ploidy. A second DNA replication and cell division of binucleated cells is supposed to be the main source of tetraploid cells, but in older animals endomitosis and proliferation of tetraploid cells may also contribute to the increasing number of polyploid cells with age (CARRIERE, 1969; BUCHER and MALT, 1971). Some experiments indicate that polyploidization of the hepatocytes depends on hormonal factors (CARRIERE, 1969). Recent data have shown the presence in the serum of adult rats of a factor inducing the formation of binucleated hepatocytes. This factor was absent in the serum of baby rats or of partially hepatectomized adult rats (NADAL, 1973).

Polyploid hepatocytes have a lower proliferating capacity. In young rats the diploid hepatocytes seem to be the chief replicating unit (POST and HOFFMAN, 1965). In adult animals incorporation of ^{3}HTdR is observed predominantly in

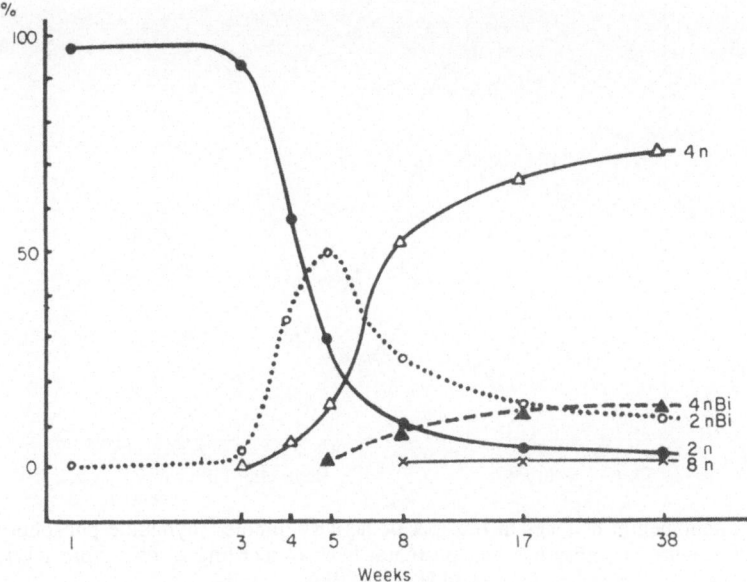

Fig. 9. Percentage of cells of different level of ploidy as a function of age (NADAL and ZAJDE-LA, 1966)

diploid cells, more rarely in tetraploid, and as an exception only in octaploid cells (BUSANNY–CASPARI, 1962; ZAVARZIN, 1967).

The decreased proliferative activity of the liver with age may be correlated with the increasing number of polyploid cells, in agreement with the idea that somatic polyploidization enables the cells to spend more time in a state of differentiated function. Such a general conclusion should be drawn with great care in view of the large variations observed in the extent of polyploidization in the liver of different species. It has been found that in fish and amphibians most of the adult hepatocytes remain as diploid cells (CARRIERE, 1969). However, the observation that some variations in the level of ploidy in Amphibia occur, depending on the breeding season, makes some authors uncertain about the real situation with polyploidization of liver cells in lower vertebrates (CARRIERE, 1969).

All of these changes in the proliferative activity of the liver cells are associated with changes of the enzymes involved in the mitotic cycle. This is well illustrated by thymidine kinase. This enzyme has a very low activity in adult liver while in embryonic liver its activity is very high (BRESNICK, 1971; BUCHER and MALT, 1971). As seen in Fig. 10, both in fetal and in postnatal development the decrease of thymidine kinase activity in the liver parallels the decrease in mitotic activity. These changes do not appear to be due to the presence of activating or inhibitory factors (KLEMPERER and HAYNES, 1968).

Some developmental changes in the properties of thymidine kinase have been also observed. The enzyme in embryonic liver was found to be much more sensitive to inhibition by TTP than the adult enzyme (BRESNICK et al., 1964; KLEMPERER and HAYNES, 1968). dCTP had little or no effect upon the partially purified enzyme from embryonic liver while it inhibited the enzyme of adult liver by

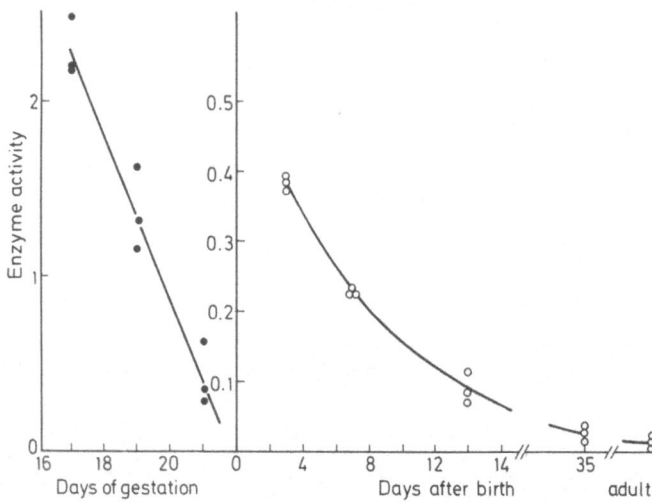

Fig. 10. Developmental changes in the specific activity (µmoles thymidine phosphorylated/min/mg of protein) in embryonic and postnatal liver. (According to data from Klemperer and Haynes, 1968)

approximately 50%. These data suggested the presence of two enzymes and the developmental replacement of one enzyme type by another. This view is supported by the reappearance of dCTP-insensitive thymidine kinase in regenerating liver (Bresnick et al., 1964). Other key enzymes involved in DNA synthesis, as dTMP kinase, dCMP deaminase, dTMP synthetase (Ferdinandus et al., 1971) and DNA polymerase (Otani and Morris, 1971) closely follow the time course of thymidine kinase.

An important feature of thymidine utilization for DNA synthesis is the balance between two opposite metabolic pathways controlling the synthesis and the degradation of thymidine. It has been shown that in parallel with the decreased biosynthetic pathway of thymidine during liver development, the catabolism of thymidine to CO_2 increased (Fig. 11). The developmental rise of the degradative pathway was completely inhibited by administration of actinomycin D, which suggested a transcriptional control of these changes (Ferdinandus et al., 1971). The inverse relationship between these two metabolic pathways again illustrates the presence of closely interrelated genes.

The appearance of a quiescent cell population in the late fetal development of the liver implies an altered ratio of functional versus proliferative activity of the liver. This is also reflected by some changes in the protein-synthesizing machinery of the hepatocytes. As shown by electron-microscopic observations the main changes consist of the development of the endoplasmic reticulum and the distribution of ribosomes into free and membrane-bound classes (Moulé, 1964; Kheissin, 1969).

The existence of a parallelism between the development of endoplasmic reticulum and differentiation has often been inferred. It seems, however, that the hepatocyte as a cellular type appears before development of endoplasmic reticulum, and the latter event is connected with functional changes. Many of the enzymic

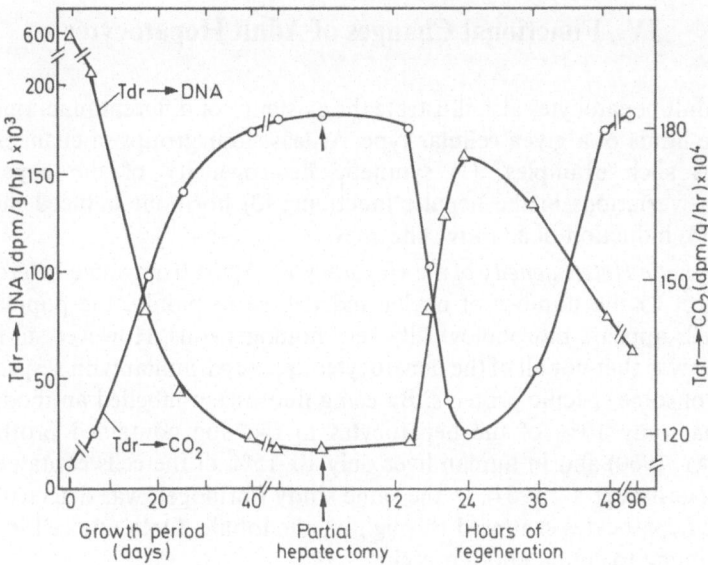

Fig. 11. Activity of synthetic and degradative pathways of thymidine (Tdr) as a function of developmental or regenerative growth (FERDINANDUS et al., 1971)

and proliferative developmental changes discussed above cannot be correlated with the development of the rough endoplasmic reticulum. A better correlation exists with the postnatal increase of albumin synthesis, which coincides with the time when the rough membranes become well developed. This is in agreement with the fact of the exclusive synthesis of serum albumin by membrane-bound polysomes (TAKAGI et al., 1969; CAMPBELL, 1970).

From this brief review the following tentative conclusions can be made:

a) The hepatocyte as a cellular type seems to appear very early during the first outburst of increased proliferation of endodermal cells in the ventral floor of the foregut. This wave of very high proliferative activity most probably includes the critical mitotic cycle which determines the hepatic cellular type.

b) The subsequent developmental changes of the hepatocytes have all the characteristics of changes in the functional state of the cellular type: they are cell-cycle-independent but depend on the inductive presence of different hormonal, nutritional, and other environmental factors.

c) The stepwise decrease in the proliferative activity of the hepatocytes during development is mainly connected with the appearance of a period of quiescence in the cell life-cycle.

d) The periods of changes in the protein pattern of the hepatocytes coincide with the periods of sharp decrease in their proliferative activity.

e) The clustered appearance of enzymes and the inverse relationship between some antigenic and proliferative changes suggest the presence of closely related genes and mutual interactions between different genes.

IV. Functional Changes of Adult Hepatocytes

The adult hepatocytes also illustrate the existence of different functional states within the limits of a given cellular type. At least four groups of changes can be quoted as such examples: (1) synthetic heterogeneity of the hepatocytes; (2) diurnal variations in the hepatic functions; (3) hormone-induced metabolic changes; (4) induction of adaptive enzymes.

1. Synthetic Heterogeneity of the Hepatocytes. Apart from some heterogeneity with respect to the number of nuclei and degree of ploidy, the population of hepatic cells appears morphologically very homogeneous. However, an interesting finding was that not all of the hepatocytes appeared similar with respect to the synthesis of some specific proteins. By using fluorescein-labelled antibody it was shown that only 10% of the hepatocytes in the dog contained prothrombin (Barnhart, 1960) and in human liver only 10–15% of the cells contained albumin (Hamashima et al., 1964). In the same study fibrinogen was detected in only 1% of the hepatocytes scattered throughout the lobule. Only one cell in a thousand was found to contain both proteins.

These data can mean either that individual hepatocytes undergo further "differentiation" or that the hepatocyte cannot perform many activites at the same time and periodically changes its synthetic program. The latter suggestion seems more probable. When albumin was detected by immunofluorescence, brightly and weakly fluorescent cells were observed among the albumin-positive cells. This suggests the presence of transition states. Tiny granules containing fibrinogen were also found in many cells, suggesting that low levels of fibrinogen synthesis may be present in many cells. Finally, it has been shown that the number of cells containing prothrombin can decrease or increase from none to practically all of them, as for instance after the administration of vitamin K (Barnhart and Anderson, 1961). This supports the view that the synthesis of different proteins by different hepatocytes reflects different functional states of individual cells which change periodically and also under the effect of inducing factors.

2. Diurnal Variations in the activity of a number of liver enzymes are well known (Black and Axelrod, 1970; Fuller, 1971; Hayashi et al., 1972). Even some structural elements of the hepatocytes such as the membranes of the endoplasmic reticulum, which are often believed to be the expression of a more stable differentiated function, undergo daily variations in their structural organization (Chedid and Nair, 1972). The proliferative activity of the hepatocytes also shows diurnal variations (Jaffe, 1954; Houbrechts and Barbason, 1972; Van Cantfort and Barbason, 1972).

In spite of the complex nature of the mechanisms controlling the variations in the protein pattern of hepatocytes, all available data clearly show that they are dependent on periodic changes of a number of nutritional, hormonal, and environmental inducing factors. Thus, they represent another typical example of reversible cell-cycle-independent changes demonstrating the lability of the functional state of a stable cellular type.

3. Hormone-Induced Changes. Striking changes in the specific proteins synthesized by the adult hepatocytes can be observed under the inducing effect of

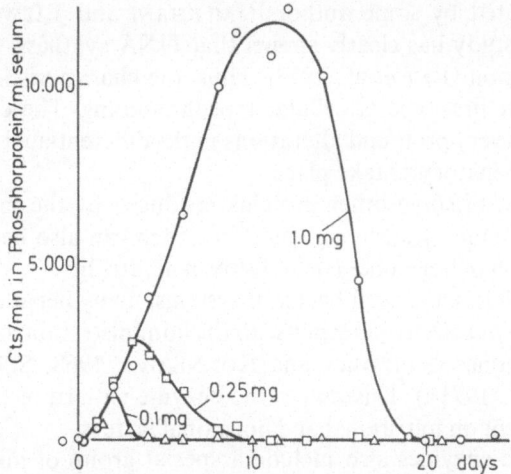

Fig. 12. Rate of serum phosphoprotein production by the liver of male Xenopus as a function of time after injection of different single doses of 17-β-estradiol (WALLACE and DUMONT, 1968)

hormones. An illustrative example is the synthesis of the yolk proteins of avian and amphibian eggs. The protein moieties of some specific yolk proteins—phosvitin and lipovitelin—are synthesized in the liver and are then transported into the egg (FOLLET et al., 1968; RUDACK and WALLACE, 1968; WALLACE and DUMONT, 1968). This specific synthetic activity of the hepatocytes is controlled by hormones: under the effect of estrogens the hepatocytes of the female Xenopus are stimulated to synthesize yolk proteins which are released into the serum and then taken up by the ovocytes under the effect of gonadotropic hormones. The liver of the male Xenopus does not synthesize these proteins, but a single injection of 17-β-estradiol can induce such a synthesis. In the absence of ovocytes these proteins accumulate in the serum. It should be stressed that the effect is transitory and dose-dependent (Fig. 12) (WALLACE and DUMONT, 1968; BEUVING and GRUBER, 1971). It is instructive that a prolonged treatment with estradiol can alter the usual function of the liver to such an extent that this organ stops the synthesis of serum albumin and globulins completely and switches on the synthesis and secretion of yolk proteins only (FOLLET et al., 1968).

The estrogen-induced synthesis of yolk proteins results in a profoundly altered cytology of the hepatocytes, which appears within a few hours after hormonal treatment. These changes are mainly expressed by the intensive formation of new granular endoplasmic reticulum which gradually fills the cytoplasm of the hepatocytes (FOLLET et al., 1968). The increased development of membrane-bound ribosomes in this case supports the idea that the granular membranes are the site of synthesis of exportable proteins and again demonstrates that they are dynamic structures rapidly changing with the functional state of the cell.

It seems very probably that these changes involve a transcriptional control (GREENGARD et al., 1964; FOLLET et al., 1968; WITTLIFF and KENNEY, 1972; JOST et al., 1973). It is also important to ascertain whether the process requires DNA synthesis, as stated by some authors (JAILKHANI and TALWAR, 1972). A more recent detailed study has clearly shown that DNA synthesis is not required for phosvitin induction (JOST et al., 1973). Thus, the characteristics of these changes are typical of the first type of cellular reprogramming. The cellular type is preserved, although very profound alterations in the differentiated function and in the cytology of the hepatocytes take place.

The synthesis of some other proteins produced by the mammalian hepatocytes such as albumin, transferrin, and fibrinogen can also be rapidly influenced by administration of hormones (JEEJEEBHOY et al., 1972).

4. Adaptive Enzymes. As in bacterial systems, many hepatic enzymes manifest marked increase in activity in response to the administration of their substrates or to certain hormones (KRITSMAN and KONNIKOVA, 1968; SCHIMKE and DOYLE, 1970; RECHCIGL, 1971a). This makes the enzyme pattern of the hepatocytes extremely dependent on nutritional and hormonal factors.

The adaptive enzymes also include a special group of microsomal enzymes induced in response to foreign pharmacological stimuli (CONNEY and BURNS, 1963; GELBOIN, 1967). The synthesis of these drug-metabolizing enzymes is the expression of one important differentiated function of the liver—detoxification.

The problem of the molecular mechanisms controlling the induction of adaptive enzymes is the same as that already mentioned in connection with the developmental and diurnal enzyme changes. Even in the best studied case—the induction of TAT—no conclusive evidence has been presented to elucidate the molecular mechanism of the enzyme induction (GELEHRTER, 1971; ROSEN and MILHOLLAND, 1971). Summarizing the main literature, it seems that we can draw the following general conclusions:

a) In many cases it has been well proved that the increased enzyme activity is due to an increased amount of enzyme;

b) The increased amount can be due both to an increased rate of synthesis and to a decreased rate of degradation (GELBOIN, 1967; RECHCIGL, 1971b; ROSEN and MILHOLLAND, 1971). In different cases the weight of these two kinetic parameters may differ;

c) The rate of synthesis can depend both on the synthesis of new mRNA (SCHMID and GALLWITZ, 1967; SEKERIS et al., 1970; BUTCHER et al., 1972; POPOV et al., 1973) and on the stabilization and translation of preexisting messages (TOMKINS et al., 1970, 1972; GELEHRTER, 1971).

It seems certain that no unifying mechanism for the induction of adaptive enzymes exists. In different cases transcriptional and/or posttranscriptional controls may play a dominant role. Even the induction of one and the same enzyme under different conditions may be regulated at different levels, as suggested for the induction of pyruvat kinase in the liver (FREEDLAND and SZEPESI, 1971).

For our discussion it is important to emphasize that the process is reversible, dependent on the continuous presence of the inducer, and does not show a dependence on previous mitotic divisions. In regard to the last point, the group of drug-metabolizing enzymes requires special comment. Several observations have shown that the induction of microsomal enzymes is associated with liver growth, involving increased mitotic activity (ARGYRIS, 1969, 1971; ARGYRIS and MAGNUS, 1968). This correlation, however, can not be interpreted as an example of the need for mitotic divisions for the expression of molecular differentiation as suggested (ARGYRIS, 1969). The rise in drug-metabolizing enzyme activities is very fast and starts as early as the first 12 hrs, concurrent with, rather than after, the induction of mitotic divisions by the drug. On the other hand, the effect depends on the presence of the inducer and is completely reversible. All of these factors make this process different from the stable cell-cycle dependent molecular differentiation.

These examples show that there is no difference, in principle, between the characteristics of the changing protein pattern in developing liver and in adult liver. In both cases the cellular reprogramming represents a process of changing the functional state of a stable cellular type.

V. Cell Population Kinetics of the Hepatocytes

The adult liver is believed to be in a steady state of cell renewal with a very low level of proliferative activity.

A very small proportion of the adult hepatocytes are in the mitotic cycle. The mitotic index of rat liver is about 0.005–0.010% (DOLJANSKI, 1960; BUCHER, 1963; LEDUC, 1964; CARRIERE, 1969; BRESNICK, 1971) and the labeling index is about 0.2–0.4%. The average life span of adult hepatocytes has been calculated to be between 150 and 500 days in rats and about 3000 days in mice (DOLJANSKI, 1960; CARRIERE, 1969; GRISHAM, 1973). These figures may be an underestimation if a DNA-repair-dependent incorporation of ^{3}HTdR is present.

Two models may be proposed for such a situation: (a) a small proliferative compartment supplying cells to a large compartment of a differentiated non-growth fraction, or (b) a large compartment of differentiated cells which enter the mitotic cycle at random.

In the case of the liver, the first model does not seem probable. Numerous data show that many quiescent liver cells are able to enter the mitotic cycle under appropriate conditions. Direct evidence against the presence of special cambial cells in the liver was presented by experiments with repeated partial hepatectomy showing that cells which have divided (^{3}HTdR-labeled) and entered a period of quiescence, reenter the mitotic cycle after a second proliferative stimulus (KLINMAN and ERSLEV, 1963).

With the second model the question arises as to whether the quiescent cells have a very long G_1 period or they represent a real quiescent population (Q-cells). Although it seems very difficult to distinguish between the two situations (BROWN, 1968; LALA, 1971; RAJEWSKY, 1972), the second possibility conforms more closely to all experimental data. Since G_1, by definition, is a period preparing the replication of DNA, it would be difficult to accept that such a long preparative period coexists with the performance of many differentiated functions. On the other hand, the synchronous entry of a large number of hepatocytes in the mitotic cycle after a lag period in partially hepatectomized rats is not compatible with a long G_1 (LAJTHA, 1963).

There is no doubt that hepatocytes which synthesize tissue specific proteins can switch on into a state of mitotic division. After partial hepatectomy almost all cells divide, especially in younger animals (see p. 207). Detailed studies show that only some cells in the central zone of the liver lobule do not divide (STÖCKER, 1966; FABRIKANT, 1969; GRISHAM, 1973). On the other hand, albumin-synthesizing cells were shown to be evenly distributed throughout the liver lobule (SCHREIBER et al., 1970). These facts, when combined, demonstrate the capacity of differentiated hepatocytes to enter the mitotic cycle.

Thus, we have to accept, that the hepatocytes can be either in a period of quiescence, performing some differentiated functions (Q-cells), or in the mitotic cycle (P-cells). Such a situation corresponds to the concept of a G_0 period (LAJTHA, 1963) in the life of differentiated cells from which the cells can enter the mitotic cycle at random and then go back into the same state of quiescence. As pointed out also by RAJEWSKY (1972) it is difficult to agree with the idea that G_0 is a phase of "decision making", from which cells go either into the mitotic cycle or into "differentiation" under the effect of specific stimuli (EPIFANOVA and TERSKIKH, 1969). As we have seen, the quiescent liver cells are already a differentiated cellular type able to produce a large spectrum of diverse proteins under the effect of different specific stimuli. The cells in the so-called G_0 compartment either represent an already differentiated cellular type (as in the case of the liver) or they have not yet reached the final differentiated cellular type (as in the case of stem cells). In the first case, different stimuli can only change the functional state of the cells or trigger them into the mitotic cycle. In the second case the cells must go through a mitotic cycle in order to reach the final cellular type. In both cases the picture is not that of cells going into a state of differentiation or into mitotic division.

In discussing the "resting" periods in the cell life, some authors (EPIFANOVA and TERSKIKH, 1969; LALA, 1971) stress the possibility of a second resting period R_2 (Q_2 in our designation) represented by cells which leave the cell cycle in G_2. Such G_2-blocked cells have been shown to exist in the epidermis (GELFANT, 1962), but there is no proof for their presence in the liver. Evidence has been presented for hepatocytes around the central vein with a delayed G_2 (EPSTEIN, 1967; PERRY and SWARTZ, 1967; FABRIKANT, 1969), but was not found to be convincing (GRISHAM, 1973). The presence of a large proportion of hepatic nuclei with a polyploid DNA content may be also an argument for G_2-blocked cells which have entered the S period again without going through mitosis (EPSTEIN, 1967; PERRY and SWARTZ, 1967). Although such a mechanism can not be ruled out, some data indicate that at least most of the polyploid nuclei in the liver are obtained by other mechanisms (see p. 208).

Bearing in mind the extremely diverse functions of the hepatocytes, we do not see any evidence for the presence of hepatocytes in a "null" state in which "a cell is able neither to proliferate nor to fulfil its specific function" (EPIFANOVA and TERSKIKH, 1969).

All these data show that the population of hepatic cells may consist of the following compartments: (1) proliferating pool or cycling cells (P-cells); (2) quiescent cells able to enter the mitotic cycle (Q_1-cells); (3) quiescent cells unable to proliferate, at least under the effect of usual proliferative stimuli (Q_0-cells); (4) possibly cells in G_2 block (Q_2-cells).

Thus, the whole picture of the cell population kinetics of the hepatocytes may be represented by the compartments and fluxes shown in Fig. 13.

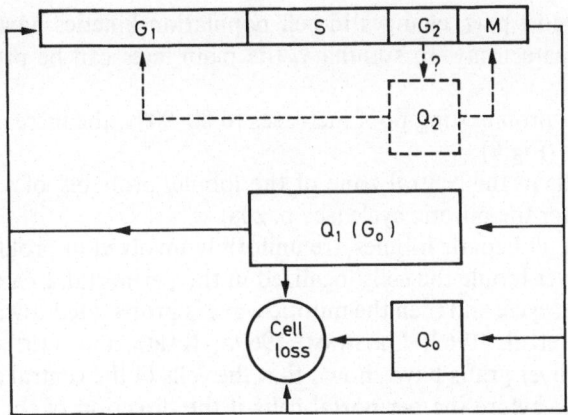

Fig. 13. Cell population kinetics of the hepatocytes

When tissues with different proliferative activities were compared, an increased length of G_2 and S was found to be correlated with a decreasing proliferation rate (LAGUCHEV et al., 1972). In this scale, the hepatocytes and the exocrine cells of the pancreas have the highest values for the length of G_2 and S.

Many hormonal, nutritional, and environmental factors can easily influence the size of the proliferating compartment of the liver. With the exception of carcinogenic agents (see p. 237) the effect of all other factors is reversible. This makes the transition into the mitotic cycle very similar to other functional transitions belonging to the reversible type of cellular reprogramming. A somewhat different situation is created by the fact that the increased proliferative activity in the adult liver is always followed by an increased polyploidization of the hepatocytes, which may lead to an irreversible loss or decrease of their proliferating capacity.

VI. Induction of Cell Proliferation in Adult Liver

The quiescent hepatocytes of adult liver can be relatively easily triggered into the cell cycle under the effect of different stimuli—altered physiological demands such as: pregnancy, lactation, nutritional, hormonal factors (additive growth); cellular loss caused by partial hepatectomy, or toxic agents such as CCL_4 (reparative growth). The morphological, ultrastructural, and biochemical changes of the hepatocytes induced to proliferate by various factors have been summarized in several reviews (DOLJANSKI, 1960; BUCHER, 1963; LEDUC, 1964; STEINER et al., 1966; CARRIERE, 1969; GRISHAM, 1969a, 1973; KHEISSIN, 1969; BRESNICK, 1971; BUCHER and MALT, 1971). We shall discuss only three points related to our main problem: A. How the cell population kinetics change after induced proliferation; B. Changes in the synthetic program of the hepatocytes that take place during the mitotic cycle; and C. The level at which these changes are regulated.

A. For the most part, changes in cell population kinetics have been studied after partial hepatectomy. In summary, the main facts can be presented as follows:

a) The daily proliferating pool increases to 80–90%, the increase being lower in older animals (Fig. 8).

b) Some cells in the central zone of the lobule, probably of a higher ploidy level, do not enter the mitotic cycle (see p. 208).

c) Although all hepatic lobules are uniformly involved in proliferating activity, within the liver lobule the cells localized in the periportal area are the first to enter the mitotic cycle and then the mitotic wave is propagated toward the central vein (OEHLERT et al., 1962; GRISHAM, 1969a; RABES and TUCZEK, 1970). Experiments with liver grafts have shown that the cells of the central zone can enter the mitotic cycle before the periportal cells if the direction of the blood flow is reversed (SIEGEL et al., 1968; GRISHAM, 1969a; SIEGEL, 1972).

d) The cell-cycle parameters of regenerating liver are not very different from those of cycling hepatocytes of normal liver. After partial hepatectomy, the first wave of DNA synthesis begins after a lag period of about 15–18 hrs (BUCHER and MALT, 1970) and has a duration of S of about 8 hrs (GRISHAM, 1969a, 1973). Accepting the concept of Q-cells (see p. 215), we should consider this lag period to be the appearance of a G_1 period, but not a prolongation of an existing G_1, as some authors (BUCHER and MALT, 1971) have stated. Some differences in the length of G_1 and of the total cycle found by different authors are well explained by the effect of age, diurnal variations, nutritional conditions, animal strain, etc. (BUCHER and MALT, 1971). After the first wave of mitotic activity comes a second wave, more or less well expressed depending on the age of the animals and on the interference with circadian rhythms (JAFFE, 1954; BUCHER and MALT, 1971; GRISHAM, 1973). In experiments with a double labeling of DNA, direct proof has been presented that many hepatocytes that have participated in the first mitotic cycle undergo a second round of DNA replication (GRISHAM, 1973).

All these data show that in respect to the proliferating pool, reparative growth in adult liver resembles developmental growth, except in the longer duration of the S period, as is characteristic of tissues with a low proliferative activity (LAGUCHEV et al., 1972).

B. The changes in the synthetic program of the hepatocytes during the mitotic cycle were studied mainly in regenerating rat liver after partial hepatectomy because of the relative synchronization of the mitotic activity of the hepatocytes and the late response of the nonparenchymal liver cells (BUCHER and MALT, 1971). These changes include, at least, the following groups:

(a) Changes in enzyme activities involved in DNA replication; (b) synthesis of proteins involved in the initiation of DNA replication and of mitosis; (c) synthesis of histones and nonhistone chromatin proteins; (d) synthetic activities connected with the reproduction of other cellular structures; (e) changes in enzyme activities and other proteins that are not directly related to cellular self-reproduction.

1. Changes in the enzymes involved in DNA synthesis have been studied thoroughly (BRESNICK, 1971; BUCHER and MALT, 1971). The main facts show that all these enzymes are either absent or have a very low activity in normal liver, but that they sharply and almost simultaneously increase after partial hepatectomy.

The pool of deoxyribonucleosidetriphosphates in normal liver is extremely small. It increases after partial hepatectomy due to the appearance of a number of enzyme activities. Thus, in normal liver the activity of CDP-reductase is undetectable, and the activities of other DNA-forming enzymes can be detected only by very sensitive techniques; however, they all increase 12 hrs after partial hepatectomy (BUCHER and MALT, 1971).

The increased utilization of thymidine in regenerating liver is coupled with its decreased catabolism (FERDINANDUS et al., 1971; BRESNICK, 1971), as shown in Fig. 11. This again shows that induction of some enzyme activities in the hepatocytes is linked with reduction of other enzyme activities and suggests the existence of mutually dependent genes.

The activity of DNA polymerase also sharply increases at the start of DNA synthesis, although this enzyme seems to be involved in the process of DNA repair. The true enzyme for the replication of DNA remains unknown (BUCHER and MALT, 1971).

2. The presence of all of these enzymes is a necessary but not sufficient condition for DNA replication, as indicated by the following facts (MUELLER, 1969; BRESNICK, 1971; BUCHER and MALT, 1971). (a) The inhibition of protein synthesis in regenerating liver stops DNA replication even though the necessary enzyme activities are present; (b) the high level of TTP and of all DNA-forming enzymes is maintained after DNA replication has been completed; and (c) in the presence of all precursors and enzymes needed, DNA replication proceeds asynchronously.

It seems that a specific initiation protein must be synthesized in order to begin replication of a given part of the genome. Also, the synthesis of a "division" protein is needed for the transition into the period of mitosis (MUELLER, 1969).

3. Many data show that the synthesis of histones takes place coordinately with the replication of DNA and all histone species are synthesized more or less synchronously (WILHELM et al., 1971). The nonhistone proteins of the chromatin were believed to turnover independently of the cell cycle, but as we have recently shown, a cell-cycle dependent fraction of chromatin nonhistone proteins is present in fibroblasts (see p. 199). Our preliminary data have shown the presence of such a fraction also in the liver.

4. Important synthetic activities of the cycling hepatocytes are related to the reproduction of all other cellular structures. This is connected with an early increase in RNA synthesis which begins only a few hours after partial hepatectomy (MCARDLE and CREASER, 1963; TSANEV and MARKOV, 1964; BUCHER and SWAFIELD, 1966; BRESNICK, 1971) and is associated with an increased RNA polymerase activity (TSUKADA and LIEBERMAN, 1964). The increased synthetic rate follows a biphasic curve, showing an early peak at 3 hrs and a second, higher increase late in G_1 (MCARDLE and CREASER, 1963; RABES and BRÄNDLE, 1969). The same biphasic changes were found in the content of nuclear RNA (TSANEV and MARKOV, 1964) and in the concentration of cAMP after partial hepatectomy (MACMANUS et al., 1972). An early peak of RNA and protein synthesis followed by a second one preceding DNA synthesis have also been reported in fibroblasts (TODARO et al., 1967) and in epidermis (TSANEV et al., 1966) which have been stimulated to proliferate. The nature of the first peak of RNA synthesis has not been established, but suggestions have been made that it represents new messen-

ger-RNA species needed for the new cellular program. In our experiments the first peak of RNA synthesis also reflected ribosomal RNA synthesis (Tsanev et al., 1966). In model experiments (see p. 227) we have obtained the same biphasic curves for RNA and protein synthesis, indicating that the first peak of RNA synthesis might represent a mixture of old and new mRNA species and of ribosomal RNA. It was due to the disturbed steady state of the system, and only the second elevation of RNA synthesis was associated with a stable establishment of a new program. That the first peak of RNA synthesis may not be related to the new program is also indicated by data showing that the early increased incorporation of RNA precursors is evenly distributed throughout the liver lobule, unlike the late peak which shows the same acinoperipheral localization as DNA synthesis (Rabes and Brändle, 1969).

A net increase in RNA was reported to take place by 20 hrs postoperatively (Bresnick, 1971); but our data show a net increase as early as 6 hrs postoperatively (Tsanev and Markov, 1964). Since the net increase primarily reflects synthesis of ribosomal RNA, and hence of ribosomes, it is important to see that during the early period of cellular reprogramming an intense formation of new ribosomes takes place.

An increased synthesis of different species of nuclear heterogeneous RNA is also found after partial hepatectomy; at the same time an increased amino acid-incorporating activity of the microsomal fraction is observed, and the overall synthesis of proteins is increased (Bucher and Malt, 1971). Most probably the larger part of these activities are connected with the synthesis of new ribosomes.

5. Some changes observed in regenerating liver do not have a direct relation to cellular reproduction. They concern a group of proteins connected with the differentiated function of the liver, and another group which is present in embryonic liver but disappears during postnatal development.

It is of great interest for our further discussion to see how the synthesis of tissue-specific liver proteins is affected by the mitotic cycle. In this respect, data are available concerning three differentiated functions of the liver: (a) the synthesis of some serum proteins, (b) the synthesis of bile acids and (c) detoxification.

a) Changes in the concentration of serum albumin observed by several authors after partial hepatectomy are meaningless, since they may depend both on altered rate of synthesis and rate of degradation. Some data point to the existence of separate control mechanisms for albumin synthesis and catabolism (Kirsch et al., 1968). On the other hand, it should be kept in mind that serum albumin has a long half-life, of several days (Ryoo and Tarver, 1968).

Few data are available on changes in the rate of albumin synthesis after partial hepatectomy. The synthesis of albumin and of fibrinogen *in vivo* was reported to be enhanced during liver regeneration (Majumdar et al., 1967). From experiments done with perfused isolated livers, it has been concluded that during liver regeneration albumin synthesis is decreased, while the synthesis of fibrinogen and of transferrin is increased (Mutschler and Gordon, 1966). The increased synthesis found in some studies may be due to radioactive contaminations of the albumin fraction. When radiochemically pure albumin is obtained by means of polyacrylamide-gel electrophoresis, a significant relative decrease of albumin syn-

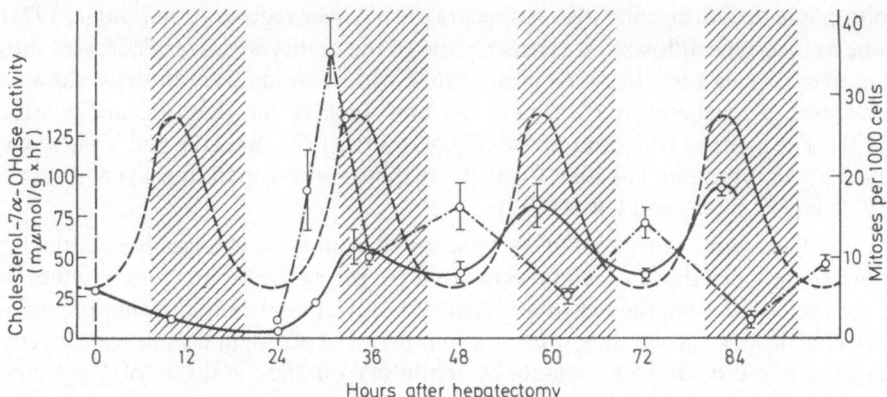

Fig. 14. Circadian rhythms of mitotic index after partial hepatectomy (—·—) and of cholesterol-7α-hydroxylase activity in normal rats (– – –) and in partially hepatectomized rats (——) as a function of time after operation performed at 10 A.M. Shaded areas represent night periods from 6 P.M. to 6 A.M. (VanCANTFORT and BARBASON, 1972)

thesis was observed in regenerating liver (MAENO et al., 1969, 1970; HILL et al., 1972). This relative decrease, however, is not relevent to the absolute rate of albumin synthesis, since an overall increase in the protein synthesis in regenerating liver takes place. The specific activity of the albumin synthesized both *in vivo* and *in vitro* was not significantly different when normal and regenerating livers are compared (MAENO et al., 1970). This is confirmed by data showing that hepatocytes proliferating *in vitro* can synthesize albumin (see p. 232).

b) Another function of the liver, connected with the synthesis of bile acids, is limited by the activity of the enzyme cholesterol 7α-hydroxylase which, unlike serum albumin, has a short half-life of about 4 hrs (VAN CANTFORT and BARBASON, 1972). When the kinetics of the activity of this enzyme and of cellular divisions were compared after partial hepatectomy, it was found that the enzyme activity dropped at once after the operation and showed no rise until the first mitotic peak (VAN CANTFORT and BARBASON, 1972) (Fig. 14). Thus in this case, a clear suppression of a specific enzyme acitivity was observed when the cells were triggered into the mitotic cycle.

c) The detoxification function of the liver, realized by the microsomal drug-metabolizing enzymes, has also been found to be affected by liver regeneration. The activity of drug-metabolizing enzymes is greatly reduced after partial hepatectomy (VON DER DECKEN and HULTIN, 1960; FOUTS et al., 1961). Their induction occurs normally or even more actively in late regenerating liver (CHIESARA et al., 1967). However, response is delayed if partial hepatectomy is performed soon after induction (HILTON and SARTORELLI, 1970).

Another interesting phenomenon pertaining to our problem is the reappearance in regenerating liver of proteins which were present in embryonic liver. Thus, the LDH_4 to LDH_3 ratio of 0.4 found in embryonic and neonatal rat livers falls to 0.025 in adult liver, but in regenerating liver temporarily increases again to the amount in the immature liver (URIEL, 1971). The immature pattern of glucose-6-

phosphate dehydrogenase also reappears during liver regeneration (URIEL, 1971). The muscle-type aldolase A, characteristic of the embryonic liver, increases during liver regeneration (RUTTER et al., 1963), while alanine transaminase shows a low level in regenerating liver as in the first week of life (NICHOL and ROSEN, 1971). The same is true of arginase (GREENGARD, 1970; SASADA and TERAYAMA, 1969). α Fetoprotein was also found to reappear in regenerating liver (ABELEV, 1971; DENÉCHAUD and URIEL, 1971).

C. The establishment of the level of control of all these changes is essential for the elucidation of the molecular mechanisms triggering the hepatocytes into the mitotic cycle. Although the existing evidence is only circumstantial, some preliminary conclusions can be drawn from a number of experimental findings concerning the effect of different metabolic inhibitors on the synthesis of RNA and proteins after partial hepatectomy. Such data have shown that in most cases the appearance of enzyme activities and proteins needed for the mitotic cycle represents a *de novo* protein synthesis. Administration of cycloheximide to partially hepatectomized rats abolishes the mitotic wave, due to a complete block of protein and DNA synthesis (VERBIN et al., 1969; BROWN et al., 1970). Puromycin has the same effect (GIUDICE et al., 1964; GOTTLIEB et al., 1964). Inhibition of RNA synthesis by actinomycin D also abolishes the mitotic wave (FUJIOKA et al., 1963; GIUDICE and NOVELLI, 1963; SCHWARTZ et al., 1965; KING and VAN LANCKER, 1969).

In this connection it should be noted that in normal liver some of the dTTP-forming enzymes are very unstable and can be stabilized by their substrates (BUCHER and MALT, 1971). Thus, dTMP kinase levels can be raised in normal liver as well as in regenerating liver by the administration of dTMP or TdR (HIAT and BOJARSKI, 1961). An important finding was, however, that this rise in normal livers can not be prevented by actinomycin D or by X-irradiation (ADELSTEIN and KOHN, 1967), while the increase induced by partial hepatectomy is prevented by these inhibitors (FAUSTO and VAN LANCKER, 1965; BAUGNET-MAHIEU et al., 1967). X-irradiation was shown to prevent the induction by partial hepatectomy of several DNA-forming enzymes (BOLLUM et al., 1960; MYERS et al., 1961; KING and VAN LANCKER, 1969). It has been suggested that such treatment would inhibit the synthesis of new enzymes without affecting the enzyme activities already present (BOLLUM et al., 1960; MYERS et al., 1961).

A transcriptional control also seems to be involved in the synthesis of chromatin proteins. Good evidence exists that the synthesis of histones is a *de novo* protein synthesis which takes place only during the cell cycle (WILHELM et al., 1971) and depends on short-lived mRNAs (SCHOCHETMAN and PERRY, 1972). The cell-cycle-dependent synthesis of some nonhistone proteins (see p. 199) suggests that transcriptional control may also be involved.

Evidence for transcriptional controls in regenerating liver has been also presented by RNA–DNA hybridization experiments. These show that new short-lived mRNAs were already synthesized 1 hr after partial hepatectomy. Other mRNAs continue to appear at various times during liver regeneration (CHURCH and McCARTY, 1967a). It is important to note that these new RNA species were found to be present in embryonic liver (CHURCH and McCARTY, 1967b). It has been concluded that partial hepatectomy makes possible the transcription of new genes which have been active in embryonic liver but were repressed during development. It could be argued that the hybridization experiments were carried out at low Cot values, thus revealing only repetetive DNA sequences which do not code for proteins. However, the possibility exists that either these sequences are involved in the activation of unique sequences, as proposed in some models (BRITTEN and DAVIDSON, 1969) or that they are part of the transcriptional unit which also contains unique sequences, as supposed in another model (GEORGIEV, 1969).

Another line of evidence concerning the levels of control in regenerating liver comes from studies on the synthesis of RNA (BUCHER and MALT, 1971). In addition to the transcription-dependent appearance of new RNA species, stabilization of labile mRNAs in regenerating liver also seems to be a real possibility (ROTH, 1964; BRESNICK, 1971). A detailed study on the stability of TdR kinase and its template has shown that the stability of the enzyme is unchanged in quiescent, regenerating, and neonatal livers as well as in Novikov ascites hepatoma, but that its message has a strongly increased stability in livers with high levels of mitotic activity (BRESNICK, 1971). These data may be interpreted as an indication for the presence of posttranscriptional controls.

It appears that posttranscriptional controls also play an important role in the increased formation of ribosomes after partial hepatectomy. This is supported again by the effect of X-irradiation. As we have shown, a preoperative whole-body irradiation suppressed DNA synthesis, but the increase in liver mass was not prevented (MARKOV et al., 1973). Such an unbalanced growth of the liver was associated with the usual increased synthesis of most of the RNA species, including the precursors of ribosomal RNA; only a fraction of the nuclear heterogeneous RNA was inhibited (MARKOV et al., 1975). The independence of new ribosome formation from the new cellular program is also confirmed by experiments showing that hydrocortisone suppresses DNA synthesis in regenerating rat liver but does not prevent the increased rate of ribosome formation (RIZZO et al., 1971).

Several data indicate that the increased overall protein synthesis following partial hepatectomy is also at least partly dependent on posttranscriptional events (TSUKADA et al., 1968a, b; NOLAN and HOAGLAND, 1971).

From all these data, the tentative conclusion can be made that after partial hepatectomy two independent groups of synthetic activities are induced in parallel: (a) The first group includes an overall increase in the rate of RNA and protein synthesis, reflected mainly by increased ribosome formation. These events are assoiated with a program already going on in quiescent cells and are regulated at different post-transcriptional levels. However, it can not be excluded that a transcriptional control may also be involved which increases the number of transcribed ribosomal RNA cistrons. (b) The second group represents the activation of a new program at a genetic level. This program is associated with the process of DNA replication and mitosis, but the presence of functional interrelations between different genes also leads to the derepression of other genes unrelated to the main events of the mitotic cycle. The second group of processes may also be subjected to some additional fine adjustments realized at different post-transcriptional levels.

VII. The Control of Q–P Transition

Different factors have been proposed as the stimulus for the transition of quiescent cells into the mitotic cycle. In order to make the problem clear it is necessary to distinguish among three groups of factors:

A. Factors which determine the capacity of the hepatocytes to proliferate (permissive factors); B. Factors which initiate the process (inducing factors) and C. Factors which realize the process (intracellular factors).

A. The data discussed here show that adult hepatocytes have the intrinsic capacity to proliferate. However, it seems that certain conditions are necessary to permit the expression of this capacity in the presence of proliferative stimuli. This is evident from the fact that adult hepatocytes are unable to proliferate in tissue cultures under conditions which permit the proliferation of embryonic hepatocytes. However, after partial hepatectomy (Hays et al., 1968, 1971) or exposure to CCl_4 (Luria et al., 1969) adult liver cells acquire the capacity to grow *in vitro*. This indicates that some factors in the organism determine the capacity of the adult hepatocytes to respond to proliferative stimuli. Different hormonal and other factors *in vivo* may be responsible for this effect, as shown by the inhibitory effect of parathyroidectomy on the proliferative response of the liver (Rixon and Whitfield, 1972). One of the permissive factors may be also the level of glucose in the blood (Weinbren and Dowling, 1972).

Our model experiments (see p. 227) have shown that different parameters of the cellular system may strongly influence its reactivity as shown, for example, by the effect of changing the cellular permeability (Fig. 16).

B. A number of inducing factors have been proposed to play a role in initiating cell proliferation in the liver after partial hepatectomy: (1) increased blood pressure, (2) increased functional demands, (3) reversible cell damage, and (4) release of humoral factors.

1. The possible significance of increased portal pressure after partial hepatectomy was emphasized by Glinos (1958). However, the regenerative reaction of liver autografts to partial hepatectomy of the nontransplanted liver (see p. 225) rejects the importance of this factor.

2. The role of functional overloading has been suggested (Goss, 1964) but there is no evidence to prove it. There are no data indicating that stimulation of a liver function can induce a hepatic proliferation comparable to that after partial hepatectomy. Only a slight increase in mitotic activity of the hepatocytes has been observed after induction of some drug-metabolizing enzymes (see p. 229). On the other hand, the stimulation of many other functions seems to exert a depressive rather than a stimulating effect on hepatocyte proliferation (see p. 229).

3. The possible role of a reversible cellular damage in inducing cellular proliferation in the liver is indicated by some cytochemical changes in the hepatocytes observed very early after partial hepatectomy. Within 30 min after operation the RNA-containing basophilic bodies in the cytoplasm already start to disappear, beginning with the cells in the periportal area and proceeding toward the center of the lobule (Glinos, 1958). These changes correspond to the appearance and progression of DNA synthesis and mitosis (Oehlert et al., 1962; Rabes and Tuczek, 1970; Grisham, 1973).

Although no quantitative changes in RNA have been observed (TSANEV and MARKOV, 1964) these structural changes correspond to a disorganization of the ergastoplasm and might be an indication of reversible cell damage. However, no evidence exists that this is the initiating factor in the start of a new transcrptional program. Similar changes were caused by other factors such as portacaval shunt (FISHER and FISHER, 1963), fasting, hypoxia, toxic agents, biliary obstruction, etc. (BADE, 1967), which do not always induce cellular proliferation.

4. In spite of some conflicting results, there is no doubt that humoral factors exist which regulate the initiation of cellular proliferation in normal liver. Their presence is demonstrated by the following experimental findings:

a) Hepatocytes grow better in plasma from partially hepatectomized animals than in normal plasma. High concentrations of normal serum inhibit the proliferation of hepatocytes *in vitro* (GLINOS and GEY, 1952).

b) Explants of regenerating liver can stimulate thymidine incorporation into DNA of explants of normal liver when the two samples are separated by a membrane filter (WRBA and RABES, 1967).

c) Dilution of the blood with saline (plasmapheresis) in normal rats increases to several times the mitotic index of the liver, while concentrating the plasma proteins by fluid restriction strongly inhibits the mitotic activity induced by partial hepatectomy (GLINOS, 1958, 1960; VIROLAINEN, 1967).

d) Injection of normal serum to partially hepatectomized animals or serum from operated animals to normal animals has given conflicting results (NADAL, 1970; BUCHER and MALT, 1971). This was explained by the extreme instability of the active substances (MOOLTEN and BUCHER, 1967; NADAL, 1973), and by some other interfering factors such as the age of the animals, diurnal variations, food intake, stress etc. (ECHAVE LLANOS, 1967; BUCHER and MALT, 1971; NADAL, 1973). By using a better standardized system, it was possible to detect the presence of substances in adult rat serum which inhibit the mitotic activity of the hepatocytes (NADAL, 1973).

e) Experiments with exchange blood transfusion (GRISHAM, 1969b) or with cross-circulation of blood between normal and partially hepatectomized rats have clearly shown a stimulation of DNA synthesis in the normal partner if the blood exchange lasted more than 10 hrs (MOOLTEN and BUCHER, 1967). Similar results were obtained, by only some of the authors, using parabiotic rats. The failure to reproduce these results in other experiments is explained by the same factors discussed under (d).

f) A strong support of humoral factors comes from experiments with liver autografts which were shown to respond to partial hepatectomy of the nontransplanted liver by a strongly increased mitotic activity, the effect being specific for liver tissue (LEONG et al., 1964; VIROLAINEN, 1967; SIEGEL, 1972).

g) A low-molecular-weight polypeptide with the characteristics of a chalone has been isolated from rabbit liver (VERLY et al., 1971).

It can be concluded from these data that humoral factors play an important role in the initiation of cell proliferation in the liver. Less certain is whether a humoral factor functions by means of the release of a substance which stimulates the quiescent cells to proliferate (positive control) or by the

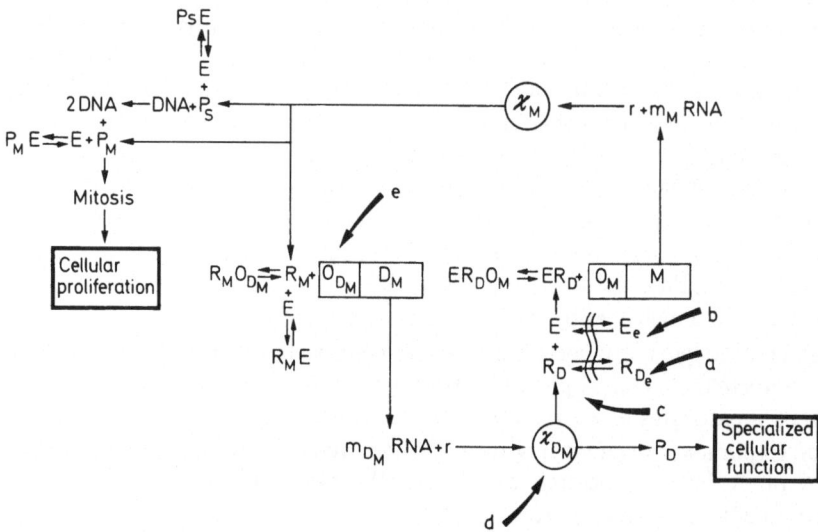

Fig. 15. Scheme of the interrelations between one functional (D_M) and one mitotic (M) operon on the basis of mutual repression. R_D and R_M – repressors; E – effector modifying the repressor; X_M and X_{D_M} – programmed ribosomes; P_S – proteins initiating DNA replication; P_M – division proteins; P_D – proteins for specialized differentiated functions. The arrows indicate some sensitive sites of the system (Sendov et al., 1970)

decrease of an inhibitory substance which normally prevents the cells from proliferating (negative control).

Many of the experimental results described above were usually interpreted in favor of the last assumption, but most of the data can be explained equally well by either a negative or positive control. It is also possible that both stimulatory and inhibitory factors are involved in the process. It seems, however, that the stimulatory factor observed in the blood of partially hepatectomized animals is due to the growth hormone released after the operation (Echave Llanos et al., 1971; Badran et al., 1972). The possibility that the inhibitor is arginase has been also investigated. One of the inhibitory factors in the cell sap of adult rat liver is indeed arginase but it was shown that a second thermolabile inhibitor was also present (Sasada and Terayama, 1969).

A major role in a negative mitotic control was ascribed to tissue-specific mitotic inhibitors called chalones (Bullough, 1965). Such substances have been isolated from at least ten differentiated tissues (Verly et al., 1971). It is not yet clear how the chalones exert their effect. They may be involved in the direct activation of some latent repressor in the cytoplasm, or initiate a chain of processes in the cellular membrane.

C. Even without knowing the exact molecular mechanisms leading to derepression of the mitotic operons, a more general model can be used to determine whether or not interrelated functional and mitotic operons can account for the regulation of cellular proliferation. We have used a simplified model of only two operons—one functional, tissue-specific, and the other mitotic—interrelated on the basis of mutual repression as shown in Fig. 15. The behavior of cellular systems

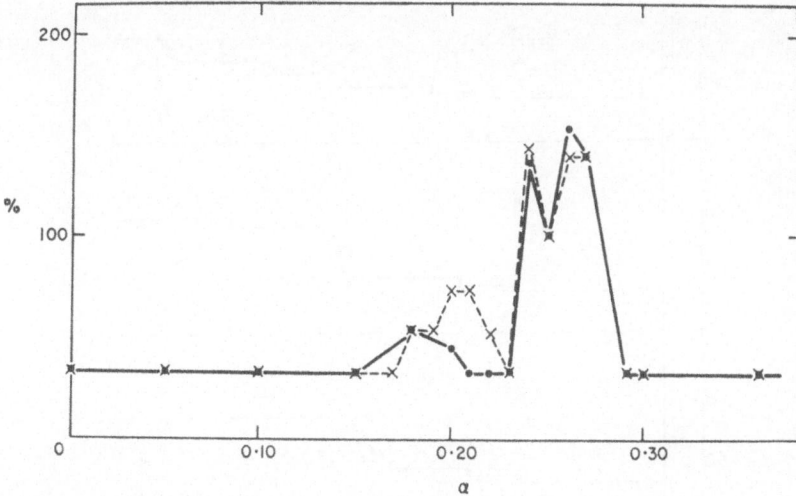

Fig. 16. Restoration of the liver in model experiments after $^2/_3$ "partial hepatectomy" as a function of a parameter determining cellular permeability (SENDOV and TSANEV, 1968)

controlled by such an intracellular mechanism has been studied by means of computer simulation and the results obtained have shown that this system can well explain the basic proliferative properties of different cellular systems (TSANEV and SENDOV, 1966; SENDOV and TSANEV, 1968; SENDOV et al., 1970).

We have applied the same mechanism to a simplified model of the liver (SENDOV and TSANEV, 1968). In the absence of cellular injury the proliferative properties of the model were found to depend strongly on some parameters such as the velocity of the blood flow, the permeability of the cellular membrane (Fig. 16), the binding constants of the repressors, etc. If the parameters of the system were properly chosen, the following behavior of the model was observed:

a) After $^2/_3$ "partial hepatectomy" a $100 \pm 20\%$ restoration of the model by mitotic divisions of the remaining cells took place (Fig. 17).

b) To obtain a regenerative response not less than 30% of the liver model had to be "excised". In real experiments it has been shown that at least 10–30% of the rat liver should be eliminated to induce cell proliferation (BUCHER and MALT, 1971).

c) In the model, the first cells to enter the mitotic cycle were the periportal cells. Mitotic activity was then propagated toward the center (Fig. 17), as in real experiments (RABES and TUCZEK, 1970).

d) The "regeneration" of the model was achieved by means of two or more mitotic waves and the time course of the regeneration was more or less protracted, depending on the parameters of the system (Fig. 17). This corresponds to the big variations in the time course of liver regeneration, depending on the animal species (MACDONALD et al., 1962; LIOSNER, 1961; STEINER et al. 1966) and age (BUCHER and MALT, 1971).

The following conclusions can be drawn from these results:

a) A system of interrelated mitotic and functional operons can explain in detail the regenerative properties of the liver after partial hepatectomy.

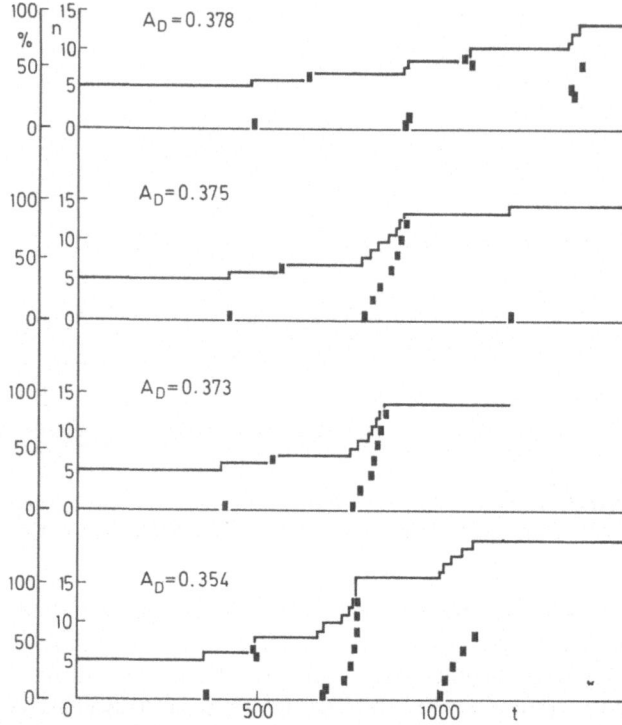

Fig. 17. Time course of liver restoration in model experiments as a function of time after $^2/_3$ "partial hepatectomy". Experiments at different values of a parameter A_D which determines the binding properties of a repressor. Abscissae—time in arbitrary units; ordinates—percentage of restoration. The black cells show the spatial distribution of mitotic cells in the course of regeneration. The direction from the bottom to the top of the figures is the direction from v. portae to v. centralis (Sendov and Tsanev, 1968)

b) No cellular damage is needed to induce cell proliferation after partial hepatectomy. With suitable parameters of the system a decreased concentration of a mitotic repressor (or a substance which activates this repressor) can account for the mitotic response.

c) A number of permissive factors is needed to determine suitable parameters of the cells in order to obtain a mitotic response.

Thus, our model experiments support the idea that a central position in the control of Q–P transitions is occupied by a system of interrelated operons in which some functional operons repress the mitotic operons. Such a system has many "sensitive sites" which, when affected, can induce cellular proliferaton; some of these factors are membrane permeability, synthetic activity of the programmed ribosomes, number of cells, concentration of metabolites, etc. This may explain why different factors such as mechanical damage, disturbed metabolism, hormones, loss of cells, etc. can lead to the same result—that is, a decreased concentration of mitotic repressors and derepression of the mitotic operons.

VIII. Mitotic Cycle
and Differentiated Function of the Hepatocyte

It has long been thought that an antagonism exists between terminal cell differentiation and cell proliferation (GROBSTEIN, 1959). This has been understood both as a decreased capacity of the terminally differentiated cells to proliferate as well as a cessation of the specific differentiated program during the mitotic cycle. Some theoretical considerations, however, show that this relationship cannot be so simple.

1. The first aspect of the problem is to determine to what extent terminally differentiated cells have a reduced mitotic activity. Theoretically, this question could be answered if we knew the functional interconnections between functional and mitotic operons. Depending on the number of functional operons involved in this interaction, a more or less pronounced effect on the mitotic activity would result. This can explain the diversity of the proliferative properties of various differentiated tissues.

The existence of differentiated liver functions repressing the mitotic activity of the hepatocytes is indicated by the developmental changes of liver enzymes. This becomes evident when the three periods in which the different enzymes appear are considered together with the periods of decreasing mitotic activity of the hepatocytes (compare Fig. 2 with Figs. 5–7). It can be seen that the three periods in which hepatocyte proliferation sharply falls (late fetal development, shortly after birth, and 2 to 3 weeks later) correlate with the appearance of three groups of liver enzymes and proteins. At the same time some other proteins and enzymes disappear. The latter group reappears when the adult hepatocytes are induced to proliferate (see p. 221–222).

The observation that the induction of some drug-metabolizing enzymes does not suppress but stimulates the mitotic activity of the hepatocytes (ARGYRIS and MAGNUS, 1968; ARGYRIS, 1969) apparently conflicts with the view of mutual repression between functional and mitotic operons. According to GRISHAM (1973) this may be due to the release of a mitotic block. There is also another possible explanation: If one group of functional operons represses another group involved in the interaction with the mitotic operons, the simultaneous stimulation of some functional activities and of cellular proliferation can be explained, as well as the additive effect of some proliferation-inducing factors (ARGYRIS, 1971). At the same time such data support the assumption that in the mutual repression between mitotic and functional operons, only some of the latter ones are directly involved.

These considerations indicate that no general rule for the proliferative properties of a differentiated tissue should exist, since different configurations of the genetic net can permit various situations.

2. The second aspect of the problem is whether the tissue-specific program is maintained during the mitotic cycle. The disturbance of the synthetic program of the cell during the mitotic cycle can result from at least three events: (a) changes in the chromatin structure, (b) disorganization of the ergastoplasm and (c) re-

pression of functional operons coupled to mitotic operons on the basis of mutual repression.

When the experimental evidence concerning the liver is interpreted, two considerations should be also kept in mind: (1) the incomplete synchrony of the cells entering the mitotic cycle and (2) the existence of long-lived mRNAs, especially for tissue-specific proteins.

Data concerning the differentiated function of the hepatocytes during the mitotic cycle are rather poor. As we have seen (p. 220) the synthesis of albumin is not inhibited in regenerating rat liver. At least three explanations of this fact are possible (1) the genes controlling albumin synthesis are repressed but the message is long-lived and is not destroyed after partial hepatectomy, thus supporting the synthesis of albumin during the mitotic cycle; (2) the albumin genes are repressed and the message is absent but due to the sequential entering of the cells into the cell cycle quiescent cells which normally do not produce albumin (see p. 212) are induced to do this as a compensatory reaction; and (3) the genes for albumin synthesis are not coupled with the mitotic operons and therefore are not repressed when the latter ones are derepressed.

It is difficult to distinguish among these possibilities. There are no data on albumin synthesis in individual hepatocytes after partial hepatectomy as in the case of normal liver. On the other hand the albumin mRNA seems to be long-lived (WILSON et al., 1967; MURTY and SIDRANSKY, 1972) but it is not known whether or not this message is destroyed after partial hepatectomy. Other proliferation-inducing agents, such as mechanical or chemical injury, may well act by destroying all messages in the cytoplasm through the activation of latent RNAses. Such a mechanism was shown to be possible by model experiments (TSANEV and SENDOV, 1966; SENDOV et al., 1970). We also found an early increase of RNAse activity in rat liver ribosomes after partial hepatectomy (BELTCHEV and TSANEV, 1966). Thus, with a long-lived message even if the albumin genes were repressed during the mitotic cycle, albumin synthesis may continue or not, depending on the kind of injury. A more direct indication that albumin genes are not repressed during the cell cycle can be deduced from experiments showing the continuous synthesis of albumin in liver cells cultivated in vitro (see p. 231).

A clear case of mutually exclusive differentiated function and proliferation in the liver is presented by the cholesterol-7α-hydroxylase. The activity of this enzyme was found to vary inversely with the mitotic activity of the hepatocytes, including both the diurnal rhythms as well as cycles induced by liver regeneration (VAN CANTFORD and BARBASON, 1972) (see Fig. 14).

From these data it seems possible to conclude that three groups of functional operons exist in the hepatocytes:

a) The first group is linked to the mitotic operons by a mutual repression realized at the level of transcription or at some post-transcriptional levels (example: cholesterol-7α-hydroxylase).

b) The second group is independent of, and does not influence, the activity of the mitotic operons (example: albumin).

c) The third group seems to be linked indirectly to the mitotic operons by repressing some of the operons of the first group (example: some drug-metabolizing enzymes).

All these considerations show that the interrelationship between differentiated function and cell cycle differs depending on the genetic net of the cell and on the metabolic stability of different mRNAs.

IX. Maintenance of the Hepatic Cellular Type upon Continuous Cellular Proliferation

Under conditions of continuous cellular proliferation changes both in the functional state and in the cellular type can occur. It may be difficult to distinguish practically between these two phenomena unless special experiments are carried out. Unfortunately no clear distinction is made between these two situations and very often the unclear term "dedifferentiation" (GROBSTEIN, 1959) is used to denote different things. It has been shown that chondrogenic cells which have been considered "dedifferentiated" may resume their lost functions under appropriate culture conditions (COON, 1966; CAHN and CAHN, 1966). However, even when grown under these optimal conditions for longer periods, chondroblasts invariably lose their typical histological features and cease synthesizing chondroitin sulfate (CHACKO et al., 1969). Such examples show that changes in phenotypic expressions may in some instances be readily reversible, in others virtually irreversible.

A prolonged cultivation of hepatic tissue *in vitro* without loss of normal morphological and biochemical features was found to be a difficult problem (LURIA et al., 1969; GUILLONZO et al., 1972). However, in many cases the observed loss of phenotypic expression is difficult to be interpreted. Because of the presence of several cellular types in the liver, the disappearance of differentiated functions can mean two different things: either a real loss of phenotypic expression of the hepatocytes or a selective outgrowth of other nonparenchymal cellular types (SANDSTRÖM, 1965). Discrimination between these two alternatives is made difficult by the fact that morphological criteria can be misleading in establishing the origin of a hepatic cell line. It seems that the "altered liver cells" observed by many authors in tissue cultures represent, in fact, cells of mesenchyme origin (SANDSTRÖM, 1965). On the other hand, clonal cell strains isolated from human liver, although resembling fibroblasts morphologically, were shown to function as hepatocytes by their ability to synthesize and secrete serum albumin (KAIGHN and PRINCE, 1971).

It has been shown that the persistence of different cellular antigens and the capacity of the liver to express differentiated functions in tissue cultures strongly depend on the culturing conditions—intercellular contacts, mechanical factors, high oxygen level etc. (SANDSTRÖM, 1965; FOGEL, 1968; LURIA et al., 1969; ALEXANDER and GRISHAM, 1970; OKADA et al., 1971).

Under suitable conditions differentiated functions of the hepatocytes can be maintained in proliferating cultures. Induction of tyrosine transaminase by corticosteroid hormones has been demonstrated in organ cultures of fetal rat liver (WICK, 1968), and in an established cell line of adult liver hepatocytes (GERSHENSON et al., 1970). When hamster embryonic liver cells were grown *in vitro* it was found that a liver-specific antigen was undetectable in cells cultivated for a few days in monolayer cultures, but if the cells were subsequently transferred into organ cultures, they formed clusters of antigen-containing cells (Fogel, 1968). These data show the importance of the contacts between cells in the maintenance of a differentiated feature.

In cultures of embryonic rat liver albumin and α fetoprotein were repeatedly accumulated in high concentrations through a *de novo* synthesis in eight subsequent changes of the culture medium (Luria et al., 1969). Fetal and newborn mouse hepatocytes under favorable culturing conditions retained many of their features of differentiation for more than 120 days of cultivation (Rose et al., 1968).

From all available data it can be concluded that only under strictly defined conditions can the hepatocytes maintain their terminally differentiated functional capacity during continuous proliferation in culture. A loss of phenotypic expressions due to alterations in the hepatic cellular type has not been directly proved, but seems very probable.

According to our model the problem of the maintenance of the cellular type upon cellular proliferation brings up the question of how the pattern of nonhistone proteins in chromatin is restored during DNA replication. Two things are obviously necessary for this restoration: (1) the same nonhistone proteins must be synthesized during the mitotic cycle, and (2) the different nonhistone species synthesized in the cytoplasm must bind to the corresponding sites of the genome.

The first point requires the derepression of the genes controlling the synthesis of the corresponding deblocking proteins during the mitotic cycle. This process can be easily achieved by suitable functional relations between operons (Tsanev and Sendov, 1971 a, b).

The second point depends on the probability of binding of these proteins to the corresponding sites in the chromatin. Whatever the exact mechanism of recognition is, the binding of these proteins in our model is supposed to be irreversible. Then, the incorporation of such a protein into the chromatin will depend only on the probability of binding p which is given by the expression

$$p = 1 - e^{-kS}$$

where k is a constant and S is the total amount of a specific deblocking protein synthesized during the mitotic cycle (Tsanev and Sendov, 1971 b). Thus, the probability always exists that a mitotic division may alter the cellular type. If kS is high enough the probability $1 - p$ of changing the pattern of nonhistone proteins will be practically negligible. We assume that this is normally the case under *in vivo* conditions. However, p exponentially decreases with decreasing S. On the other hand, the probability of altering the cellular type is roughly proportional to the number of mitotic divisions that a cell undergoes (Tsanev and Sendov, 1971 b). This shows that any factor which interferes with the synthesis of the nonhistone proteins will exponentially increase the probability of changing the deblocked region of the genome. This results either in cellular death or in alteration of the cellular type. The disturbed normal metabolism *in vitro* and the increased rate of proliferation are factors which can cause such changes.

X. Mitotic Cycle and Hepatocarcinogenesis

Three points related to the process of malignant transformation of liver cells are relevant to the problem discussed:

1. What is the role of the cell cycle in hepatocarcinogenesis?

2. How is the differentiated function of liver cells affected by malignant transformation? and 3. How do the proliferative properties of liver cells change in hepatomas?

We shall discuss the experimental evidence concerning these three points in the light of a possible mechanism of carcinogenesis based on the probability of inaccurate reproduction of the cellular type during the mitotic cycle (TSANEV and SENDOV, 1971 b). As it has been pointed out in the previous section, a decreased synthesis of some deblocking proteins would exponentially increase the probability of inaccurate reproduction of the pattern of deblocked operons. The consequences differ, depending on the region of the genome affected. Disturbances in the mitotic region or in the region coding for universal proteins would be of no importance for carcinogenesis since such cells will die. However, if tissue-specific operons are affected there is a chance that the cell can survive. On the other hand, if the tissue-specific operons affected are interconnected with the mitotic operons on the basis of mutual repression, a more or less pronounced increase in the mitotic activity will arise. Thus, malignant transformation appears as an alteration of the cellular type and the theoretical answers to the above questions can be formulated as follows:

1. Mitotic divisions should play a double role in carcinogenesis. First of all, they appear as a necessary condition for malignant transformation, since no changes in the deblocked region of the genome can take place without a mitotic division. From this point of view mitotic divisions play the same role in malignant transformation as in cytodifferentiation, the role of embryonic inducers being played by the carcinogenic agents. The second role of mitotic divisions consists in increasing the number of initially transformed cells until the tumor reaches a critical size, whereupon the new cellular type escapes the control of the neighboring normal cells (BULLOUGH, 1965; TSANEV and SENDOV, 1969).

2. The differentiated function of the cells should be altered since the essence of malignancy is assumed to be a random blocking of tissue-specific operons. However, the resulting changes can be extremely variable depending on the number and combinations of tissue-specific operons affected. It can be predicted that if n operons control the differentiated function of a cellular type, the number of different malignant transformations will be 2^n-1 or less if some combinations prove lethal.

3. Concerning the proliferative capacity of neoplastic cells big variations can be expected depending on the number of tissue specific-operons affected and on their mutual relations with the mitotic operons. It can be expected that a higher degree of loss of phenotypic expressions will lead to a higher mitotic activity.

All experimental evidence agrees with these predictions. The requirement for mitotic divisions in malignant transformation of the hepatocytes has been shown in two aspects: the dependence of the effectiveness of some carcinogens on the phases of the mitotic cycle, and the need of mitotic divisions to reveal an already initiated carcinogenic lesion.

The first aspect is illustrated by experiments showing that partial hepatectomy increases the effect of some carcinogens. Thus, when acetylaminofluoren was given concomitantly with partial hepatectomy, a decreased latent period for tumor production was observed (LAWS, 1959). A higher rate of hepatoma production was observed in partially hepatectomized rats treated with diethylnitrosamine (GRÜNTHAL et al., 1970). Urethane becomes a powerful hepatocarcinogen when given at the late prereplicative period after partial hepatectomy (CHERNOZEMSKI and WARWICK, 1970). The noncarcinogen 2-methyl-4-dimethylaminoazobenzene can induce hepatomas if administered to partally hepatectomized rats (WARWICK, 1967). 7,12-Dimethylbenz(a) anthracene was also found to induce tumors in regenerating but not in intact liver (MARQUARDT et al., 1972).

According to our model, the greater effectiveness of some carcinogens at particular phases of the mitotic cycle may be due to either their effect on the synthesis of deblocking nonhistone proteins, or to the interference of the carcinogen with the process of binding these proteins to the corresponding sites of the genome. It seems very probable that different carcinogens affect different points in the chain of events leading to the restoration of the deblocked region of the genome upon mitotic division. Such an opportunity is presented by the binding of hepatic carcinogens to different macromolecules of the chromatin (O'BRIEN et al., 1969; WARWICK, 1971; SUGIMOTO and TERAYAMA, 1972; JUNGMANN and SCHWEPPE, 1972).

The second aspect is demonstrated by the "unmasking effect" of hepatic proliferation on latent carcinogenic lesions. Thus, ionizing radiations can induce hepatomas if such treatment is followed by regeneration of the liver (COLE and NOWELL, 1965; CURTIS et al., 1968). The damage can remain in a latent form for as long as 9 months until a stimulation of hepatic proliferation comes to reveal it (CURTIS et al., 1968).

There are abundant data showing extremely large variations in the alterations of the protein pattern in malignantly transformed hepatocytes. These alterations have several general features: (a) lack of tumor specificity, (b) tendency to resemble fetal or neonatal liver, (c) correlation with tumor growth, and (d) irreversibility.

a) No alterations in the protein pattern of hepatomas have been found that could be considered tumor-specific. Different hepatomas may exhibit quite different synthetic patterns (EMMELOT, 1971; REYNOLDS et al., 1971). Thus, glucose-6-phosphatase activity is absent in some hepatomas (WEBER and CANTERO, 1955) but in others it remains unchanged (HADJIOLOV, 1969). Different agents used to induce tyrosine-aminotransferase in different hepatomas showed a wide diversity in the inductive response. There were no two hepatoma lines responding similarly to all inducing agents (REYNOLDS et al., 1971). Studies on different liver-specific antigens in chemically induced hepatomas have shown that each individual hepatoma had lost different antigens. Even in the same animal different hepatomas had different antigen structures (GUELSTEIN, 1962, 1966; KHRAMKOVA and GUELSTEIN, 1965). The same was observed with α fetoprotein: even a single liver was found to contain both α FP-positive and -negative hepatoma nodules (MCINTIRE et al., 1972).

Thus, the hepatomas illustrate well the important biological phenomenon that many different neoplastic types may correspond to one normal cellular type (GREENSTEIN, 1954), as predicted by our model. The hepatocytes appear as a cellular type characterized by an extremely high number of deblocked functional operons, as indicated by the high number of different functional states of the liver. Thus, it is predicted that the hepatic cellular type exhibits a very high number, if not the highest, of different neoplastic transformations. This seems to be confirmed by the fact that more than 40 different hepatomas have already been obtained (POTTER, 1968).

b) Many authors have observed that the changes in the protein pattern of hepatomas tend to resemble the picture found in "immature" liver (EMMELOT, 1971). As we have seen, the so-called immature liver represents an already established cellular type and the developmental changes referred to as "maturation" or "differentation" are nothing but changes in the functional state of the hepatic cellular type, induced by hormonal and environmental factors. On the other hand, fetal manifestations, present in many hepatomas, are not obligatory (WEINHOUSE et al., 1972). As already mentioned, α fetoprotein is present in many hepatomas as well as in embryonic liver, but it is absent in other hepatoma lines.

The reappearance in hepatomas of some antigens that are present only in fetal or neonatal liver can be explained if we accept the idea that mutual repression and derepression exist between different operons and that in carcinogenesis some tissue-specific operons are blocked. Such a disturbance of the genetic net will lead to derepression of genes which had been repressed by functional operons inactivated in the process of carcinogenesis. In most cases such changes appear as a reversal of the cell into a previous functional state.

c) Different hepatomas also show large variations in growth rate, due mainly to variations in their proliferating fraction, which is usually increased. The cell-cycle parameters of hepatomas are not essentially different from those observed in normal liver. Thus, in chemically induced rat hepatomas the cell generation time was even longer than in 3-week-old rat liver, but the growth fraction was increased to 15% as compared to 10% in young rats (POST and HOFFMAN, 1964b). The percentage of cells that divide during 24 hrs is much higher in hepatomas (10–30%) than in normal adult liver (0.1–0.5%) but is lower than in regenerating liver (64%) (EMMELOT, 1971). In one of the most slowly growing hepatomas the labeling index was much higher (4%) than in adult liver, while the length of the S period remained the same (FRANKFURT, 1967). These data show that changes in cell-cycle parameters are unimportant for malignant growth—this corresponds to data on other malignant cells (BRESCIANI, 1968). The altered proliferating fraction and the establishment of unlimited growth point to disturbances in the mechanisms controlling the Q–P transition.

Good correlations have been observed between the degree of loss of liver-specific functions and growth or proliferation rates. Thus, hepatomas with liver-like enzyme pattern exhibit a very low growth rate; this rate increases if some deletions in the normal enzyme pattern occur, while in hepatomas with many such deletions a rapid growth rate is observed (OTANI and MORRIS, 1971; WEINHOUS et al., 1972). Changes in the isozyme pattern of glucose phosphorylating enzymes, in fructose aldolase, and in pyruvate kinase have shown also that the liver-specific and feedback-regulated enzymes disappeared, whereas hexokinase

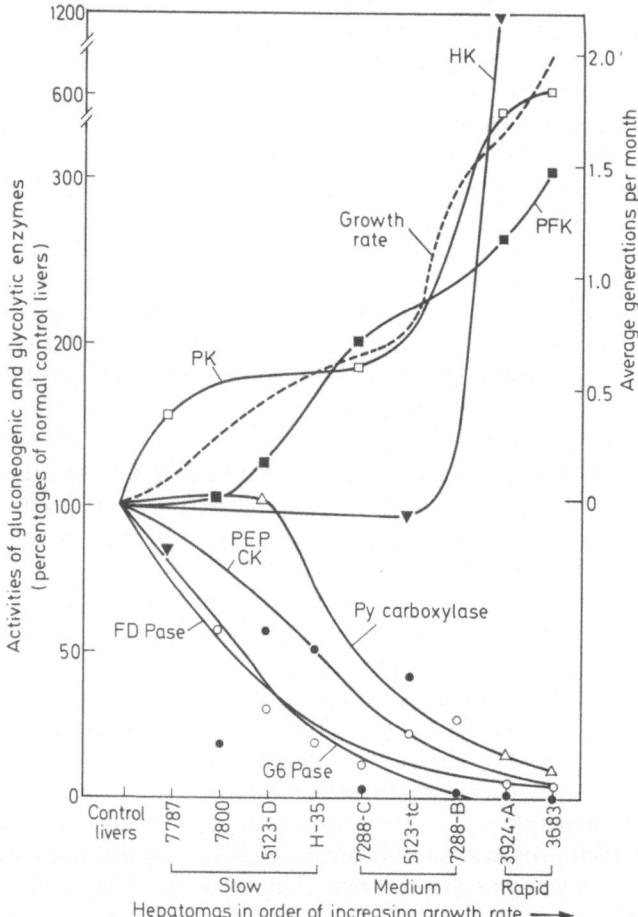

Fig. 18. Behavior of carbohydrate-metabolizing enzymes in relation to hepatoma growth rate
(Weber, 1971)

and pyruvate kinase of the muscle type increased in parallel with the increasing
growth rate (Fig. 18). The degree of decreased glucose-6-phosphatase activity in
hepatomas also showed a good correlation with growth rate (Weber, 1963). The
correlation between proliferation and loss of functions is also reflected by the
diurnal variations in the mitotic activity of hepatomas. In a slow growing "differentiated" hepatoma the circadian rhythm of DNA synthesis and of mitosis were
comparable to that of normal growing liver, while in a fast-growing "undifferentiated" hepatoma the rhythmic character of the proliferative activity was seen
only in the curve of mitosis, the amplitude of the variations being much lower
(Nash and Echave Llanos, 1971). Some type of circadian rhythm of DNA
synthesis and of several enzymes was also observed in two minimal deviation
hepatomas (Pitot, 1966).

The observed correlation between the extent of loss of liver-specific functions
and the increase in proliferation rate in hepatomas favors the assertion that tissue

homeostasis is maintained by a mutual repression between tissue-specific and mitotic operons. Such data also support a model in which not only one but a number of functional operons are involved in this control. It should be mentioned that the correlation discussed may not always be very strict since some functional operons may not be interconnected with the mitotic operons, as in the case of albumin. This explains why some biochemical changes can be correlated with the growth rate of hepatomas, while others can not (PITOT, 1968; OVE et al., 1970; WEBER, 1971).

d) The last but very important feature of the changes observed in hepatomas is their irreversibility. Many identical changes taking place in embryonic or in adult liver are reversible, or at least each individual step is reversible (see p. 203). Some changes observed during the early stage of hepatocarcinogenesis are also reversible and seem to be due to non-specific transient disturbances of the steady state. Thus, administration of hepatocarcinogens in the diet of rats led to the prompt appearance of α fetoprotein in the serum of all animals within a few weeks (WATABE, 1971; KITAGAWA et al., 1972; KROES et al., 1972), the effect being reversible and unrelated to the obligatory appearance of tumors.

Unlike these early precancerous stages, the properties of already developed hepatomas are irreversibly altered, as in a new cellular type. It can be argued that this irreversibility may also be due to a sequence of many reversible steps. However, the fact that many functional changes induced in normal hepatocytes are similar to those observed in hepatomas, but do not lead to neoplastic expression, suggests that in hepatomas some irreversible changes take place which prevents the reestablishment of normal relations between functional and mitotic operons. As already discussed, a possible mechanism is the failure of some functional operons to be deblocked during a mitotic cycle. Thus, according to this concept the molecular mechanisms of normal cytodifferentiation and of carcinogenesis are identical, the difference being that in the first case the process takes place as a regular event occurring at a fixed time and affecting strongly determined genes, while in the second case it is a random process. This is in agreement with the suggestion of several authors that cancer is a disease of cytodifferentiation and that the neoplastic changes may have an epigenetic nature (MARKERT, 1968; PITOT, 1968; TSANEV and SENDOV, 1971b; EMMELOT, 1971; DUSTIN, 1972).

XI. General Conclusions

1. Clarifying the relationship between cell proliferation and differentiated function in the hepatic cellular type presents some difficulties due to the multiple functions of the liver and the lack of suitable markers of cytodifferentiation. Study of this problem is facilitated if two different forms of cellular reprogramming are clearly distinguished. The first form represents functional changes within the limits of the cellular type. These changes are reversible, dependent on the presence of different inducing factors, and are cell-cycle-independent. The second form represents alterations of the cellular type. Such processes are normally irreversible

and cell-cycle-dependent. It is probable that two different molecular mechanisms are involved in the control of these two types of cellular reprogramming.

2. The cell-cycle-dependent step in the emergence of the hepatic cellular type seems to be a very early event, occurring during the intense proliferation of endodermal cells in the ventral floor of the foregut which form the hepatic primordium. All further developmental changes in the protein pattern of the hepatocytes have the characteristics of functional changes.

3. The developmental and all other functional changes in the protein pattern and in the proliferative activity of the hepatocytes suggest the presence of closely linked genes. A mutual repression between tissue-specific and mitotic genes is indicated by metabolic and proliferative changes observed during development, during liver regeneration, and in hepatomas. In all these cases the changes in the proliferative activity are associated with changes in enzyme activities and other antigens which are not involved in the process of self reproduction. A model of functionally interconnected mitotic and tissue-specific genes can well explain the proliferative properties of the liver.

4. The interrelations between mitotic cycle and differentiated function may differ, depending on the particular function studied. It appears that there are at least three types of such interrelations: mutual repression (e.g. cholesterol-7α-hydroxylase), independence (e.g. albumin), and simultaneous stimulation (e.g. some drug-metabolizing enzymes). These various types of interrelations depend on the functional links between different genes.

5. The only changes that affect the hepatic cellular type seem to take place during malignant transformation of the hepatocytes. These changes are cell-cycle-dependent and many data indicate that they result from a random blocking of tissue specific genes. This can explain the existence of many malignant transformations corresponding to one normal cellular type.

References

ABELEV, G. I.: Alpha-fetoprotein in oncogenesis and its association with malignant tumors. Advan. Cancer Res. **14**, 295–358 (1971).

ABELEV, G. I., PEROVA, S. D., KHRAMKOVA, N. I., POSTNIKOVA, Z. A., IRLIN, I. S.: Embryonic serum alpha-globulin and its synthesis by transplantable mouse hepatomas. Biokhimiya **28**, 625–634 (1963) (In Russian).

ADELSTEIN, S. J., KOHN, H. I.: Failure of X-rays and actinomycin D to prevent the response of renal and hepatic thymidilate kinase treatment. Biochim. Biophys. Acta **138**, 163–168 (1967).

ALEXANDER, R. W., GRISHAM, J. W.: Explant culture of rat liver. I. Method, morphology, and cytogenesis. Lab. Invest. **22**, 50–62 (1970).

ARGYRIS, TH. S.: Enzyme induction and the control of growth. In: DUNPHY, E., VANWINKLE, JR., W. (Eds.): Repair and Regeneration, pp. 201–216. New York: McGraw-Hill 1969.

ARGYRIS, TH. S.: Additive effects of phenobarbital and high protein diet on liver growth in immature male rats. Develop. Biol. **25**, 293–309 (1971).

ARGYRIS, TH. S., MAGNUS, D. R.: The stimulation of liver growth and demethylase activity following phenobarbital treatment. Develop. Biol. **17**, 187–201 (1968).

BADE, E. G.: Biological and ultrastructural aspects of early stages of liver regeneration. In: TEIR, H., RYTÖMAA, T. (Eds.): Control of Cellular Growth in Adult Organisms, pp. 260–269. London: Academic Press 1967.

BADRAN, A. F., SURUR, J. M., BALDUZZI, R., ECHAVE LLANOS, J. M.: Assay of plasma from intact and hepatectomized mice on mouse regenerating liver DNA synthesis. Virchows Arch. Abt. B. Zellpath. **10**, 176–178 (1972).

BARNHART, M. I.: Cellular site for prothrombin synthesis. Am. J. Physiol. **199**, 360 (1960).

BARNHART, M. I., ANDERSON, G. F.: A cellular study of prothrombin synthesis altered by coumadin and vitamin K. Federation Proc. **20**, 50 (1961).

BAUGNET-MAHIEU, L., GOUTIER, R., SEMAL, M.: Effects of total-body X-irradiation and AET on the activity of thymidine phosphorylating kinases during liver regeneration. Radiation Res. **31**, 808–825 (1967).

BELTCHEV, B., TSANEV, R.: Ribonucleic acid degrading activity of rat liver microsomes following partial hepatectomy. Nature **212**, 531–532 (1966).

BEUVING, G., GRUBER, M.: Induction of phosvitin synthesis in roosters by estradiol injection. Biochim. Biophys. Acta **232**, 529–536 (1971).

BLACK, I. B., AXELROD, J.: The regulation of some biochemical circadian rhythms. In: LITWACK, G. (Ed.): Biochemical Actions of Hormones, pp. 135–155. New York-London: Academic Press 1970.

BOLLUM, F. J., ANDEREGG, J. W., MCELYA, A. B., POTTER, V. R.: Nucleic acid metabolism in regenerating rat liver. VII. Effect of X-radiation on enzymes of DNA synthesis. Cancer Res. **20**, 138–143 (1960).

BRESCIANI, F.: Cell proliferation in cancer. European J. Cancer **4**, 343–366 (1968).

BRESNICK, E.: Regenerating liver: an experimental model for the study of growth. Methods Cancer Res. **6**, 347–397 (1971).

BRESNICK, E., TOMPSON, U. B., MORRIS, H. B., ZIEBELT, A. G.: Inhibition of thymidine kinase activity in liver and hepatomas by TTP and dCTP. Biochem. Biophys. Res. Commun. **16**, 278–284 (1964).

BRITTEN, R., DAVIDSON, E.: Gene regulation for higher cells: a theory. Science **165**, 349–357 (1969).

BROWN, J. M.: Long G_1 or G_0 state: a method of resolving the dilemma for the cell cycle of an in vivo population. Exptl. Cell Res. **52**, 565–570 (1968).

BROWN, R. F., UMEDA, T., TAKAI, S.-I., LIBERMAN, I.: Effect of inhibitors of protein synthesis on DNA formation in liver. Biochim. Biophys. Acta **209**, 49–53 (1970).

BRUGAL, G.: Relations entre la prolifération et la différenciation cellulaires: étude autoradiographique chez les embryons et les jeunes larves de Pleurodeles waltlii Michah (Amphibien Urodèle). Develop. Biol. **24**, 301–321 (1971).

BUCHER, N. L. R.: Regeneration of mammalian liver. Intern. Rev. Cytol. **15**, 245–300 (1963).

BUCHER, N. L. R., MALT, R.: Regeneration of liver and kidney. Boston: Little, Brown & Co. 1971.

BUCHER, N. L. R., SWAFFIELD, M. N.: Nucleotide pools and (6-^{14}C) orotic acid incorporation in early regenerating rat liver. Biochim. Biophys. Acta **129**, 445–459 (1966).

BULLOUGH, W. S.: Mitotic and functional homeostasis: a speculative review. Cancer Res. **25**, 1683–1727 (1965).

BUSANNY-CASPARI, W.: Autoradiographische Untersuchungen mit ^{3}H-Thymidin über die DNS-Synthese in Leberzellen verschiedener Ploidiestufen. Frankfurter Z. Pathol. **72**, 123–134 (1962).

BUTCHER, F. R., BUSHNELL, D. E., BECKER, J. E., POTTER, V. R.: Effects of cordycepin on induction of tyrosine aminotransferase employing hepatoma cells in tissue culture. Exptl. Cell Res. **74**, 115–123 (1972).

CAHN, R. D., CAHN, M. B.: Heritability of cellular differentiation: clonal growth and expression of differentiation in retinal pigment cells in vitro. Proc. Natl. Acad. Sci. **55**, 106–114 (1966).

CAMERON, I. L., JETER JR., J. R.: Relationship between cell proliferation and cytodifferentiation in embryonic chick tissues. In: Developmental Aspects of the Cell Cycle, pp. 191–222. New York: Academic Press 1971.

CAMPBELL, P. N.: Functions of polyribosomes attached to membranes of animal cells. FEBS Letters **7**, 1–7 (1970).

CARRIERE, R.: The growth of liver parenchymal nuclei and its endocrine regulation. Intern. Rev. Cytol. **25**, 201, 277 (1969).

Chacko,S., Abbott,J., Holtzer,S., Holtzer,H.: The loss of phenotypic traits by differentiated cells. VI. Behavior of the progeny of a single chondrocyte. J. Exp. Med. 130, 417–442 (1969).

Chedid,A., Nair,V.: Diurnal rhythm in endoplasmic reticulum of rat liver: electron microscopic study. Science 175, 176–179 (1972).

Chernozemski,I., Warwick,G.P.: Liver regeneration and induction of hepatomas in B6AF mice by uretan. Cancer Res. 30, 2685–2690 (1970).

Chiesara,E., Clementi,F., Conti,F., Meldolesi,J.: The induction of drug-metabolizing enzymes in rat liver during growth and regeneration. Lab. Invest. 16, 254–267 (1967).

Church,R.B., McCarty,B.J.: Ribonucleic acid synthesis in regenerating and embryonic liver. I. The synthesis of new species of RNA during regeneration of mouse liver after partial hepatectomy. J. Molec. Biol. 23, 459–475 (1967a).

Church,R.B., McCarty,B.J.: Ribonucleic acid synthesis in regenerating and embryonic liver. II. The synthesis of RNA during embryonic liver development and its relationship to regenerating liver. J. Molec. Biol. 23, 477–486 (1967b).

Cole,L.J., Nowell,P.C.: Radiation carcinogenesis: the sequence of events. Science 150, 1782–1786 (1965).

Conney,A.H., Burns,J.J.: Induced synthesis of oxidative enzymes in liver microsomes by polycyclic hydrocarbons and drugs. In: Advan. Enzyme Regulation 1, 189–214 (1963).

Coon,H.G.: Clonal stability and phenotypic expression of chick cartilage cells in vitro. Proc. Natl. Acad. Sci. 55, 66–73 (1966).

Curtis,H.J., Czernik,C., Tilley,J.: Tumor induction as a measure of genetic damage and repair in somatic cells of mice. Radiation Res. 34, 315–319 (1968).

Djondjurov,L., Markov,G., Tsanev,R.: Distribution of tryptophan-containing proteins and of newly synthesized RNA in metaphase chromosomes. Exptl. Cell Res. 75, 442–448 (1972).

Doljanski,F.: The growth of the liver with special reference to mammals. Intern. Rev. Cytol. 10, 217–241 (1960).

Dustin Jr.,P.: Cell differentiation and carcinogenesis: a critical review. Cell Tissue Kinet. 5, 519–533 (1972).

Echave Llanos,J.M.: Liver tissue growth factors and circadian rhythms in liver regeneration. In: Teir,H., Rytömaa,T. (Eds.): Control of Cellular Growth in Adult Organisms, pp. 209–220. London: Academic Press 1967.

Echave Llanos,J.M., Gomez Dumm,C.I., Nessai,A.C.: Ultrastructure of STH cells of the pars distalis of hepatectomized mice. Z. Zellforsch. 113, 29–38 (1971).

Emmelot,P.: Some aspects of the mechanisms of liver carcinogenesis. In: Liver Cancer, IARC Sci. Publ. 1, pp. 94–109. Lyon 1971.

Epifanova,O.I., Terskikh,V.V.: On the resting periods in the cell life cycle. Cell Tissue Kinet. 2, 75–93 (1969).

Epstein,C.: Cell size, nuclear content and the development of polyploidy in the mammalian liver. Proc. Natl. Acad. Sci. 57, 327–334 (1967).

Fabrikant,J.I.: Size of proliferating pools in regenerating liver. Exptl. Cell Res. 55, 277–279 (1969).

Fausto,N., van Lancker,J.: IV Thymidylic kinase and DNA polymerase activities in normal and regenerating liver. J. Biol. Chem. 240, 1247–1255 (1965).

Ferdinandus,J., Morris,H., Weber,G.: Behaviour of opposing pathways of thymidine utilization in differentiating, regenerating, and neoplastic liver. Cancer Res. 31, 550–556 (1971).

Fisher,E.R., Fisher,B.: Ultrastructural hepatic changes following partial hepatectomy and portacaval shunt in the rat. Lab. Invest. 12, 929–942 (1963).

Fogel,M.: Phenotypic changes in cultured hamster embryonic liver cells as studied by immunofluorescence. Develop. Biol. 17, 85–100 (1968).

Follet,B.K., Nicholls,T.J., Redshow,M.R.: The vitellogenic response in the south african clawed toad (Xenopus laevis Daudin). J. Cell Physiol. 72, 91–102 (1968).

Fouts,J., Dixon,R., Shultice,R.: The metabolism of drugs by regenerating liver. Biochem. Pharmacol. 7, 265–270 (1961).

Frankfurt,O.: The kinetics of cell populations in the process of carcinogenesis and tumor growth. Tsytologia 9, 1103–1120 (1967) (In Russian).

FREEDLAND, R.A., SZEPESI, B.: Control of enzyme activity: nutritional factors. In: RECH-
CIGL JR., M. (Ed.): Enzyme Synthesis and Degradation in Mammalian Systems, pp. 103–
140. Basel: Karger 1971.

FRIEDEN, E., JUST, J.J.: Hormonal responses in amphibian metamorphosis. In: Biochemical
Actions of Hormones, Vol. 1, pp. 1–52. New York: Academic Press 1970.

FUJIOKA, M., KOGA, M., LIEBERMAN, I.: Metabolism of ribonucleic acid after partial hepatectomy. J. Biol. Chem. **238**, 3401–3406 (1963).

FULLER, R. W.: Rhythmic changes in enzyme activity and their control. In: RECHCIGL JR., M.
(Ed.): Synthesis and Degradation in Mammalian Systems, pp. 311–338. Basel: Karger
1971.

GELBOIN, H. V.: Carcinogens, enzyme induction, and gene action. Advan. Cancer Res. **10**, 1–
81 (1967).

GELEHRTER, TH. D.: Regulatory mechanisms of enzyme synthesis: enzyme induction. In:
RECHCIGL JR., M. (Ed.): Enzyme Synthesis and Degradation in Mammalian Systems,
pp. 165–199. Basel: Karger 1971.

GELFANT, S.: Initiation of mitosis in relation to the cell division cycle. Exptl. Cell Res. **26**, 395
(1962).

GEORGIEV, G.: On the structural organization of operon and the regulation of RNA synthesis
in animal cells. J. Theoret. Biol. **25**, 473–490 (1969).

GERSHENSON, L. E., ANDERSSON, M., MOLSON, J., OKIGAKI, T.: Tyrosine transaminase induction by dexamethasone in a new rat liver cell line. Science **170**, 859–861 (1970).

GIUDICE, G., KENNEY, F.T., NOVELLI, G. D.: Effect of puromycin on DNA synthesis by regenerating rat liver. Biochim. Biophys. Acta **87**, 171–173 (1964).

GIUDICE, G., NOVELLI, G. D.: Effect of actinomycin D on the synthesis of DNA polymerase in
hepatectomized rats. Biochem. Biophys. Res. Commun. **12**, 383–387 (1963).

GLINOS, A. D.: The mechanisms of liver growth and regeneration. In: MCELROY, W.D.,
GLASS, B. (eds.): The Chemical Basis of Development, pp. 813–839. Baltimore: Johns Hop-
kins Press 1958.

GLINOS, A. D.: Environmental feedback control of cell division. Ann. N.Y. Acad. Sci. **90**, 592–
602 (1960).

GLINOS, A. D., GEY, G. O.: Humoral factors involved in the induction of liver regeneration in
the rat. Proc. Soc. Exptl. Biol. Med. **80**, 421–425 (1952).

GOSS, R. J.: Adaptive Growth. New York: Academic Press 1964.

GOTTLIEB, L. I., FAUSTO, N., VAN LANCKER, J.: Molecular mechanism of liver regeneration. The
effect of puromycin on deoxyribonucleic acid synthesis. J. Biol. Chem. **239**, 555–559 (1964).

GRACHEVA, N. D.: Autoradiographic studies on the proliferative processes during histogenesis of rat liver. Arch. Hist. Embryol. **50**, 38–46 (1966) (In Russian).

GREENGARD, O.: Enzymic Differentiation in mammalian liver. Science **163**, 891–895 (1969).

GREENGARD, O.: The developmental formation of enzymes in rat liver. In: Biochemical Actions of Hormones, Vol. 1, pp. 53–87. New York: Academic Press 1970.

GREENGARD, O., FEDERMAN, M., KNOX, W. E.: Cytomorphometry of developing rat liver and
its application to enzyme differentiation. J. Cell Biol. **52**, 261–272 (1972).

GREENGARD, O., GORDON, M., SMITH, M. A., ACS, G.: Studies on the mechanism of diethylstil-
bestrol-induced formation of phosphoprotein in male chicken. J. Biol. Chem. **239**, 2079–
2082 (1964).

GREENSTEIN, J. P.: Biochemistry of the Cancer. New York: Academic Press 1954.

GRISHAM, J. W.: A morphologic study of deoxyribonucleic acid synthesis and cell proliferation in regenerating rat liver: autoradiography with thymidine-^{3}H. Cancer Res. **22**, 842–
849 (1962).

GRISHAM, J. W.: Cellular proliferation in the liver. In: FRY, R.J.M., GRIEM, M.L., KIR-
STEN, W.H. (Eds.): Normal and Malignant Growth, pp. 28–43. Berlin-Heidelberg-New
York: Springer 1969.

GRISHAM, J. W.: Hepatocytic proliferation in normal rats after multiple exchange transfusions
with blood from partially hepatectomized rats. Cell Tissue Kinet. **2**, 277–282 (1969).

GRISHAM, J. W.: Effects of drugs on hepatic cell proliferation. In: ZIMMERMAN, A.M., PADIL-
LA, G. M., CAMERON, I. Z. (Eds.): Drugs and the Cell Cycle, pp. 95–136. New York: Academic Press 1973.

GROBSTEIN, C.: Differentiation of vertebrate cells. In: BRACHET, J., MIRSKY, A. E. (Eds.): The Cell, vol. 1, pp. 437–496. New York: Academic Press 1959.

GRÜNTHAL, D., HELLENBROICH, D. O., SÄNGER, P., MAASS, H.: Der Einfluß von partiellen Hepatectomien auf die Hepatomrate nach Diäthylnitrosamin-Gaben. Z. Naturforsch. **25 b**, 1277–1281 (1970).

GUELSTEIN, V.: Morphological and immunological liver changes in the course of experimental carcinogenesis in mice. Acta Uni. Intern. contra Cancr. **18**, 60–62 (1962).

GUELSTEIN, V.: Production and progression of experimental tumors in liver. Thesis, Moscow 1966 (In Russian).

GUILLONZO, A., OUDEA, P., LEGUILLY, Y., OUDEA, M. C., LENOIR, P., BOUREL, M.: An ultrastructural study of primary cultures of adult human liver tissue. Exp. Molec. Path. **16**, 1–15 (1972).

HADJIOLOV, D.: Glucose-6-phosphatase activity in primary rat liver tumors induced by high doses of 4-dimethylaminoazo-benzene. Z. Krebsforsch. **72**, 43–46 (1969).

HAMASHIMA, Y., HARTER, J. G., COONS, A. H.: The localization of albumin and fibrinogen in human liver cells. J. Cell Biol. **20**, 271–279 (1964).

HARRIS, H.: Nucleus and Cytoplasme. Oxford: Clarendon Press 1968.

HAYASHI, SH., ARAMAKI, Y., NOGUCHI, T.: Diurnal changes in ornithine decarboxylase activity of rat liver. Biochem. Biophys. Res. Commun. **46**, 795–800 (1972).

HAYS, D. M., MATSUSHIMA, Y., TEDO, I., TSUNODA, A.: Liver regeneration: influence of interval postextirpation on *in vitro* growth of cells from the remnant liver. Proc. Soc. Exp. Biol. Med. **138**, 658–660 (1971).

HAYS, D. M., TEDO, I., OKUMURA, S., NAGASHIMA, K.: Hepatic regeneration: effect of size of resection on *in vitro* growth of remnant liver cells. Nature **220**, 286–287 (1968).

HIATT, H. H., BOJARSKI, T. B.: The effects of thymidine administration on thymidylate kinase activity and on DNA synthesis in mammalian tissues. In: Cold Spring Harbor Symp. Quant. Biol. **26**, 367–369 (1961).

HILL, H. Z., WILLSON, S. H., HOAGLAND, M. B.: Patterns of albumin and general protein synthesis in rat liver as revealed by gel electrophoresis. Biochim. Biophys. Acta **269**, 477–484 (1972).

HILTON, J., SARTORELLI, A. C.: Induction by phenobarbital of microsomal mixed oxidase enzymes in regenerating liver. J. Biol. Chem. **245**, 4187–4192 (1970).

HOLTZER, H., WEINTRAUB, H., BIEHL, J.: Cell cycle-dependent events during myogenesis, neurogenesis and erythrogenesis. In: MONROY, A., TSANEV, R. (Eds.): Biochemistry of Cell Differentiation, pp. 41–53. London-New York: Academic Press 1973.

HOUBRECHTS, N., BARBASON, H.: Apparition du rythme nycthémèral des mitoses chez le très jeune rat. Compt. Rend. Serie D, **274**, 3457–3460 (1972).

JAFFE, J. J.: Diurnal mitotic periodicity in regenerating rat liver. Anat. Rec. **120**, 935 (1954).

JAILKHANI, B. L., TALWAR, G. P.: Induction of phosvitin by oestradiol in rooster liver needs DNA synthesis. Nature New Biol. **239**, 240–241 (1972).

JEEJEEBHOY, K. N., BRUCE-ROBERTSON, A., HO, J., SODTKE, U.: The effect of cortisol on the synthesis of rat plasma albumin, fibrinogen and transferrin. Biochem. J. **130**, 533–538 (1972).

JOHNSON, G. R., JONES, R. O.: Differentiation of the mammalian hepatic primordium *in vitro*. J. Embryol. Exp. Morphol. **30**, 83–96 (1973).

JOST, J.-P., KELLER, R., D.-VENTLING, C.: Deoxyribonucleic acid synthesis during phosvitin induction by 17 β-estradiol in immature chicks. J. Biol. Chem. **248**, 5262–5266 (1973).

JUNGMANN, R. A., SCHWEPPE, J. S.: Binding of chemical carcinogens to nuclear proteins of rat liver. Cancer Res. **32**, 952–959 (1972).

KAIGHN, M. E., PRINCE, A. M.: Production of albumin and other serum proteins by clonal cultures of normal human liver. Proc. Natl. Acad. Sci. (Wash.) **68**, 2396–2400.

KAUFMAN, S. A.: Metabolic stability and epigenesis in randomly constructed genetic nets. J. Theoret. Biol. **22**, 437–467 (1969).

KHEISSIN, E. M. (Ed.): Normal and Pathological Cytology of Liver Parenchyma. Leningrad: Nauka 1969 (In Russian).

KHRAMKOVA, N. I., GUELSTEIN, V. I.: Antigenic structure of mouse hepatomas V. Organospecific liver antigens and embryonic α-globulin in hepatomas of mice induced with ortoaminoazotoluene. Neoplasma 12, 251–260 (1965).

KING, C. D., VAN LANCKER, J. L.: Molecular mechanisms of liver regeneration. VII. Conversion of cytidine to deoxycytidine in rat regenerating livers. Arch. Biochem. Biophys. 129, 603–608 (1969).

KIRSCH, R., FRITH, L., BLACK, E., HOFFENBERG, R.: Regulation of albumin synthesis and catabolism by alteration of dietary protein. Nature 217, 578–579 (1968).

KITAGAWA, T., YOKOCHI, T., SUGANO, H.: α-Fetoprotein and hepatocarcinogenesis in rats fed 3'methyl-4(dimethylamino)azobenzene or N-2-fluorenylacetamide. Int. J. Cancer 10, 368–381 (1972).

KLEMPERER, H. G., HAYNES, G. R.: Thymidine kinase in rat liver during development. Biochem. J. 108, 541–546 (1968).

KLINMAN, N. R., ERSLEV, A. J.: Cellular response to partial hepatectomy. Proc. Soc. Exp. Biol. Med. 112, 338–340 (1963).

KRITSMAN, M. G., KONNIKOVA, A. S.: Induction of Enzymes. Moscow: Medicina 1968 (In Russian).

KROES, R., WILLIAMS, G. M., WEISBURGER, J. H.: Early appearance of serum α-fetoprotein during hepatocarcinogenesis as a function of age of rats and extent of treatment with 3'methyl-4-dimethylaminoazobenzene. Cancer Res. 32, 1526–1532 (1972).

LAGUCHEV, S. S., BARDIK, YU. V., GOLOLOBOVA, M. T., MARKELOVA, I. V., SOKOLOVA, L. V.: Duration of mitotic cycle periods in organs with different levels of proliferation. Tsitologia 14, 619–624 (1972) (In Russian).

LAJTHA, L. G.: On the concept of the cell cycle. J. Cell. Comp. Physiol. 62, Suppl. 1, 143–145 (1963).

LALA, P. K.: Studies on tumor cell population kinetics. In: BUSCH, H. (Ed.): Methods in Cancer Research 6, 3–95 (1971).

LAWS, J. O.: Tissue regeneration and tumor development. Brit. J. Cancer 13, 669–674 (1959).

LEDUC, E. H.: Regeneration of the liver. In: ROUILLER, CH. (Ed.): The Liver: Morphology, Biochemistry, Physiology, Vol. 2, pp. 63–89. New York: Academic Press 1964.

LEONG, G. F., GRISHAM, J. M., HOLE, B. V., ALBRIGHT, M. L.: Effect of partial hepatectomy on DNA synthesis and mitosis in heterotopic partial autografts of rat liver. Cancer Res. 24, 1496–1501 (1964).

LIOSNER, L. D.: Regeneration of the liver in anura. Bull Exp. Biol. Med. 8, 98–101 (1961) (In Russian).

LURIA, E. A., BAKIROV, R. D., YELISEYEVA, T. A., ABELEV, G. I., FRIEDENSTEIN, A. Y.: Differentiation of hepatic and hematopoietic cells and synthesis of blood serum proteins in organ cultures of the liver. Exptl. Cell Res. 54, 111–117 (1969).

MACDONALD, R. A., GUINEY, T., TANK, R.: Regeneration of the liver in Triturus Viridescens. Proc. Soc. Exp. Biol. Med. 111, 277–280 (1962).

MACMANUS, J. P., FRANKS, D. J., YOUDALE, T., BRACELAND, B. M.: Increase in rat liver cyclic AMP concentration prior to the initiation of DNA synthesis following partial hepatectomy or hormone infusion. Biochem. Biophys. Res. Commun. 49, 1201–1207 (1972).

MAENO, H., SCHREIBER, G., WEIGAND, K., WEINSSEN, U., ZÄHRINGER, J.: Impairment of albumin synthesis in cell-free systems from rat liver. FEBS Letters 6, 137–140 (1970).

MAENO, H., SCHREIBER, G., WEIGAND, K., WEINSSEN, U., ZÄHRINGER, J., ROTERMUND, H.-M.: Zur Synthese von Albumin in zellfreien Systemen. Hoppe-Seyler's Z. Physiol. Chem. 350, 1165–1166 (1969).

MAJUMDAR, C., TSUKADA, K., LIEBERMAN, I.: Liver protein synthesis after partial hepatectomy and acute stress. J. Biol. Chem. 242, 700–704 (1967).

MARKERT, CL. L.: Neoplasia: a disease of cell differentiation. Cancer Res. 28, 1908–1914 (1968).

MARKOV, G., DESSEV, G. N., RUSSEV, G. C., TSANEV, R. G.: Effects of γ-irradiation on biosynthesis of different types of RNA in normal and in regenerating liver. Biochem. J. 146, 41–51 (1975).

MARKOV, G., PETROV, P., DESSEV, G., TSANEV, R.: Effect of ionizing radiations on rat liver regeneration. Effect of irradiation applied before partial hepatectomy. Bull. Inst. Biochem. **4**, 215–225 (1973) (In Bulgarian).

MARQUARDT, H., PHILIPS, F., BENDICH, A.: DNA binding and inhibition of DNA synthesis after 7,12-dimethylbenz(a)anthracene administered during the early prereplicative phase in regenerating rat liver. Cancer Res. **32**, 1810–1813 (1972).

MCARDLE, A. H., CREASER, E. H.: Nucleoproteins in regenerating rat liver. I. Incorporation of $^{32}P_i$ into ribonucleic acid of liver during early stages of regeneration. Biochim. Biophys. Acta **68**, 561–568 (1963).

MCCLURE, M. E., HNILICA, L. S.: Nuclear proteins in genetic restriction. III. The cell cycle. Sub-cell Biochem. **1**, 311–332 (1972).

MCINTIRE, K. R., VOGEL, C. L., PRINCLER, G. L., PATEL, I. R.: Serum α-fetoprotein as a biochemical marker for hepatocellular carcinoma. Cancer Res. **32**, 1941–1946 (1972).

MONDER, C., COUFALIK, A.: Influence of cortisol on glycogen synthesis and gluconeogenesis in fetal rat liver in organ culture. J. Biol. Chem. **247**, 3608–3617 (1972).

MOOLTEN, F. L., BUCHER, N. L. R.: Regeneration of rat liver: transfer of humoral agent by cross circulation. Science **158**, 272–274 (1967).

MOULE, Y.: Endoplasmic reticulum and microsomes of rat liver. In: LOCKE, M. (Ed.): Cellular Membranes in Development, pp. 97–133. New York: Academic Press 1964.

MUELLER, G.: Biochemical events in the animal cell cycle. Federation Proc. **28**, 1780–1789 (1969).

MURTY, C. N., SIDRANSKY, H.: Studies on the turnover of mRNA in free and membrane-bound polyribosomes in rat liver. Biochim. Biophys. Acta **281**, 69–78 (1972).

MUTSCHLER, L. E., GORDON, A. H.: Plasma protein synthesis by the isolated perfused regenerating rat liver. Biochim. Biophys. Acta **130**, 486–492 (1966).

MYERS, D. K., HEMPHILL, C. A., TOWNSEND, C. M.: Deoxycytidylate levels in regenerating rat liver. Can. J. Biochem. Physiol. **39**, 1043–1054 (1961).

NADAL, C.: Contrôle de la multiplication et de la polyploïdie des hepatocytes du rat. Etude des perturbations de la régulation physiologique causées par l'injection de produits irritants. Wilhelm Roux' Arch. **166**, 136–149 (1970).

NADAL, C.: Synchronization of baby rat hepatocytes: a new test for the detection of factors controlling DNA synthesis in rat hepatic cells. Cell Tissue Kinet. **6**, 437–446 (1973).

NADAL, C., ZAJDELA, F.: Polyploïdie somatique dans le foie de rat. I. Le rôle des cellules binucléées dans la genèse des cellules polyploïdes. Exptl. Cell Res. **42**, 99–116 (1966).

NASH, R. E., ECHAVE LLANOS, J. M.: Twenty-four-hour variations in DNA synthesis of a fast-growing and a slow-growing hepatoma: DNA synthesis rhythm in hepatoma. J. Natl. Cancer Inst. **47**, 1007–1012 (1971).

DE NECHAUD, B., URIEL, J.: Antigènes cellulaires transitoires du foie de rat. I. Sécrétion et synthèse des protéines sériques au cours du développement et de la régénération hépatiques. Int. J. Cancer **8**, 71–80 (1971).

NICHOL, CH. A., ROSEN, F.: Changes in alanine transaminase activity related to corticosteroid treatment or capacity for growth. In: WEBER, G. (Ed.): Advances in Enzyme Regulation, Vol. 1, pp. 341–361. Oxford: Pergamon Press 1963.

NOLAN, R. D., HOAGLAND, M. B.: Cytoplasmic control of protein synthesis in rat liver. Biochim. Biophys. Acta **247**, 609–620 (1971).

O'BRIEN, R. L., STANTON, R., CRAIG, R. L.: Chromatin binding of benzo(a)pyrene and 20-methylcholanthrene. Biochim. Biophys. Acta **186**, 414–417 (1969).

OEHLERT, W., HÄMMERLING, W., BÜCHNER, F.: Der zeitliche Ablauf und das Ausmaß der Desoxyribonukleinsäure-Synthese in der regenerierenden Leber der Ratte nach Teilhepatektomie. Beitr. Pathol. Anat. **126**, 91–112 (1962).

OKADA, T. S., EGUCHI, G., TAKEICHI, M.: The expression of differentiation by chicken lens epithelium in *in vitro* cell cultures. Develop. Growth Differentiation **13**, 323–336 (1971).

OTANI, T., MORRIS, H. P.: Comparison of activity and isozyme patterns of four enzymes from hepatomas of different growth rates. J. Natl. Cancer Inst. **47**, 1247 (1971).

OVE, P., JENKINS, M. D., LASZLO, J.: DNA polymerase patterns in developing rat liver. Cancer Res. **30**, 535–539 (1970).

PERRY, L. D., SWARTZ, F. J.: Evidence for a subpopulation of cells with an extended G_2 period in normal adult mouse liver. Exptl. Cell Res. **48**, 155–157 (1967).

PITOT, H.: Some biochemical aspects of malignancy. Ann. Rev. Biochem. **35**, 335–368 (1966).

PITOT, H.: Some aspects of the developmental biology of neoplasia. Cancer Res. **28**, 1880–1887 (1968).

POPOV, P. G., KAVRAKIROVA, S. V., MALEEV, A. CH.: Effect of olivomycin on the induced synthesis of tyrosine aminotransferase and tryptophan oxygenase in rat liver. Biochem. Pharmacol. **22**, 1526–1529 (1973).

POST, J., HOFFMAN, J.: Changes in the replication times and patterns of the liver cell during the life of the rat. Exptl. Cell Res. **36**, 111–123 (1964a).

POST, J., HOFFMAN, J.: The replication time and pattern of carcinogen-induced hepatoma cells. J. Cell Biol. **22**, 341–350 (1964b).

POST, J., HOFFMAN, J.: Further studies on the replication of rat liver cells *in vivo*. Exptl. Cell Res. **40**, 333–339 (1965).

POST, J., HUANG, C., HOFFMAN, J.: The replication time and pattern of the liver cell in the growing rat. J. Cell Biol. **18**, 1–12 (1963).

POTTER, V. R.: Summary of discussion on neoplasms. Cancer Res. **28**, 1901–1907 (1968).

RABES, H. M., BRÄNDLE, H.: Synthesis of RNA, protein, and DNA in the liver of normal and hypophysectomized rats after partial hepatectomy. Cancer Res. **29**, 817–822 (1969).

RABES, H., TUCZEK, H. V.: Quantitative autoradiographische Untersuchung zur Heterogenität der Leberzellproliferation nach partieller Hepatectomie. Virchows Arch. Abt. B. Zellpath. **6**, 302–312 (1970).

RAFTELL, M., PERLMANN, P.: Antigen development in neonatal rat liver. Exptl. Cell Res. **49**, 317–331 (1968a).

RAFTELL, M., PERLMANN, P.: Synthesis of stage specific serum proteins and tissue antigens by early neonatal rat liver. Exptl. Cell Res. **49**, 332–339 (1968b).

RAJEWSKY, M. F.: Proliferative parameters of mammalian cell systems and their role in tumor growth and carcinogenesis. Z. Krebsforsch. **78**, 12–30 (1972).

RECHCIGL, M., JR. (Ed.): Enzyme synthesis and degradation in mammalian systems. Basel: Karger 1971a.

RECHCIGL, M., JR.: Intracellular protein turnover and the role of synthesis and degradation in regulation of enzyme levels. In: RECHCIGL, M., JR. (Ed.): Enzyme Synthesis and Degradation in Mammalian Systems, pp. 236–310. Basel: Karger 1971b.

REYNOLDS, R. D., SCOTT, D. F., POTTER, V. R., MORRIS, H. P.: Induction of tyrosine aminotransferase and amino acid transport in Morris hepatomas and in adult and neonatal rat liver. Cancer Res. **31**, 1580–1589 (1971).

RIXON, R. H., WHITFIELD, J. F.: Parathyroid hormone—a possible initiator of liver regeneration. Proc. Soc. Exp. Biol. Med. **141**, 93 (1972).

RIZZO, A. J., HEILPERN, P., WEBB, T. E.: Temporal changes in DNA and RNA synthesis in the regenerating liver of hydrocortisone-treated rats. Cancer Res. **31**, 876–881 (1971).

ROSE, G. G., KUMEGAWA, M., CATTONI, M.: The circumfusion system for multipurpose culture chambers. II. The protracted maintenance of differentiation of fetal and newborn mouse liver *in vitro*. J. Cell Biol. **39**, 430–497 (1968).

ROSEN, F., MILHOLLAND, R. J.: Control of enzyme activity by glucocorticoids. In: RECHCIGL, M., JR. (Ed.): Enzyme Synthesis and Degradation in Mammalian Systems, pp. 77–102. Basel: Karger 1971.

ROTH, J. S.: Increased stability of templates for some enzymes of DNA synthesis in tumors: deoxycytidylate deaminase, thymidine and thymidylate kinase. Life Sci. **3**, 1145–1149 (1964).

RUDACK, D., WALLACE, R. A.: On the site of phosvitin synthesis in Xenopus laevis. Biochim. Biophys. Acta **155**, 299–301 (1968).

RUTTER, W. J., BLOSTEIN, R. E., WOODFIN, B. M., WEBER, C. S.: Enzyme variants and metabolic diversification. In: WEBER, C. (Ed.): Advances in Enzyme Regul., Vol. 1, pp. 39–56. Oxford: Pergamon Press 1963.

RUTTER, W. J., PICTET, R. L., MORRIS, P. W.: Toward molecular mechanisms of developmental processes. Ann. Rev. Biochem. **42**, 601–646 (1973).

Ryabinina, Z. A., Benyush, V. A.: Cell polyploidization and hypertrophy in growth and reparative processes. Moscow: Medicina 1973 (In Russian).

Ryoo, H., Tarver, H.: Studies on plasma protein synthesis with a new liver perfusion apparatus. Proc. Soc. Exp. Biol. Med. **128**, 760–772 (1968).

Sandström, B.: Studies on cells from liver tissue cultivated *in vitro*. I. Influence of the culture method on cell morphology and growth pattern. Exptl. Cell Res. **37**, 552–568 (1965).

Sasada, M., Terayama, H.: The nature of inhibitors of DNA synthesis in rat-liver hepatoma cells. Biochim. Biophys. Acta **190**, 73–87 (1969).

Scherrer, K., Marcaud, L.: Messenger RNA in avian erythroblasts at the transcriptional and translational levels and the problem of regulation in animal cells. J. Cell Physiol. **72**, Suppl. 1, 181–212 (1968).

Schimke, R. T., Doyle, D.: Control of enzyme levels in animal tissues. Ann. Rev. Biochem. **39**, 929–976 (1970).

Schmid, W., Gallwitz, D., Sekeris, C. E.: On the mechanism of hormone action. VIII. Further studies on the effect of cortisol on isolated rat-liver nuclei. Biochim. Biophys. Acta **134**, 80–84 (1967).

Schochetman, G., Perry, R. P.: Early appearance of histone messenger RNA in polyribosomes of cultured L cells. J. Molec. Biol. **63**, 591–596 (1972).

Schreiber, G., Lesch, R., Weinssen, U., Zähringer, J.: The distribution of albumin synthesis throughout the liver lobule. J. Cell. Biol. **47**, 285–289 (1970).

Schultze, B., Maurer, W.: Proliferation of liver epithelia in three-week-old rats. Cell Tissue Kinet. **6**, 477–487 (1973).

Schwartz, H. S., Sodergreen, J. E., Garofalo, M., Sternberg, S. S.: Actinomycin D effects on nucleic acid and protein metabolism in intact and regenerating liver of rats. Cancer Res. **25**, 307–317 (1965).

Sekeris, C. E., Niessing, J., Seifart, K. H.: Inhibition by α-amanitin of induction of tyrosine transaminase in rat liver by cortisol. FEBS Letters **9**, 103–104 (1970).

Sendov, Bl., Tsanev, R.: Computer simulation of the regenerative processes in the liver. J. Theoret. Biol. **18**, 90–104 (1968).

Sendov, Bl., Tsanev, R., Mateeva, E.: A mathematical model of the regulation of cellular proliferation in the epidermis. Bull. Mathem. Inst. Bulg. Acad. Sci. **11**, 221–246 (1970) (In Bulgarian).

Siegel, B.: Extracellular regulation of liver regeneration. In: Goss, R. J. (Ed.): Regulation of Organ and Tissue Growth, pp. 271–282. New York: Academic Press 1972.

Sigel, B., Baldia, L. B., Brightman, S. A., Dunn, M. R., Price, R. I. M.: Effect of blood flow reversal in liver autotransplants upon the site of hepatocyte regeneration. J. Clin. Invest. **47**, 1231–1237 (1968).

Steiner, J. W., Perz, Z. M., Taichman, L. B.: Cell population dynamics in the liver. Exp. Molec. Pathol. **5**, 146–181 (1966).

Stöcker, E.: Der Proliferationsmodus in Niere und Leber. Verhandl. Deut. Ges. Pathol. **50**, 53–74 (1966).

Stöcker, E., Heine, W.-D.: Regeneration of liver parenchyma under normal and pathological conditions. Beitr. Path. **144**, 400–408 (1971).

Stöcker, E., Schultze, Br., Heine, W.-D., Liebscher, H.: Wachstum und Regeneration in parenchymatösen Organen der Ratte. Z. Zellforsch. **125**, 306–331 (1972).

Sugimoto, T., Terayama, H.: Pattern of 2-methyl-4-dimethyl-aminoazobenzene-binding proteins in the liver of partially hepatectomized rats and of continuously dye-fed rats in comparison with 3'-methyl-4-dimethyl-amino-azobenzene. Cancer Res. **32**, 1878–1883 (1972).

Takagi, M., Tanaka, T., Ogata, K.: Evidence for exclusive biosynthesis *in vivo* of serum albumin by bound polysomes of rat liver. J. Biochem. **65**, 651–653 (1969).

Tata, J. R.: Regulation of protein synthesis by growth and developmental hormones. In: Biochemical Actions of Hormones, Vol. 1, pp. 89–133. New York: Academic Press 1970.

Todaro, G., Matsuya, Y., Bloom, S., Robbins, A., Green, H.: Stimulation of RNA synthesis and cell division in resting cells by a factor present in serum. In: Defendi, V., Stoker, M. (Eds.): Growth Regulating Substances for Animal Cells in Culture, pp. 87–101. Philadelphia: Wistar Institute Press 1967.

TOMKINS, G. M., LEVINSON, B. B., BAXTER, J. D., DETHLETSEN, L.: Further evidence for post-transcriptional control of inducible tyrosine aminotransferase synthesis in cultured hepatoma cells. Nature New Biol. **239**, 9–14 (1972).

TOMKINS, G. M., MARTIN, D. W., JR., STELLWAGEN, R. H., BAXTER, J. D., MAMONT, P., LEVINSON, B. B.: Regulation of specific protein synthesis in eukaryotic cells. Cold Spring Harbor Symp. Quant. Biol. **35**, 635–640 (1970).

TSANEV, R.: Cellular reprogramming and cellular differentiation. In: MONROY, A., TSANEV, R. (Eds.): Biochemistry of Cell Differentiation, pp. 177–179. London-New York: Academic Press 1973.

TSANEV, R., MARKOV, G.: Early changes in liver nucleic acids after partial hepatectomy. Folia Histochem. Cytochem. **2**, 233–242 (1964).

TSANEV, R., RUSSEV, G.: Distribution of newly synthesized histones during DNA replication. European J. Biochem. **43**, 257–263 (1974).

TSANEV, R., SENDOV, BL.: A model of the regulatory mechanism of cellular multiplication. J. Theoret. Biol. **12**, 327–341 (1966).

TSANEV, R., SENDOV, BL.: Possible molecular mechanism for cell differentiation. J. Theoret. Biol. **30**, 337–393 (1971 a).

TSANEV, R., SENDOV, BL: An epigenetic mechanism for carcinogenesis. Z. Krebsforsch. **76**, 299–319 (1971 b).

TSANEV, R., DJONDJUROV, L., IVANOVA, E.: Cell-cycle-dependent turnover of non-histone proteins. Exptl. Cell Res. **84**, 137–142 (1974).

TSANEV, R., IORDANOVA, L., KOLEVA, R.: Incorporation du ^{32}P dans les différentes fractions de l'ARN de la peau après lésion mécanique. Arch. Bioch. Cosmetol. **85**, 9–15 (1966).

TSUKADA, K., LIEBERMAN, I.: Synthesis of ribonucleic acid by liver nuclear and nucleolar preparations after partial hepatectomy. J. Biol. Chem. **239**, 2952–2956 (1964).

TSUKADA, K., MORIYAMA, T., DOI, O., LIEBERMAN, I.: Ribosomal change in liver after partial hepatectomy and acute stress. J. Biol. Chem. **243**, 1152–1159 (1968).

TSUKADA, K., MORIYAMA, T., UMEDA, T., LIEBERMAN, I.: Relationship between the ribosomal alteration after partial hepatectomy and increase in liver protein synthesis *in vivo*. J. Biol. Chem. **234**, 1160–1165 (1968).

URIEL, J.: Transitory liver antigens and primary hepatoma in man and rat. In: Liver Cancer, IARC, pp. 58–68. Lyon 1971.

VAN CANTFORT, J. J., BARBASON, H. R.: Relation between the circadian rhythms of mitotic rate and cholesterol-7α-hydroxylase activity in the regenerating liver. Cell Tissue Kinet. **5**, 325–330 (1972).

VERBIN, R. S., SULLIVAN, R. J., FARBER, E.: The effects of cycloheximide on the cell cycle of the regenerating rat liver. Lab. Invest. **21**, 179–182 (1969).

VÉRLY, W. G., DESCHAMPS, Y., PHUSHPATADAM, J., DESROSIERS, M.: The hepatic chalone. I. Assay method for the hormone and purification of the rabbit liver chalone. Canad. J. Biochem. **49**, 1376–1383 (1971).

VIROLAINEN, M.: Humoral factors in liver cell proliferation. In: TEIR, H., RYTÖMAA, T. (Eds.): Control of Cellular Growth in Adult Organisms, pp. 232–249. London: Academic Press 1967.

VON DER DECKEN, A., HULTIN, T.: The induced synthesis and transport of yolk proteins and their accumulation by the oocyte in Xenopus laevis. J. Cell. Physiol. **72**, 73–90 (1968).

WARWICK, G. P.: The covalent binding of metabolites of ^{3}H-2-methyl-4-dimethylaminoazobenzene to rat liver and the carcinogenecity of the unlabelled compound in partially hepatectomized rats. European. J. Cancer **3**, 227–233 (1967).

WARWICK, G. P.: Metabolism of liver carcinogens and other factors influencing liver cancer induction. In: Liver Cancer, IARC, pp. 121–157. Lyon 1971.

WATABE, H.: Early appearance of embryonic α-globulin in rat serum during carcinogenesis with 4-dimethylaminoazobenzene. Cancer Res. **31**, 1192–1194 (1971).

WEBER, G.: Behavior and regulation of enzyme systems in normal liver and in hepatomas of different growth rates. In: Advances in Enzyme Regulation, vol. 1, pp. 321–340. Oxford: Pergamon Press 1963.

WEBER, G.: Biochemistry of liver cancer. In: Liver Cancer, IARC, pp. 69–93. Lyon 1971.

Weber, G., Cantero, A.: Glucose-6-phosphatase activity in normal, precancerous and neoplastic tissues. Cancer Res. **15**, 105–108 (1955).

Weinbren, K., Dowling, F.: Hypoglycaemia and the delayed proliferative response after
subtotal hepatectomy. Brit. J. Exp. Path. **53**, 78–84 (1972).

Weinhouse, S., Shatton, J. B., Criss, W. E., Morris, H. P.: Molecular forms of enzymes in
cancer. Biochimie **54**, 685–693 (1972).

Wicks, W. D.: Induction of tyrosine-α-ketoglutarate transaminase in fetal rat liver. J. Biol.
Chem. **243**, 900–906 (1968).

Wilhelm, J. A., Spelsberg, T. S., Hnilica, L. S.: Nuclear proteins in genetic restriction. I. The
histones. Sub-Cell Biochem. **1**, 39–65 (1971).

Wilson, S. H., Hill, H. Z., Hoagland, M. B.: Protein synthesis by stable polysomes. Biochem.
J. **103**, 567–572 (1967).

Wittliff, J. L., Kenney, F. T.: Regulation of yolk protein synthesis in amphibian liver. I.
Induction of lipovitellin synthesis by estrogen. Biochim. Biophys. Acta **269**, 485–492
(1972).

Wrba, H., Rabes, H.: Humoral regulation of liver regeneration. In: Teir, H., Rytömaa, T.
(Eds.): Control of Cellular Growth in Adult Organisms, pp. 221–231. London: Academic
Press 1967.

Zavarzin, A. A.: DNA synthesis and cell population kinetics in the ontogenesis of mammal-
ia, pp. 47–53, 141–155. Leningrad: Nauka 1967 (In Russian).

Histones, Differentiation, and the Cell Cycle

THADDEUS W. BORUN

Fels Research Institute, Temple University School of Medicine
Philadelphia, PA, USA

I. Introduction

ALBRECHT KOSSEL discovered a new class of basic proteins known as histones among the nuclear proteins of goose erythrocytes in 1884 (155). In 1900 TRAUBE reported the chemical synthesis and definitive identification of a white crystalline alkaloid called theophylline, which is one of the active ingredients of tea. According to HNILICA (122) theophylline was also discovered in KOSSEL's laboratoy. We shall see by the end of this paper that it is possible to use theophylline in cells to determine one of the important functions of the histones and that it is quite intriguing that both of these apparently unrelated discoveries came from the same source.

In a short letter to the editor of Nature, dated August 21, 1950, EDGAR and ELLEN STEDMAN proposed that the histones, which were still a rather undefined class of acid-soluble nuclear protein, might have gene suppression as one of their physiological functions (258). The STEDMANS' "one gene-one histone" hypothesis suggested that there might be very many different kinds of histone polypeptides which were specifically bound to and repressed different specific regions of the eucaryotic genome. It was implied that this hypothetical multitude of species and tissue specific histone molecules was somehow responsible for the gene modulations which caused the appearance of different cellular phenotypes during embryonic development and under various physiological conditions in the adult (259). At the time there did not seem to be much further speculation about what might be controlling the appearance of the histones as they controlled changes in gene expression (186). However, this proposal stimulated an enormous number of experiments., including some (7, 129) whose data were initially interpreted to be consistent with the STEDMAN's hypothesis.

But by 1968, after eighteen years of intense research in many different laboratories, it was becoming apparent that the "one gene–one histone" idea was not correct. In place of the STEDMANS' hypothesis, the following generalizations about the histones were well on their way toward being accepted as established facts.

1. There are five main classes of histone polypeptide in the somatic cells of most eucaryotes (45, 64, 65, 75, 121, 138, 139, 142, 171, 172, 217, 229). Members of a given histone class, derived from different tissues and species, have very similar,

but not identical, chemical and physical properties (29, 62, 78, 119, 181, 195, 207). In addition, nucleated red blood cells contain an erythrocyte-specific histone as well as the usual five main kinds of histone (120, 193, 194, 258, 259).

2. DNA replication and the net synthesis of the five main histone classes are normally tightly coupled processes during the S phase of the somatic cell cycle (128, 269) in a wide variety of organisms and cell types (6, 24, 26, 39, 67, 83, 101, 113, 180, 222–225, 242, 243, 245, 246, 257, 271, 286, 292).

3. In a given somatic cell type, approximately constant relative amounts of the five main histone classes are accumulated in complexes with DNA in the chromosomes during S phase; and at other parts of the cell cycle these relative amounts do not change (30, 47, 75, 101, 113, 163, 257, 271, 275).

4. Once synthesized and incorporated into chromosomes; both DNA and histone polypeptides have very low metabolic turnover rates (53, 54, 57, 218).

5. The histones, like almost all other known proteins, are synthesized on cytoplasmic polyribosomes (233, see also 25, 49, 64, 97, 169, and 182 for additional background) and are encoded in messenger RNA molecules (135). These histone mRNA species appear to be subject to extensive translational and transcriptional controls that insure the coupling of DNA replication to histone synthesis (30).

6. The histones, unlike very many other proteins, are subject to an extremely extensive array of post-synthetic amino-acid modifications, which in 1968 were known to include lysine and N-terminal amino-acid acetylation (8, 72, 91, 216), lysine methylation (190, 204), and the phosphorylation of serine and threonine (154, 162, 201, 265).

In 1968 GEOFFREY ZUBAY proposed a theory of eucaryotic cell differentiation (291) which contains the elements of many more recent theories (61, 274, 278) and reflected the impact of the data leading to the preceeding generalizations, the discoveries of microbial genetics (135), and some of his own experimental observations (256) on ideas about the role of histones in differentiation.

ZUBAY's theory proposed that:

1. DNA replication and histone synthesis are coupled processes in most eucaryotes because the DNA: histone complex (nucleohistone) is responsible for the gross structural organization of the chromosomes and both components of the complex are thus necessary for chromosome replication. Histones function in such complexes to facilitate packaging of large amounts of DNA within the structure of the chromosome and are also responsible for a nonspecific, non-inducible repression of genes which lasts from one round of chromosome replication until the next in a given cell.

2. There are at least two different kinds of potential histone-binding sites in eucaryotic DNA. Histone binding to "operator-like" regions of the DNA during chromosome replication prevents subsequent net RNA transcription of the "structural genes" controlled by the "operator-like" sites. DNA: histone interactions at other regions of the genome have no effect on RNA transcription.

3. Differentiation, or release of nonspecific, histone-induced genetic inactivation, occurs only during chromosome replication and is caused by an interaction between specific non-histone chromosomal components and the histones at specific "operator-like" control regions of the DNA. It was suggested that the non-histone chromosomal components responsible for differentiation might be repressible or co-repressible proteins analogous to those participating in the control of operons in procaryotes (135). What controlled the appropriately timed produc-

tion of either the histones or the non-histone chromosomal differentiators was not specified.

Experimental support for the idea that differentiation occurs during chromosome replication by mechanisms like those proposed in ZUBAY's theory comes from two different kinds of data, also appearing in 1968. The first data came from ISHIKAWA et al. (133) who observed that differentiating cells had to undergo cell divisions requiring DNA replication (125) before the progeny could express new phenotypes that were different from those of the progenator (124). The second set of data came from the *in vitro* reconstitution experiments of JOHN PAUL and STEWART GILMOUR (212). In such experiments, chromosome components are dissociated by treatment with high salt and urea concentrations, the components are then separated and combinations of components from different sources are reconstituted by a gradual gradient dialysis. Experiments of this sort approximate the conditions obtaining during chromosome replication, at least with respect to the interaction of histones and non-histone components at DNA binding sites, and indicate that DNA reconstituted solely with histones is completely inactive as a template for RNA synthesis. In these and in subsequent experiments, which may be definitive (14, 93, 94), it was shown that the non-histone components of the chromosome apparently reverse the inhibitory effects of the histones and permit the production of specific messenger RNA transcripts in eucaryotic cells.

Recently it has been suggested that "acidic" non-histone chromosomal proteins may be responsible for the production of specific messenger RNA transcripts and cellular differentiation (20, 32, 90, 260), and there is currently much interest in these proteins and their metabolism. However the mechanisms by which appropriate schedules of "acidic" non-histone protein synthesis and processing are controlled during development remain a complete mystery.

Since the renaissance of histone studies in the laboratoy of EDGAR and ELLEN STEDMAN in the 1940's, the putative responsibility for specific gene regulation and the control of differentiation thus seems to have passed from the histone to the non-histone components of the chromosome. Evidence that the histone composition does not vary appreciably during differentiation (12, 13, 23, 52, 77, 79, 177, 178, 195, 220, 241) while the composition of the non-histone protein component of the chromosomes changes a great deal (59, 166, 167, 221, 240) is accumulating and seems consistent with such a conclusion. However, evidence is also accumulating that suggests that at least some of the histones do play important roles in cellular differentiation. In this paper we shall examine those roles by reviewing the literature as it seems relevant to the participation of the histones in various aspects of development. We shall use the six generalizations detailed above as an outline and shall include some of the author's work, conducted in collaboration with others, which seems relevant to various issues as they arise in the course of discussion.

II. Histone Multiplicity

Determining the number of different kinds of histone polypeptide in eucaryotes was one of the central problems in testing the Stedmans' "one gene–one histone" hypothesis (259). After a few reports that seemed consistent with the idea of many different kinds of histone (168, 192, 194, 259), work from a large number

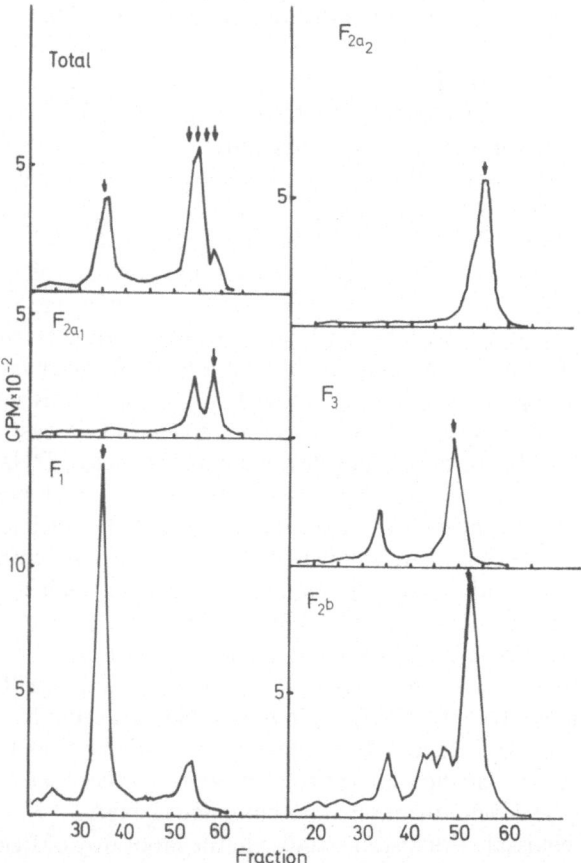

Fig. 1. Gel radioelectrophorograms of histones extracted from ^{14}C lysine–labeled Hela S-3 cells as described in Ref. (30), according to the method of Johns et al. (140). Resolved on 7.5% SDS gels containing 1% SDS and 0.5 M urea according to the procedures described in Ref. (30)

of laboratories accumulated and suggested that there might be only a few main histone fractions. One of the interesting and confusing aspects of the histone literature is the unfortunate fact that there are actually more different histone fraction nomenclature systems (at least 7; see Table 3, page 10 of Hnilica (122)) than principal types of histone polypeptide. In 1960 Johns et al. (138), using a technique first reported by P. F. Davidson in 1957 (68, 69), initiated use of the "F" (for fraction) nomenclature. Since this system is simple, very extensively used and appears to have the oldest continuous history, we shall use it throughout the rest of this paper.

The Stedmans initially discriminated between the main arginine-rich and the subsidiary lysine-rich histone fractions (122, 259). In light of what we know about histones F_{2a1}, F_{2a2}, F_{2b}, and F_3 in the arginine-rich fraction and histone F_1 in the lysine-rich fraction, this initial distinction remains one of the most important, as we shall presently see.

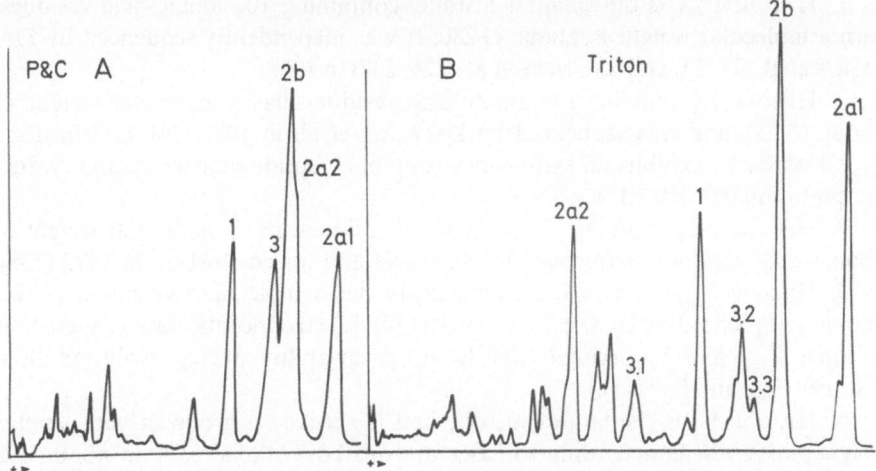

Fig. 2. (A) Histones extracted from Hela S-3 cells according to the procedure outlined in Ref. 32 and run in a Panyim and Chalkley gel for 16 hrs at 75 volts. This gel contained 2.5 M urea. (B) The same histone preparation resolved according to the method outlined in Ref. (5) on 12% acrylamide gels containing 6 mM Triton X-100 at pH 3 (5% HAC). The gel was run for 11 hrs at 100 volts. Both gels were then stained with Amido Black, destained electrically, and scanned at 600 nanometers. Absorbance is the ordinate

While the older complex chemical (119, 138, 139) and chromatographic (64, 65, 171, 162, 229) methods pointed to the existence of five main histone fractions, the evolution of simple methods of histone electrophoresis in acrylamide gels under non-aggregating conditions firmly established the principle of limited histone heterogeneity and finally allowed the standardization of different histone investigations (70, 174, 179, 230, 244). Figure 1 shows radioactive Hela-cell histone fractions prepared according to the method of JOHNS (140) partially resolved by electrophoresis on SDS-urea gels (233), a technique for analyzing histones first reported in 1967. It is evident in these preparations that the differentially extracted and selectively precipitated histone fractions from the Hela cells are cross-contaminated to various extents with different amounts of the same five histone polypeptides as well as some non-histone protein. Another approach to non-aggregating conditions was published by BONNER et al. (29) in 1967, in which histones were separated in acrylamide gels containing urea and acetic acid. This latter method was further prefected by PANYIM and CHALKLEY in 1969 (209) and has become in its various modifications (15) the standard means of resolving the five main histone polypeptide groups. Fig. 2 A shows Hela-cell histone resolved on a Panyim and Chalkley gel and stained with Amido Black. Histone F_{2a1} polypeptides migrate at the fastest rate, followed in order by histones F_{2a2}, F_{2b}, F_3, and finally F_1, which migrates at the slowest rate under these electrophoresis conditions. Figure 2 B shows the resolution of the same acid soluble nuclear extract into further subfractions on Triton, acetic acid–urea gels (5). Some salient facts about the principle classes of histone polypeptides are summarized below and are discussed in greater detail by HNILICA (122).

1. Histone F_{2a1} is the smallest histone, containing 102 amino-acid residues, with a molecular weight of about 11280. It was idependently sequenced by DE-LANGE et al. (71, 73) and SAUTIERE et al. (226, 237) in 1968.

2. Histone F_3 contains 135 amino-acid residues, has a molecular weight of about 15320, and was sequenced by DELANGE et al. in 1970 (74). Like histone F_{2a1}, histone F_3 exhibits an extremely strong polypeptide sequence stability during evolution (80, 210, 211).

3. Histone F_{2a2} contains 131 amino-acid residues, has a molecular weight of about 15000, and was sequenced by SAUTIERE and his co-workers in 1972 (238).

4. Histone F_{2b} contains 125 amino acids, has a molecular weight of 13774, and was sequenced by IWAI et al. in 1970 (134). Electrophoretic data suggest that histones F_{2a2} and F_{2b} exhibit slightly more variability during evolution than histones F_{2a1} and F_3 (211).

5. Histone F_1 is the largest histone and is actually a group of very similar polypeptides which, according to COLE and his co-workers, contain about 212 amino-acid residues (46) and have molecular weights of about 19500 to 22000 (150, 151, 272). There is good evidence that there are groups of different histone F_1 molecules in different tissues and species (151–153, 60, 267). Although no single type of histone F_1 molecule has yet been completely sequenced, sequencing studies are at present being carried out in COLE's laboratory (263).

6. Histone F_{2c} has a molecular weight of 16000 to 20000 (58, 76) and occurs only in nucleated erythrocytes and some of their precursors (12, 23, 120, 193, 194, 258, 259). It is somewhat similar in chemical composition to histone F_1 (see p. 32 of HNILICA [122]), and has an electrophoretic mobility on Panyim and Chalkley gels, intermediate between that of histone F_1 and the monomer of histone F_3. This histone appears to have a complex relationship to the genetic inactivity of mature nucleated erythrocytes (12, 23). It is somewhat ironic that KOSSEL discovered the histones in the atypical nucleated erythrocyte and that the STEDMANS were probably most misled in their generalization about histone variability because they included in their studies the nucleated erythrocyte histones with their peculiar amino-acid composition.

Most of the eucaryotic nuclear proteins that are soluble in dilute mineral acids fall into one or another of the preceeding histone classes. However, we shall see in the next section that there is definitive evidence that such crude acid-soluble nuclear "histone" preparations contain many other species of "non-histone" proteins whose metabolism is very different from that of the five main histone polypeptide classes.

III. The Relationship of DNA Replication to Histone Synthesis

DNA replication and net histone synthesis are very probable inextricably coupled processes in the somatic cells of almost all known higher eucaryotes. There is evidence that the two processes may not be coupled in yeast (117, 284) and that histone synthesis can proceed if DNA replication is partially inhibited in Physarum polycephalum (188). The principal apparent exception to the general rule of coupled DNA replication and histone synthesis in higher eucaryotes occurs in male gametes during meiosis (11, 27, 28, 219, 220). We shall first duscuss the general phenomena and then the exceptions.

A. The Concurrent Synthesis of DNA and Histones during the Somatic Cell Cycle

In 1950 HEWSON SWIFT (269) and later in 1953 HOWARD and PELC (128) showed that DNA is replicated during a discrete period of pre-mitotic interphase called the S (for synthetic) phase of plant and animal mitotic cell cycles. The period after mitosis which precedes S phase is called G_1 (for gap), and the period following S phase which precedes mitosis is called G_2. In 1955 DAVID BLOCH and GABRIEL GODMAN published cytochemical evidence that both DNA and histones accumulated during S phase in animal cells. They suggested that the two processes might be inextricably coupled (24). While the precise mechanism by which *in vivo* DNA replication is coupled to cytoplasmic histone synthesis is still unknown today, the initial observations of BLOCH and GODMAN and their inference from those observations have since come to be recognized as generally true. In 1955 MAX ALFRET demonstrated that the apparent coupling of DNA replication and histone synthesis also occurred in plant cells (6). Since these initial investigations, the cytochemical literature has been replete with observations suggesting that DNA and histone synthesis are coupled processes in a wide variety of organisms and cell types including: protozoans (83, 222–225); echinoderms (67); plant cells (26, 180, 286, 292) and mammalian cells (242, 243, 245, 246). These cytochemical studies are based on observations of parallel increases in:

a) Feulgen-stained DNA and fast green-stained "histone".

b) ^{3}H-thymidine into DNA and ^{3}H-amino acids, usually lysine and arginine, into nuclear proteins.

By their nature, cytochemical studies are not usually definitive on the molecular level. These studies were also subject to an important objection which I shall call the "blotter theory" of UMAÑA et al. (275). The advocates of the "blotter theory" argued that it was possible that net histone synthesis could only be observed when there was newly synthesized DNA present to "blot" it up and make it visible by staining reactions or keep histone radioactivity concentrated in the nucleus. Thus, it was postualted that the histones might be synthesized in all parts of the cell cycle but their net synthesis was apparent only in S phase. This objection was also pertinent to biochemical investigations where it was argued that observations of concurrent net synthesis of DNA histones did not *prove* that the processes were actually coupled in the cell. Unfortunately, there was no way of the "blotter theory" paradox if one studied only DNA and histone. However, after the discovery of histone messenger RNA (BORUN et al., 1967; 30) it became possible to prove that DNA and histone synthesis were in fact coupled.

Almost all biochemical studies of the relationship between DNA and histone synthesis in a wide variety of experimental systems indicate that net DNA replication is coupled to the net synthesis of the five main classes of histone polypeptide (12, 23, 31, 32, 101, 106, 113, 123, 177, 188, 233, 257, 260, 264, 271). In contrast, a number of studies of regenerating rat liver (50, 123, 273, 275) and one study of the Hela cell cycle (235) are frequently cited as evidence that DNA replication and histone synthesis are not coupled in these systems. It is probable that the apparent synthesis of "histones" prior to DNA replication in these studies was due to the increased pre-replicative synthesis of non-histone contaminants of the histone preparations (20, 32, 90, 260). The existence of such non-histone contamination in histone preparations was reported by MIRSKY and POLLISTER in 1947 (182), BUT-LER et al. in 1954 (48) and has subsequently been extensively discribed by others (141, 143, 163, 261, 264).

Since DNA replication and histone synthesis have underlying molecular mechanisms, it is certainly possible that drugs, mutations, or evolution could modify the coupling mechanism and produce either histone synthesis in the absence of DNA replication (as occurs *in vitro*) or DNA replication in the absence of histone synthesis. These possibilities are examined in the following sections.

B. The Effect of Inhibiting DNA Replication on Histone and Non-Histone Synthesis

In two early and often quoted studies, FLAMM and BIRNSTIEL (82) used effective concentrations of FUdR to inhibit DNA replication in cultured tobacco cells and ONTKO and MOOREHEAD (199) used X-irradiation to inhibit DNA replication in Ehrlich ascites tumor cells. Both of these studies reported only 14–16% decreases in amino-acid incorporation into crude acid-soluble nuclear extracts and each concluded that DNA and histone synthesis could therefore be dissociated. Subsequent work has demonstrated that crude acid-soluble nuclear extracts are badly contaminated with high specific activity non-histone proteins whose synthesis is not coupled to DNA replication, and whose high rate of amino-acid incorporation obscures the fact that histone synthesis is rapidly terminated if DNA replication is inhibited in S phase of the cell cycle (30, 32, 233, 257, 260, 290). Histone polypeptide synthesis has been shown to be inhibited when DNA replication is prematurely terminated by drugs such as hydroxyurea (51, 288, 290), FUdR (126), cytosine arabinoside (30, 32, 55, 233, 260), and by hormone treatment (170, 270) in a wide variety of cell types including rat liver and brain (270, 288), mouse salivary glands (31) Chinese hamster cells (106), mouse mammary epithelium (177), and in malignant human Hela cells (233, 260, 290). It is probable that recent reports (19, 21, 144) of the apparent dissociation of DNA replication and histone synthesis are due to unsuspected contamination of the histone preparations with non-histone proteins and the incomplete inhibition of DNA replication in these studies (107). Although MOHBERG and RUSCH (188) have demonstrated that histone synthesis could apparently continue after DNA replication was partially inhibited in the primative syncitial slime mold, Physarm polycelphalum, there are no convincing demonstrations of appreciable net DNA replication without net synthesis of the five main histone polypeptide classes or net histone synthesis in the absence of DNA replication in the somatic cells of the higher eucaryotes. In contrast to the other histones, the erythrocyte-specific histone F_{2c}

is synthesized after the cessation of DNA replication in nucleated red blood cells. It will be interesting to determine what characteristic of this histone's messenger RNA allows its translation in the absence of DNA replication (12, 23).

C. The Effect of Inhibiting Protein Synthesis on DNA Replication

HENRY HARRIS was probably the first investigator to report that eucaryotic DNA replication required protein synthesis (114). Since his initial report, an important but complex literature on the subject has accumulated. While a complete discussion of this literature is outside the scope of this paper, (see 88, 118, 127, 132, 189, 281) currently available data suggest that in higher eucaryotes, proteins synthesized in G_1 and S phase are required for DNA replication. It has been speculated that the required proteins facilitate the initiation (127) and/or the elongation (281) of DNA replication units or "replicons" (130, 208) during synthesis of eucaryotic chromosome segments (89). Recently, WEINTRAUB has collected arguments suggesting that the histones are the chain elongation proteins required for DNA replication during S phase (281). These arguments are consistent with some observations that I made in ELLIOTT ROBBINS' laboratory in 1967 (126). When cycloheximide was added to a log phase culture of Hela S-3 cells, protein synthesis stopped at 3 min and DNA replication essentially ceased at 7 min after the addition of the drug. We independently determined that the 4-min time differential in these inhibitions is just about the time it takes for most of the histones to "chase" from the cytoplasmic site of their synthesis into nuclear chromosomes, and probably reflects the histone pool size in these cells. None of these data however, *prove* that DNA replication ceases solely because of a lack of histones; and we shall see in a subsequent section that the G_1 proteins required for the initiation of DNA replication in S phase may include the kinases which phosphorylate histone F_1.

Recently, HEREFORD and HARTWELL (117) and WILLIAMSON (284) have reported that all the proteins required for yeast chromosome replication are synthesized prior to the initiation of DNA replication in this primitive eucaryote. These findings may imply an important distinction between the replication of the small yeast chromosomes and the more complex chromosomes of higher eucaryotes. Alternatively, the chromosomes in higher eucaryotes may contain many components or replicons, each of which is functionally similar to a small yeast chromosome. Since mammalian replicons are known to be initiated at different times during S phase, the synthesis of initiating proteins would be required throughout S phase to permit the eventual replication of all the replicons.

D. The Apparent Lack of Coupling between DNA Replication and Histone Synthesis in Meiotic Male Gamates

BLOCH and BRACK (25) were the first authors to note that the synthesis of histone-like proteins occurred in the cytoplasm of developing grasshopper spermatides and that their synthesis and subsequent transfer to the nucleus occurred some weeks after DNA replication had ceased in these cells. Subsequent cytochemical and biochemical studies by Bloch and Teng (27) and others (11, 28, 219, 220) have confirmed these observations in other developing male gamate systems. At the present time it is not known whether these observations actually represent an uncoupling of DNA and histone synthesis. This is because during meiosis there is first an apparent increase in the relative amount of DNA in these systems, which is followed by an apparent return in the relative amount of histone to normal DNA: histone ratios. There are three ways that such observations could be accounted for.

a) DNA and histones are synthesized concurrently but the histones are partially degraded and independently resynthesized later.

b) DNA and histones are synthesized concurrently but part of the histone is sequestered in some unobservable form for a period of time and then released. The data of Bloch and Teng (27) suggest that anomalously staining forms of histone F_1 could be sequestered in the spermatide during synapsis and then become available later to bind the dye used to observe the histones. These anomalous histone forms may be similar to the keratinoid meshes observed in spermatazoa by Bril-Peterson and Westerbrink (43) and by Henricks and Mayer (116).

c) DNA and histone synthesis are actually uncoupled in this kind of system and these cells represent the only known case in the higher eucaryotes where DNA can be replicated without histones. We shall see later that this possibility may be unlikely and that the second possibility mentioned above may be the best explanation we have at the present.

In light of the preceeding data it is probably safe to conclude that whenever any somatic cell in a higher eucaryote is making DNA in appreciable amounts in its nucleus, it is probably also synthesizing histones in appreciable amounts to complex with the newly synthesized DNA.

IV. Relative Amounts of Different Histone Polypeptides Synthesized and Retained on Chromosomes during the Somatic Cell Cycle and Differentiation

There have been many reports that during the proliferation of a given cell type, nearly the same relative amounts of the five main classes of histone polypeptide are synthesized in S phase and that during the other portions of the cell cycle the relative amounts of these histones classes do not vary significantly (47, 75, 101, 113, 164, 257, 271, 275). During the whole Hela S-3 cell cycle the relative percentages of the different histone classes recovered from nuclei do not vary more than 2 to 6% from their average values, which is about the level of error in this kind of determination (176). This general invarience in relative proportion of total histone mass is probably the rule for the amounts of histones F_{2a1}, F_{2a2}, F_{2b}, and F_3 synthesized and accumulated on chromosomes throughout most of the differentiation process in erythropoetic cell lines (23), Physarum polycephalum (188), Xenopus laevis (52), Triturus phyrrhogaster (13), Drosophila melanogaster (77), Drosophila virilis (197), mouse mammary gland (262) and sea urchin (243, 241). The histone F_1 group is the exception to this generalization since it has been determined that various components of this group are species- and tissue-specific

(5, 60, 151–153, 267). On the basis of this determination alone it is apparent that at some point in the development of an organism, various components of the histone F_1 group must undergo changes in their relative rates of synthesis. In all of the embryological studies cited above except one (52), members of the histone F_1 group were observed to undergo apparent changes, usually increases in amount, during the course of development from zygote to adult. RUDDERMAN et al. (234) recently showed that the messenger RNA for a new histone F_1 molecule (F_g) appears and is translated as sea urchin embryos make the transition from the morula to the gastrula.

V. Histone Turnover after Incorporation into Chromosomes

In the classical studies of histone turnover, BYVOET (53) showed that histone F_1, the other histone fractions and DNA all turned over at the same low rates in a variety of quiescent rat tissue. BYVOET went on to demonstrate that these turnover rates were not affected in quiescent tissues during the induction of enzymes (54). PIHA et al. (218) conducted similar studies in mouse liver and brain and found that the whole histone fraction and DNA had similar turnover rates in these tissues. In subsequent studies of rapidly proliferating Hela and mouse mastocyte cells, HANCOCK (112) found that unfractionated whole histones and DNA had nearly the same low turnover rate. In their extensive studies of rapidly dividing Chinese hamster cells, GURLEY and HARDIN (102, 103, 105) were able to determine that while the F_{2a1}, F_{2a2}, F_{2b}, and F_3 histone fractions did not turnover relative to DNA, approximately one-half of the histone F_1 fraction appeared to be turned over every 2 to 3 cell generations. It may be concluded from these studies that all the histone classes (except the F_1 group) have very stable, if not permanent, associations with DNA once they have been incorporated into the chromosomes. The F_1 histones apparently have stable associations with DNA in the chromosomes of non-proliferating rat liver, even if such cells are induced to produce new enzymes and presumably the RNA templates of those enzymes by drug treatment (54). However, the F_1 histone group may be metabolically labile in proliferating tissues. In a subsequent section we shall see other data connecting the metabolism of the F_1 histones and especially its phosphorylation with cellular proliferation.

VI. The Molecular Basis of Histone Synthesis and Its Control during the Somatic Cell Cycle and Differentiation

A. Mammalian Histone Messenger RNA Transcription and Translation during the Somatic Cell Cycle

In 1947 MIRSKY and POLLISTER observed that as they purified histone preparations, the tryptophan content of the preparations decreased indicating that the histones lacked tryptophan (182). In 1955 CRAMPTON et al. (64) confirmed this finding with their extensive amino-acid analyses of the histones. These observations and the already popular idea that proteins were synthesized on cytoplasmic polyribosomes and presumably had messenger RNA molecules directing their synthesis (135) ultimately led to the discovery of histone messenger RNA in 1967 by BORUN et al. (30). In 1965 ELLIOTT ROBBINS and the author began a series of experiments which demonstrated that histone synthesis was tightly coupled to DNA replication and occurred on small cytoplasmic polyribosomes during S phase of the Hela S-3 cell cycle (30, 233). In those experiments complete and nascent histone polypeptide chains were recognized by their lack of tryptophan (64, 182), molecular weights by electrophoresis in acrylamide gels containing SDS (244, 279) and the apparently absolute dependence of their synthesis on DNA replication during the Hela cell cycle (see Fig. 3). This DNA-dependent histone synthesis in Hela cells seems not to be similar to the cytoplasmic synthesis of histone-like polypeptides in developing grasshopper spermatides which was reported by BLOCH and BRACK in 1964 (25) and which occurs long after the cessation of DNA replication in that system. In 1967 BORUN et al. demonstrated the specific association of rapidly labeled, 7–9S RNA species with histone synthesizing polyribosomes. (We now know that the histone messages have mobilities on gels which suggest that they have S values ranging from about 7 to 12 S). This specific association and the apparently absolute dependence of their association with polyribosomes upon DNA replication during the Hela S-3 cell cycle led us to conclude that the 7–9 S species are the messenger RNA templates of the histones (see Fig. 4). This conclusion was definitively proven to be true in 1972 by JACOBS-LORENA et al. (136), who used a mouse ascites tumor protein-synthesizing system to translate isolated Hela histone messenger RNA preparations *in vitro* into the five main classes of histone polypeptide. Subsequent to these observations, GALLWITZ and MUELLER (84–86) and GALLWITZ and BREINDL (42–87) have independently observed, isolated and translated RNA fractions containing Hela histone messenger RNA. In 1972 SCHOCHETMAN and PERRY (239) outlined the following general properties of mammalian histone messenger RNA.

a) Their molecular weights are approximately 150000 to 220000 (86).

b) Their functional association with polyribosomes *in vivo* is dependent upon DNA replication (1, 30, 86, 149, 239, 289.

c) Their half-life on polyribosomes is short compared with other mRNA species (30, 63, 86, 215).

d) Their processing and transport from nucleus to cytoplasm is more rapid than that of other RNA species (239).

e) They lack the poly (A) sequences at their 3′OH termini which characterize most other mRNA species (1, 2, 95, 239).

At the present time it is not known whether the transcription of histone mRNA is as closely coupled to DNA replication *in vivo* as is its translation. Preliminary evidence suggests the possibility of such a tight transcriptional couple with DNA replication in the Hela S-3 cell on the following grounds (38). Although histone mRNA could be recovered from the supernatant remaining after sedimentation of the ribosomes of S-phase Hela cells, greatly diminished amounts of this RNA were recovered from similar supernatants of S-phase cells treated for 30 min with ara C to inhibit DNA and histone synthesis; and none could be recovered from G_1 supernatants. In Fig. 5A it can be seen that although there are many

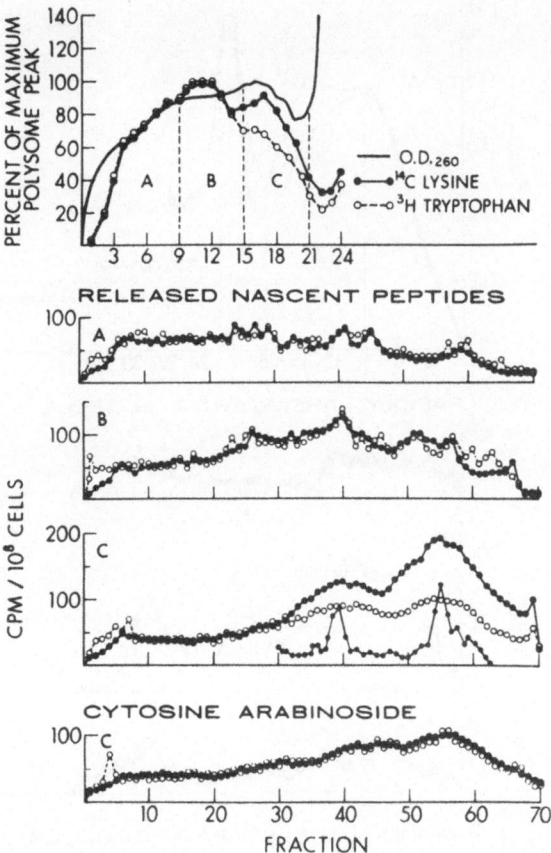

Fig. 3. Top panel: Polyribosome profile (OD_{260}) and incorporation of acid-precipitable H^3-tryptophan and C^{14}-lysine into nascent polypeptides. Synchronized Hela cells were pulse-labeled with these amino acids for 1 min, 2.5 hrs after the beginning of DNA synthesis. Polyribosomes were analyzed on 15–30% sucrose gradients centrifuged at 24000 rpm for 2 hrs. Middle three panels: Gel electropherograms (7.5% acrylamide gels, electrophoresed at 90 volts for 12 hrs) showing the distribution of tryptophan and lysine radioactivity in different sizes of polypeptides released from (A) large, (B) medium, and (C) small polyribosomes of gradient as indicated in top panel. In panel C marker electropherogram of Hela cell nuclear histones, labeled with C^{14}-lysine, has been superimposed for comparison. Bottom panel: Electropherogram of polypeptides from (C) small polyribosomes, isolated from cells pre-treated for 1 hr with cytosine arabinoside (40 µg/ml). [See Ref. (30) for complete details]

different species of putative mRNA in the G_1 supernatant, most of these species appear to contain poly (A) and are absorbed in oligo dT-cellulose columns, Fig. 5 B. Species of RNA in S-phase supernatants similar in electrophoretic mobility to polyribosomal histone mRNA can be seen by comparing Fig. 5 C and 6 C. Resolution of the question of the couple between DNA replication and histone mRNA transcription will probably have to be resolved by the *in vitro* translation of the mRNA in the G_1 supernatant. Such studies are nearly complete and suggest that there is no translatable histone mRNA in the cytoplasm of G_1 cells (38).

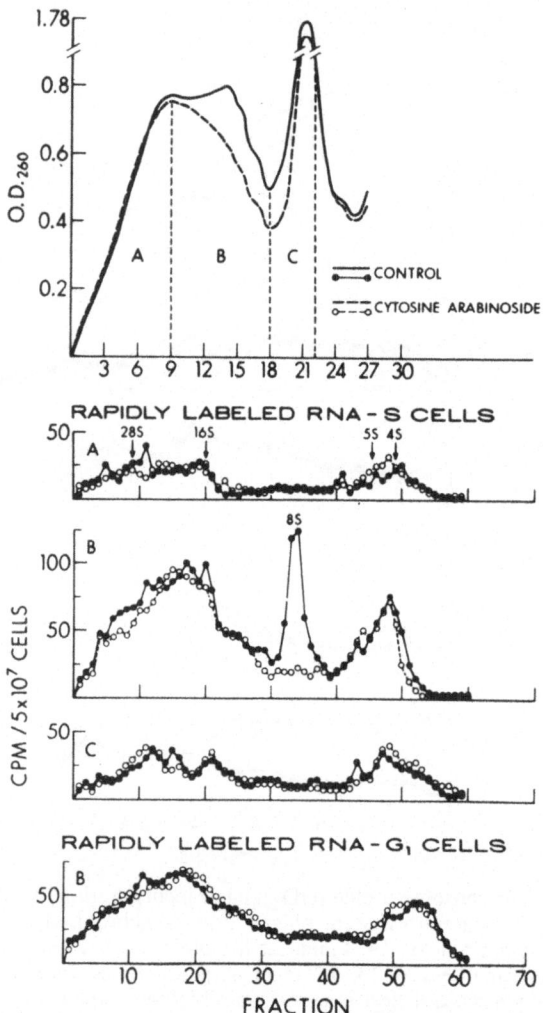

Fig. 4. Top panel: Polyribosome OD$_{260}$ of Hela cells pulsed with H$_3$-uridine for 30 min, 2.5 hrs after the beginning of DNA synthesis. Polyribosomes were analyzed on 7.5–45% sucrose gradients centrifuged at 24000 rpm for 3.25 hrs. Control cells, ———; cells pretreated 1 hr with cytosine arabinoside (40 μg/ml), ---. Middle three panels: Gel electropherograms (2.4% acrylamide gels, electrophoresed at 50 volts for 16 hrs) showing the distribution of H^3-uridine radioactivity in different species of rapidly labeled RNA extracted from the (A) large plus medium-sized polyribosomes, (B) small polyribosomes, and (C) single ribosomes of gradients as indicated in top panel. In each panel cytosine arabinoside (○—○) and control cells (●—●) are compared. Bottom panel: Rapidly labeled RNA species associated with the small polyribosomes of similarly prepared postmitotic (G$_1$) cells. [See Ref. (30) for details]

GREENBERG and PERRY (95) and ADESNIK and DARNELL (1, 2) have reported that histone mRNA species identified by radioisotopic labeling did not contain significant sequences of poly (A) at their 3′OH termini. My colleagues and I have confirmed these observations using amounts of histone mRNA that could be identified by their absorbance at 260 nm (37). The RNA species shown in Fig. 6 have

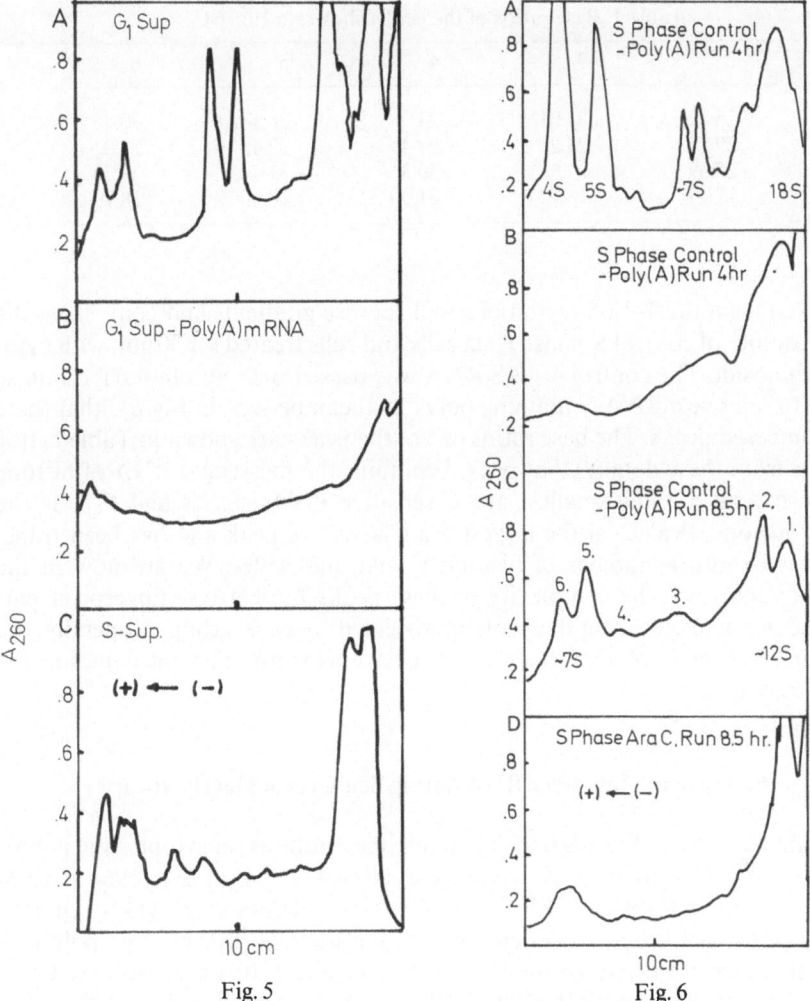

Fig. 5   Fig. 6

Fig. 5. (A) RNA from the 102000 g cytoplasmic supernatant of Hela S-3 cells in G_1 synchronized as described in Ref. (32). The RNA was extracted and run on 6% acrylamide gels
containing 2% SDS and other components as described in Ref. (38). Electrophoresis was for
8 hrs at 250 volts. (B) The same RNA preparation after adsorption of poly A containing RNA
species on oligo dT cellulose columns in 0.1 M KCl as described in Ref. (38) and run on 6%
gels as in 6A. (C) RNA from the 102000 g cytoplasmic supernatant of S-phase Hela cells
synchronized as described in Ref. (32) and resolved in 6% acrylamide gels run for 8 hrs at 250
volts as described in (38)

Fig. 6. (A) Polyribosomal 4-18S RNA prepared from synchronized S-phase Hela S-3 cells as
described in Ref. (37) after absorpiton of poly A containing RNA on oligo dT cellulose in
0.1 M KCl, 0.01 M Tris, pH 7.2. This RNA was run on 6% acrylamide gels in 6 mm I.D.
quartz tubes at 250 volts for 4 hrs as described in (37). (B) The RNA from the preparation in
(A) which contains poly (A) and is absorbed on oligo dT cellulose. These RNA species were
run as in (A). (C) The same RNA as in (A) run for 8.5 hrs at 250 volts. (D) Polyribosomal 4-18S
RNA prepared from synchronized S-phase Hela cells after treatment for 3 hrs in 80 mg/ml ara
C at a concentration of 10^6 cells/ml as described in (37) and run on 6% acrylamide gels at 250
volts for 8.5 hrs

Table 1. Base ratios of the peaks shown in Fig. 6 C

Peak	2	3	4	5	6
C	26.72	31.01	24	26.11	30.45
A	29.79	23.2	23.4	22.48	21.0
G	25.69	27.66	30.36	31.4	29.74
U	17.79	18.12	21.30	20.00	18.79

been taken from the 4–18 S region of zonal sucrose gradients and come from the polyribosomes of control S-phase Hela cells and cells treated for 30 min with cytosine arabinoside. The control 4–18 S RNA was passed over an oligo dT-cellulose column to remove mRNA containing poly (A). It can be seen in Fig. 6 C that there are 6 numbered peaks. The base ratios of 5 of the peaks are shown in Table 1. It is probable from these data (9) that peak 5 contains the messenger RNA of histone F_{2a1} since peak 5 is the smallest ara C-sensitive RNA species and F_{2a1} is the smallest histone. Peak 2 ist the largest ara C-sensitive peak and has been translated *in vitro* into a number of histone F_1-like molecules. We are now in the process of analyzing the complexity of these peaks by RNAase fingerprint patterns and are characterizing the proteins specified by each group. Hopefully, the histone messenger RNA molecules will not have as many different names as the histone fractions.

B. Histone Messenger RNA during Sea Urchin Development

The first evidence of DNA-dependent histone synthesis on cytoplasmic polyribosomes in the sea urchin was a report by Nemer and Lindsay (196) in 1969. Later that year, Kedes and P.R.Gross (146) and Kedes et al. (147) reported further evidence of DNA-dependent histone synthesis on cytoplasmic polyribosomes and described newly synthesized messenger-like 9–10 S RNA species whose association with histone-synthesizing polyribosomes was also dependent upon DNA replication. From these initial observations there has evolved a large literature concerned with the metabolism of histone mRNA during the development of echinoderm embryos. In 1971 Moav and Nemer (187) published nearly conclusive evidence that the histone-like polypeptides found associated with sea urchin polyribosomes were histones. In that year Kedes and Birnstiel began publishing a series of investigations which suggest that the DNA complements of sea urchin histone mRNA are closely clustered on the chromosome and are repeated about 400 to 1200 times (148, 280). Soon after Jacobs-Lorena et al. (136) and Gallwitz and Breindl (42, 87) independently demonstrated the *in vitro* translation of isolated mammalian histone mRNA, K.W.Gross et al. (98) definitively showed that isolated RNA fractions from the sea urchin could be translated *in vitro* into sea urchin histones. Kedes and his co-workers have partially fractionated sea urchin histone mRNA and identified the message for sea urchin histone F_{2a1} (100). Recently, Fraquhar and McCarthy (81) observed that unfractionated histone mRNA could be identified by hybridization techniques in both the unfertilized eggs and in developing embryos of *Strongylocentrotus purpuratus*. These hybridization techniques were extended to *Arbacia punctulata* by SKOULTCHI and P.R.GROSS (252) and recently K.W.GROSS et al. (99) isolated histone mRNA from the post-polyribosomal supernatant of unfertilized *Lytechinus pictus* eggs and translated it *in vitro* into histone. These important observations demonstrate that histone mRNA can be stored in untranslated form in the sea urchin egg. It remains to be determined whether the transcription which leads to the histone mRNA stockpile is coupled to DNA replication. RUDDERMAN et al. (234) recently reported that a new mRNA for histone F_{1g} appears and is translated (241) as sea urchin embryos begin gastrulation. In the following section, we shall see that these observations are consistent with the idea that histone F_1 may play a critical role in controlling the proliferation of different cell types during differentiation and adulthood. In conclusion, the data of this section seems to justify the following observations:

a) Mammalian and sea urchin histone mRNA species appear to have similar sizes and metabolism.

b) The mechanism which couples histone messenger RNA translation to DNA replication in sea urchins and mammals is a complete mystery. It has not been possible to isolate a soluble inhibitor of *in vitro* histone synthesis from the post-polyribosomal cytoplasmic supernatant of cells not making DNA (137, 214, 231). However, BUTLER and MUELLER (51) have observed that the continued translation of Hela histone mRNA after the inhibition of DNA replication is apparently responsible for the abnormally rapid destruction of histone mRNA at such times. If translation was inhibited with cycloheximide, histone mRNA was not degraded in the absence of DNA replication.

c) It is now possible to dismiss completely the objections of the "blotter" theorists. Since the definitive identification of histone mRNA molecules (87, 98, 136) it is possible to look back at our initial description of these RNA species in the Hela S-3 cell (30) and see that by using the criterion of message on the polyribosomes as an index of histone synthesis (Fig. 4) it is evident that DNA replication and histone synthesis are indeed coupled in this and perhaps in all higher eucaryotic somatic cell types.

VII. The Post-Translational Modifications and Their Relationship to the Control of RNA Transcription, Histone mRNA Translation, and Cellular Proliferation

It is now known that the histone polypeptides are probably subject to more post-synthetic modifications than any other group of functionally related proteins. These modifications include:

a) The N-terminal amino-acid acetylations first observed by PHILLIPS in 1963 (216).

b) The ε-N-monomethylation of lysine residues in histones F_{2a1}, F_{2a2}, and F_3 discovered by MURRAY in 1964 (190). PAIK and KIM discovered ε-N-dimethyllysine in 1967 (204) and HEMPEL et al., discovered ε-N-trimethyllysine in 1968 (115).

c) The ε-N-acetylation of internal lysine residues of histones F_{2a1} and F_3 was discovered by ALLFREY and his co-workers in 1968 (91, 276).

d) 3-Methylhistidine was found by GERSHEY et al. in 1959 (92).

e) ω-N-methylarginine was observed in histones by PAIK and KIM in 1970 (205).

f) The earliest investigations of protein phosphorylations were concerned with phosphate storage proteins (BURNETT and KENNEDY, 1954 (44). Early studies of histone serine and threonine phosphorylation were conducted by LANGEN (162), KLEINSMITH et al. (154) and STOCKEN et al. (201–203, 265).

The biological functions of most of these modifications have not yet been determined and have been the subjects of some controversy. While a number of their conclusions have stimulated part of this controversy, the continuing investigations of histone modifications by ALLFREY, MIRSKY and their co-workers have probably been the largest factor responsible for focusing attention on the potential importance of this subject area.

In this section, we shall briefly discuss histone acetylation reactions, review some aspects of histone methylation, and conclude the data portion of this paper with a number of recent observations which suggest that histone F_1 phosphorylation may be related to the control of cellular proliferation during differentiation, malignant transformation processes, and to the somatic cell cycle.

A. Reactions Involved in Histone Acetylation

The histone acetylation literature is complex and often contradictory and is best kept outside the scope of this paper. The interested reader will find an excellent introduction to this topic in HNILICA (122, pp. 79–84).

The *in vivo* study of histone acetylation is complicated by an extraordinary number of intervening variables. These include:

a) Changes in the size of the acetate pool during the course of the experiment.

b) Use of acetate for amino-acid synthesis and acetylation via acetyl CoA.

c) Changes in the acetyl CoA pool during the course of the experiment.

d) Changes in the activity levels of acetylating enzymes.

e) Variations in the availability of histone substrates which may be acetylated both at their N-terminal amino acids and internally.

f) Variations in the conformation of the substrates which may affect acetylation rates.

g) Turnover of different acetate moieties at different rates in the same molecule.

It is thus evident that the interpretation of labeled acetate incorporation data is exceedingly complex. Even if one determines the quantity of each type of acetylated histone species by sensitive methods such as the Triton gel system developed by ZWEIDLER and COHEN (5), it is still very difficult to estimate accurate acetate turnover rates in a given quantitiy of acetylated histone at different times during the course of an experiment. For these reasons, the role of histone acetylation in cellular processes has not yet been determined. ALFRED ZWEIDLER and the author have recently examined the acetylation levels of different histone polypeptides throughout the Hela S-3 cell cycle (293). In this study we determined the relative amounts of the acetylated forms of histones F_{2a1} and F_3 at different times during the cell cycle in control and ara C-treated cells. The relative amounts of all the acetylated histones remain essentially constant throughout the cell cycle in the control and the treated cells. It is thus apparent that in this cell type, at least, profound variations in the rate of DNA, histone, RNA, and total protein synthesis can occur without observable changes in histone F_{2a1} and F_3 acetylation levels. An interesting increase in the proportion of histone molecules that are acetylated with an increase in the time since the last round of histone synthesis in the cell was observed in this study. Similar increases in methylated histone levels have also been observed [see BORUN et al. (33)].

B. Histone Methylation during the Somatic Cell Cycle

Since histone methyl groups have very stable associations with the histone polypeptide chains (33) many of the objections to the study of acetylation do not apply to studies of this kind of modification. There have been a number of studies of histone methylation during the somatic cell cycle in attempts to determine the function of these stable modification reactions which seem to affect all the histones except the F_1 group. In the earlier studies (248, 273) the data appeared to suggest that histone methylation reactions might be causally related to chromosome condensation, since the rate of methylation was highest as the cells approached mitosis. In a more recent study (33) however, BORUN et al. showed that lysine methylation (the predominant form of histone residue methylation) occurred throughout the entire Hela S-3 cell cycle and increased its rate in mid-S to methylate newly synthesized histones. We further noted in that study that histone methyl groups had such low turnover rates that G_1 cells had histones which were essentially as methylated as the histones of the pre-mitotic parent. Since the pre-mitotic chromosomes were condensing and the G_1 chromosomes were unwinding, histone methylation by itself does not appear to have any simple relationship to pre-mitotic chromosome condensation. Since the times of maximal histone lysine methylations (late S and G_2) were not coincident with the times of maximal DNA or RNA synthesis in the Hela cell, we concluded that histone methylations may function as polypeptide finishing reactions which confer some useful three-dimensional property that may be related to the metabolic stability of the histone polypeptides in chromosomes. Because very high percentages (60% to nearly 100%) of the histone molecules other than F_1 in the Hela cell are methylated, it isn't likely that this type of modification is related to highly selective regulatory processes affecting small fractions of the histone-DNA chromosomal complex in its cell type. Finally, we observed (33) but did not then comment upon an interesting correlation of unknown significance between the apparent evolutionary stability of a histone polypeptide sequence and the methylation levels in these proteins. Thus histones F_{2a1} and F_3, which have the highest evolutionary stability, are methylated at the highest rates and to the highest levels. Histones F_{2a2} and F_{2b}, which vary somewhat in evolution, have lower methylation rates but lose these methyl groups at about half the rate of the F_{2a1} and F_3 histones. Histone F_1, which varies considerably in different species and cell types, was not methylated at all. For a more detailed discussion of these and other aspects of histome methylation, the interested reader is directed to other sources [HNILICA (122) pp. 89–90; PAIK and KIM (206); LEE et al. (165); and BORUN et al. (34)].

C. Histone Phosphorylation and the Control of Cell Proliferation

The study of protein phosphorylation was extended to the investigation of histone serine and threonine residue phosphorylation by LANGAN (162), KLEIN-SMITH et al. (154) and the nearly prophetic research of STOCKEN, ORD and their co-workers at Oxford (66, 201–203, 213, 249, 253, 254, 265). Since these initial studies of histone phosphorylation, it has been proposed that these phosphorylations were related to processes which control:
    a) Selective RNA transcription (55, 56, 154, 162, 175, 183–185, 175, 250, 251).
    b) Histone transport from cytoplasm to nucleus (105, 198).
    c) DNA replication (16–18, 66, 109, 145, 170, 184, 198, 201–203, 213, 228, 232, 247, 249–250, 253, 265).
    d) Mitotic processes (41, 108, 109, 111, 145, 157–160).
    e) Polyribosome function (104, 105, 107, 110).
We shall present evidence which suggests that the phosphorylations which affect histone F_1 molecules are different from the phosphorylations affecting other Hela histone molecules. While the function of the other histone phosphorylations is unknown, it appears that histone F_1 phosphorylations may be related to both DNA replication and mitotic chromosome processing in the Hela and other cell types. It further appears that histone F_1 phosphorylation may be part of the reactions which control cellular proliferation.

In 1971, DAWN MARKS, WOON KI PAIK, and the author began studying histone phosphorylation as part of a larger systematic investigation of the relationship of histone modifications to the processes which control the periodic synthesis of DNA and histones during the Hela S-3 cell cycle. Figures 7–10 illustrate some results of that study (176). It can be seen in Fig. 7 that there are at least two different sorts of histone phosphorylation during the Hela cell cycle. Histones F_{2a1} and F_{2a2} (109) have much lower apparent phosphorylation rates than histone F_1. Other experiments, not shown here but discussed in detail elsewhere (36, 176), indicated that the apparent turnover rates of the phosphate groups of histones F_{2a1} and F_{2a2} were about 3 times greater than histone F_1 phosphate group turnover in both G_1 and S phase. A comparison of Figs. 7A and B shows that changes in the apparent phosphorylation rates of histones F_1, F_{2a1}, and F_{2a2} are not obviously coupled to increases in the rate of RNA synthesis during mid-G_1 of the Hela cell cycle. In contrast, histone F_1 phosphorylation rates do parallel increases in DNA replication rates occurring through G_1 into early and mid-S phase. However, it can be seen in Fig. 7B and 8 that this apparent coupling between changes in the rates of DNA replication and histone F_1 phosphorylation is not maintained in late S and G_2, or in cells treated with ara C in S phase. In Fig. 7B it may be seen that DNA replication has declined to about 30% of its maximal rate by 16 hrs after mitosis (G_2), but the apparent rate of histone F_1

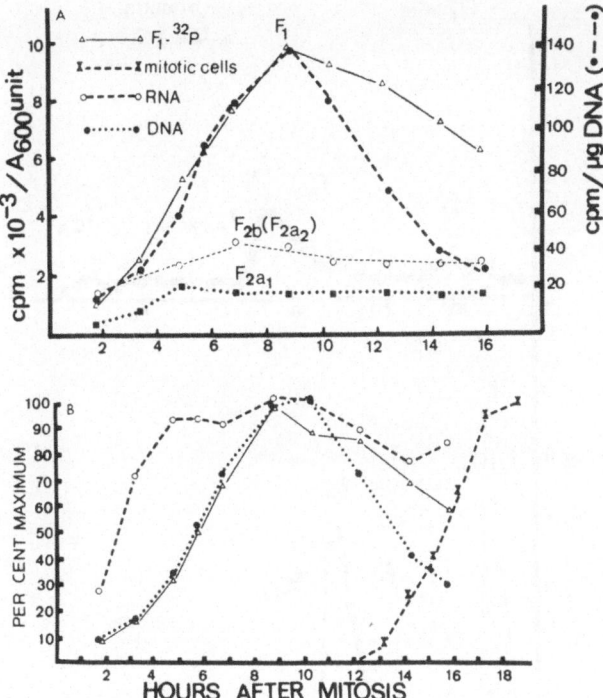

Fig. 7. (A) Specific activity of Hela histone fractions after ^{32}P pulses at various times during the cell cycle. 50-ml aliquots were taken at the indicated times from a culture containing 6.2×10^5 synchronized cells per ml. The aliquot was sedimented, resuspended for 1/2 hr in 10 ml of Medium A containing 1/10 the normal amount of phosphate and 730 μ Ci of ^{32}P and the histones were isolated. The histones (corresponding to 3.4×10^6 cells per gel) were subjected to electrophoresis (6 M urea, 16 hrs, 1 ma per gel). The absorbance at 600 nm of the stained histone bands was determined and the specific activity was calculated (cpm per A_{600} unit) after the amount of radioactivity in gel slices was determined. The specific activity of DNA was determined as described under "Experimental Procedure" (176). △—△ f_1 histone; ○---○ f_{2a2}; ■---■ f_{2a1} and ●---● DNA. (B) The temporal relationship of F_1 histone phosphorylation, nucleic acid synthesis, and per cent of cells accumulating in metaphase after colcamide treatment during the Hela cell cycle. Aliquots of a synchronized culture were pulsed with ^{32}P as described in Fig. 5A. The histones were isolated, subjected to electrophoresis, and the amount of radioactivity in histone F_1 was determined by counting of gel slices. DNA and RNA were isolated from nuclear pellets remaining after histone extraction, and separated as dexscribed under "Experimental Procedure" (176). Aliquots were counted to determine the extent of ^{32}P incorporation, and the results are presented as per cent of maximum incorporation. Colcamide was added to a concentration of 0.05 μg per ml to a part of the culture at 12 hrs after mitosis and the per cent of cells accumulating in metaphase was determined by phase contrast microscopy. X---X, metaphase cells; △—△ f_1 histone; ○·····○ DNA; ○---○ RNA. [See Ref. (176)]

phosphorylation is still at 60% of its maximum value. In Fig. 8 it can be seen that treatment of G_1 cells with ara C has no effect on apparent rates of histone phosphorylation, but ara C treatment in S phase reduces histone F_1 phosphorylation by about 50% without affecting the apparent phosphorylation of histones F_{2a1} or F_{2a2}. At present we do not know the function of histone F_{2a1} or F_{2a2}

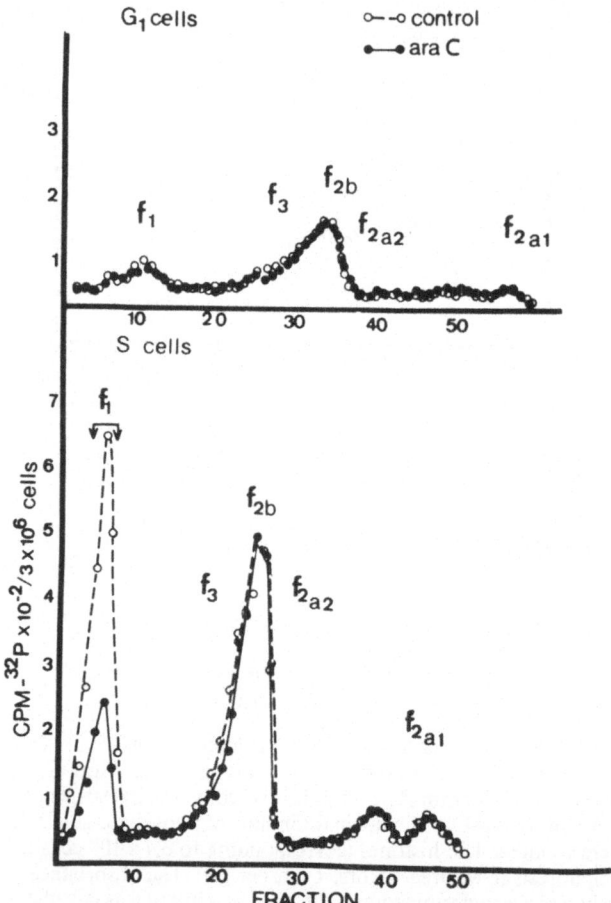

Fig. 8. The effect of cytosine arabinoside on the phosphorylation of various histone fractions during G_1 and S. At 1 (G_1), and 7 hrs (S) after mitosis, cytosine arabinoside (40 µg per ml of cells) was added to aliquots of cells. 1 hr later control and cytosine arabinoside-treated cells were pulsed for 30 min with ^{32}P (50 Å Ci per ml). Histones were then isolated and subjected to electrophoresis (2 M urea, 26 hrs, 190 volts). The amount of radioactivity in each histone fraction was determined by counting gel slices. ○---○ control cells; ●—● cells treated with cytosine arabinoside. [See Ref. (176)]

phosphorylation reactions. The present data suggest that while they have no obvious relationship to changes in RNA transcription they may have some relationship to chromosome replication processes and probably have functions different from histone F_1 phosphorylation reactions.

Just as the interpretation of *in vivo* acetylation experiments is difficult, the interpretation of *in vivo* phosphorylation data based on the incorporation of ^{32}P may be subject to a large number of complicating variables which are very difficult to control. Some of these include:

a) Changes in the phosphate pool size during the course of the experiment.

b) Changes in the amount of histone polypeptide.

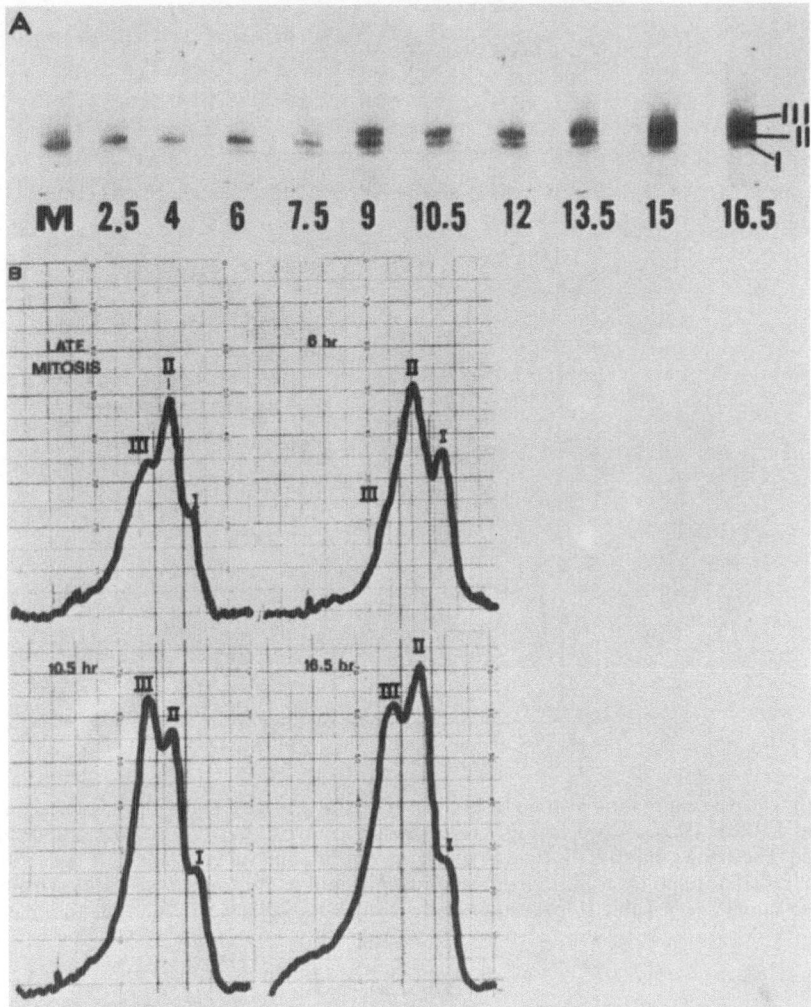

Fig. 9 A and B. The three forms of histone F_1 resolved on high resolution polyacrylamide gels. The stained gels (25 cm in length, containing 2.5 M urea, run at 200 volts for 68.5 hrs at 4°) were aligned, and the F_1 regions were cut out, photographed, and scanned. Forms I, II, and III are indicated as well as the time (in hrs after mitosis, M) at which the histones were isolated. (A) Photographs of the F_1 region. Electrophoresis is from top to bottom. (B) Patterns of the absorbance at 600 nm of the F_1 region of 4 of the gels shown in (A). Electrophoresis is from left to right. [See Ref. (176)]

    c) Conformational changes in the histone substrates.
    d) Variations in the level of histone kinase activity.
    e) Variations in the level of histone phosphatase activity.
    f) Histone phosphate exchange reactions which do not change the amount of a phosphorylated histone derivative but which would increase the specific activity of the derivative during labeling with [32]P.

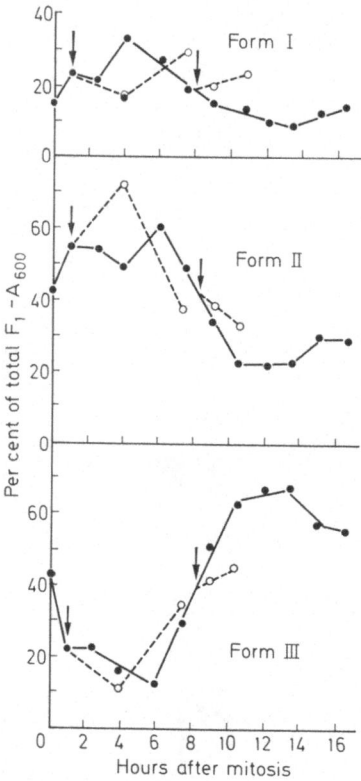

Fig. 10. The per cent of total histone F_1 found during the cell cycle under the areas designated Forms I, II, or III described in Fig. 7 (176) ●—● control; ○---○ cytosine arabinoside–treated. The arrows indicate the time at which cytosine arabinoside was added to a portion of the culture. The inhibitor was present from this time until the histones were isolated from the treated samples [see Table II for further description of the cytosine arabinoside treatment in Ref. (176)]

In 1971 Balhorn et al. published an important method for determining the amount of different phosphorylated histone F_1 molecules by means of electrophoresis of the histones on long gels for about 3 days (15). It is evident that this means of estimating histone F_1 phosphorylation is not subject to the complications of labeling experimentation outlined above, and the results of these determinations reflect the net result of the phosphorylations affecting histone F_1 molecules. We therefore used the technique of Balhorn et al. (15) to examine the amount of different phosphorylated histone F_1 molecules at different points in the mitotic cycle of control and cytosine arabinoside treated cells. In Fig. 9 it is evident that there are three principal forms of histone F_1 in the Hela S-3 cell; Form I may not be phosphorylated, Form II is probably more phosphorylated than Form I and Form III is probably more phosphorylated than Form II (36, 176). All three forms are present in Hela cells just after mitosis, but Form III is rapidly degraded during G_1 to Forms I and II. Form III appears in the nucleus when DNA replication begins in S phase and it increases throughout S and G_2 until just before mitosis

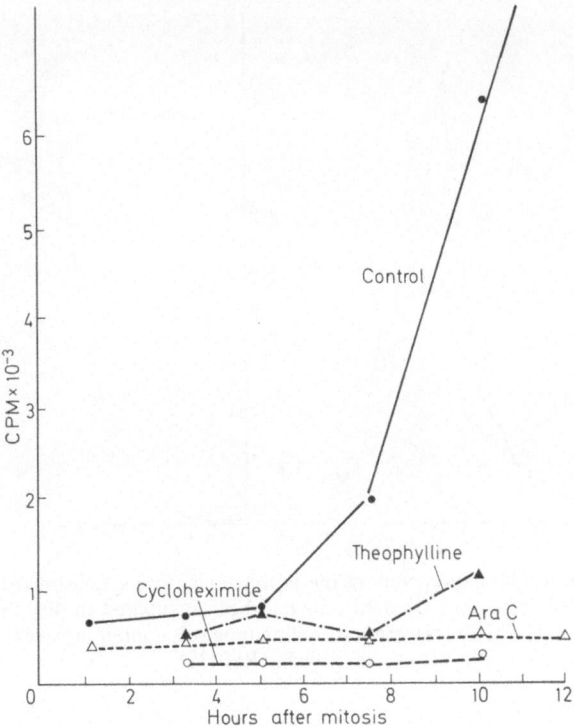

Fig. 11. The rate of ^{14}C thymidine incorporation into control and drug-treated Hela S-3 cells at different points in the cell cycle. Cells were synchronized and treated as described in (35). Cytosine arabinoside (40 µg/ml) was added to one part of the culture at 2 hrs after mitosis and was allowed to remain thereafter. Another part of the culture was harvested and resuspended at the same cell concentration in warm complete medium (35) containing 5 mM theophylline at 2 hrs after mitosis. Another aliquot was treated with 100 µg/ml of cycloheximide at 2 hrs after mitosis

when it comprises about 60% of the total mass of histone F_1 molecules. 30% of the F_1 is in Form II and 10% is in Form I at that time. This histone F_1 phosphorylation cycle begins at the usual time in cells which had been treated with ara C in G_1 to inhibit the initiation of DNA replication (Fig. 10), indicating that previously synthesized histone F_1 molecules can be phosphorylated to the Form II and III levels in the absence of DNA replication. If DNA and histone synthesis are prematurely terminated by ara C treatment in S phase, histone F_1 phosphorylation does continue but at a considerably reduced rate suggesting that newly synthesized histone F_1 molecules are normally phosphorylated as well as the old. Recently, DAWN MARKS and the author have extended these observations as shown in Fig. 11–13 (35). In Fig. 11 it can be seen that various drug treatments in G_1 inhibit the initiation of DNA replication in Hela cells. It is evident in Fig. 12 that while the ara C treatment inhibits the initiation of DNA replication, it does not stop the phosphorylation of previously synthesized histone F_1 molecules. These results indicate that histone F_1 phosphorylation does not require concurrent DNA replication to proceed. However, in Fig. 13 it can be seen that DNA

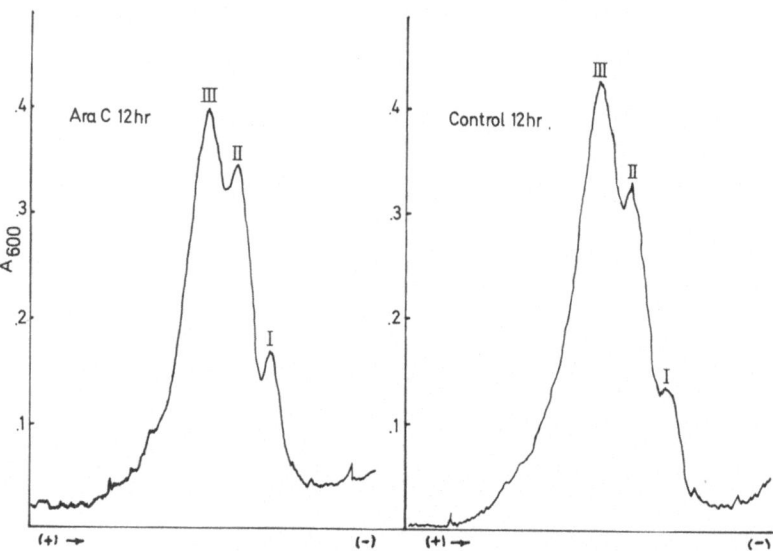

Fig. 12. Gel scans at 600 nanometers of the F_1 histones from synchronized Hela S-3 cells at 12 hrs after mitosis. The ara C-treated cells had been incubated in 40 μg/ml of ara C since 2 hrs after mitosis. The histones were resolved on long gels containing acetic acid and urea, as described in Ref. (35)

replication apparently needs histone F_1 phosphorylation in order to proceed. Treatment of G_1 Hela S-3 cells with 5 mM theophylline leads to an inhibition of histone F_1 phosphorylation and DNA replication in S phase. Theophylline (and here we return to KOSSEL) presumably produces its effect by increasing cellular levels of cyclic AMP (10) through its inhibition of phosphodiesterase which degrades cyclic AMP. Presumably the high cyclic AMP levels apparently prevent histone F_1 phosphorylation to Form III. This may in turn prevent the initiation of DNA replication. Control experiments showed that theophylline at these concentrations did not affect rates of total protein synthesis. Treatment of G_1 Hela S-3 cells with effective concentrations of cycloheximide also prevents the initiation of DNA replication and the phosphorylation of histone F_1 to the putative multiphosphorylated state. These experiments suggest that certain kinds of histone F_1 phosphorylations are necessary but not sufficient causes of the initiation of DNA replication in the Hela cell. The fact that histone F_1 phosphorylation is not tightly coupled with DNA replication in late S, G_2, and mitosis of the Hela cell cycle suggests that histone F_1 phosphates may participate in both DNA replication and chromosome processing in the following fashion:

1. Histone F_1 molecules are phosphorylated prior to the initiation of DNA replication in a given region of the eucaryotic chromosome to allow an opening of the DNA supercoil (3,40, 251). This relaxation of the supercoiling allows access of the replicating enzymes to initiation sites.

2. Once the sister chromatides have been replicated they are separated in reactions which may involve an interaction of phosphorylated histone F_1 molecules with other molecules, which may be proteins (77, 277, 253), poly ADPR (96,

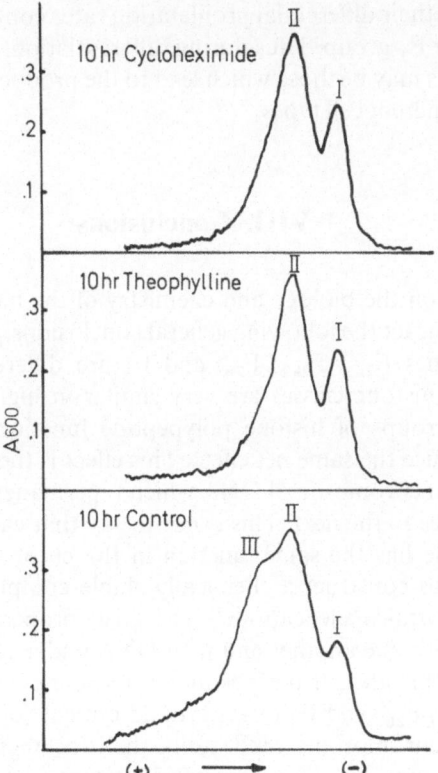

Fig. 13. The F_1 histones from control, 5 mM theophylline and 100 µg/ml cycloheximide-treated synchronized Hela S-3 cells at 10 hrs after mitosis. The theophylline- and cycloheximide-treated cells were prepared as described in Fig. 11 and in detail in Ref. (35). Histone F_1 was resolved on long acrylamide gels containing urea and acetic acid as described in Ref. (35)

232, 254, 287) polynucleotides (96) or unknown chromosomal components. Ultimately, these interactions would lead to separated, replicated, and condensed chromosomes during metaphase.

3. It is evident that control of histone F_1 phosphorylation (see 10, 191) may be one of the principal means by which eucaryotes control cellular proliferation and mitotic chromosome processing. If a cell loses the capacity to inhibit the histone F_1 phosphorylation cycle while retaining the enzymatic capacity to replicate its chromosomes, then chromosome replication and cell division would occur continuously. If that same cell regained the capacity to inhibit the phosphorylation of histone F_1, then it could stop proliferating. It is certainly possible from the preceeding to postulate that pathologies in the feedback mechanisms which integrate control of histone F_1 phosphorylation with cell surface-mediated proliferative stimuli (4, 200) and their cellular effects (10, 191, 228) lead to loss of growth control and malignant transformation (see 36 for a more complete discussion of the preceeding). During development, the different cell types appearing in the

embryo may have their differential proliferation rates controlled by various members of the histone F_1 group. Thus during differentiation, one of the most critical sequences of events may be those which lead to the production of different histone F_1 molecules in different cell types.

VIII. Conclusions

From the data on the biology and chemistry of the histones reviewed here, I believe we may abstract the following general conclusions.

1. While histones F_{2a1}, F_{2a2}, F_{2b}, and F_3 are different from one another, members of these histone classes are very similar in higher eucaryotes because these four main groups of histone polypeptide function together in all these organisms to produce the same net effect. This effect is the nonspecific repression of critical regions of chromosomal DNA which is permanent from one replication of the chromosomes to the next. This is not to say that each of the four kinds of histone polypeptide has the same function in the chromosome, rather they all function together to construct a chemically stable complex with DNA perhaps similar to certain viral nucleocapsids. The great propensity of the histones to form aggregates with one another and with DNA under physiological conditions is consistent with this idea. If the postulated nucleohistone complex containing histones F_{2a1}, F_{2a2}, F_{2b}, and F_3 forms around critical "operator-like" regions of the DNA during chromosome replication, then the regions controlled by the "operator" (which may be a few structural genes or whole batteries of genes) would not be transcribed because RNA polymerase molecules could not attach to the DNA. If the nonspecific repression of the DNA control regions is prevented during chromosome replication by the specific interaction of a non-histone chromosome component with the DNA and the histones near the conrol region, then the genetic components controlled by the critical DNA segment possess at least the potential to become activated as outlined in Zubay's theory in the introduction to this paper. The critical ideas here are that the bulk of histones F_{2a1}, F_{2a2}, F_{2b}, and F_3 form a kind of "fail-safe" repressor structure in complexes with DNA which may interact with non-histone components of the chromosome but do so only during chromosome replication to produce progeny which are different from the pregenitor cells. The dependence of phenotype development upon DNA replication (124, 125, 133) during differentiation suggests that this type of interaction between the histone, non-histone, and critical control regions of the genome can occur *only* during chromosome replication. The lack of much structural variation in these four main histone classes in different eucaryoties, the tight coupling of histone synthesis with DNA replication, the constancy in relative amount of these histones in chromosomes, and the metabolic stability of histones F_{2a1}, F_{2a2}, F_{2b}, and F_3 once they have been incorporated into chromosomes together with the observation of the necessity of cell divisions requiring DNA replication during differentiation (124, 125) are all consistent with this data. Furthermore, the *in vitro* chromatin reconstitution experiments which show that histones will completely repress DNA template capacity unless the chromatin also contains "non-

histone" components which prevent the histone repression and which are responsible for the production of specific RNA transcripts add yet another dimension to the argument in favor of this idea.

If the preceeding is true, what directs the appearance of the appropriate non-histone-specific gene regulators during differentiation? Ultimately this control element may lie outside of the chromosomes in the cortex of the oocyte. There is ample evidence at the present time that the oocyte is highly structured and that cleavage differentially partitions components of the oocyte (22, 156, 161, 227, 236, 255, 268, 282, 283, 285). It is possible that the non-histone components (20), or their messenger RNA species (156), if they are proteins, could be sequestered in the egg cortex and activated at the appropriate times. Alternatively, patterns of macromolecular organization (131) on or near the oocyte surface may by themselves without the help of nucleic acid mediators have some role in the orchestration of developmental operations.

2. At the present time it is clear histone F_1 phosphorylation may play an important role in the control of Hela cell proliferation. Recently, BORUN, MARKS, AUGENBLICH, and BASERGA have extended observations of histone F_1 phosphorylation to the WI-38 cell, stimulated to proliferate by serum. In this system histone F_1 was observed to undergo changes in phosphorylation, that are essentially similar to those in the Hela cell but not identical in every detail. Data of this sort suggest that histone F_1 phosphorylation cycles may be important components in the reaction chains which transduce environmental stimuli to remain quiescent or to proliferate from the cell's surface to the cell's chromosomes. The idea outlined here and developed elsewhere in greater detail (36) suggests that in order for a cell to proliferate, it must not only have all the required replicating enzymes in active form and the sub-units of the chromosomal macromolecules but it must also be able to phosphorylate its histone F_1 molecules to the appropriate levels in order to open up the nucleoprotein supercoils stabilized by unphosphorylated histone F_1 molecules. It seems probable from the data we have reviewed in this paper that different histones F_1 molecules found in different cell types may be required in order to specifically regulate the proliferation of that cell type during differentiation and in the adult. The observation of RUDDERMAN, BAGLIONI, and P. R. GROSS that a new histone F_{1g} mRNA appears and is translated at the transition of the sea urchin from a mass of cells into a tissue forming gastrula are consistent with this idea.

At the present time, the role of histone F_1 phosphorylation in processes regulating RNA transcription is not clear (175, 56). In addition to such a possible role in RNA transcription control processes, interesting studies by GURLEY and his co-workers suggest that histone F_1 and its phosphorylated derivatives may play some role in the processes which control messenger RNA translation by polyribosomes (104, 110).

Clearly, much more work is going to be required before we obtain a complete molecular description of histone function. I hope that this paper has shown that such a description would not only be interesting in itself but may contain key information about the processes which control cellular differentiation, cellular proliferation, and mitotic chromosome processing.

Acknowledgements

The author would like to acknowledge the excellent technical assistance of BARBARA KELLER, SALLY NG, RAE SPENCER, JOSEPH ORGANTINI, ALICE LU, and ERLINDA CABACUNGAN.

Most of the later experiments in this paper were supported by a grant from the National Cancer Institute, N.I.H.N.C.I. Grant # CA11463.

I am indebted to Dr. ROBERT PERRY and Ms. DAWN KELLY for use of their oligo dT cellulose column, Dr. HUGH ROBERTSON for some of the base ratio data, and I have had many conversations with Drs. ALFRED ZWIEDLER, LEONARD COHEN, and WOON KI PAIK.

References

1. ADESNIK, M., DARNELL, J. E.: The biogenesis and characterization of histone mRNA in Hela cells. J. Molec. Biol. **67**, 397–406 (1972).
2. ADESNIK, M., SALDITT, M., THOMAS, W., DARNELL, J. E.: Evidence that all messenger RNA molecules (except histone messenger RNA) contain poly (A) sequences and that poly (A) has a nuclear function. J. Molec. Biol. **71**, 21–30 (1972).
3. ADLER, A. J., LANGAN, T. A., FASMAN, G. D.: Complexes of DNA with lysine-rich (F_1) histone phosphorylated at two seperate sites: circular dichroism studies. Arch. Biochem. Biophys. **153**, 769–777 (1972).
4. ALEXANDER, P., HORNING, E. S.: Observations on the Oppenheimer method of inducing tumors by subcutaneous implantation of plastic films. In: Carcinogenesis: Mechanisms of Action (WOLSTENHOLM and O'CONNER, Eds.), CIBA Foundation Symposium, pp. 12–22. London: Churchill 1959.
5. ALFAGENE, C. R., ZWEIDLER, A., MAHOWALD, A., COHEN, L. H.: Histones of Drosophila Embryos: Electrophoretic Isolation and Structural Studies. J. Biol. Chem. Submitted (1974).
6. ALFRET, M.: Quantitative cytochemical studies on patterns of nuclear growth. In: Fine Structure of Cells, pp. 137–163. New York: Interscience 1955.
7. ALLFREY, V. G., LITTAU, V. C., MIRSKY, A. E.: On the role of histonesin regulating ribonucleic acid synthesis in the cell nucleus. Proc. Natl. Acad. Sci. U.S. **49**, 414–421 (1963).
8. ALLFREY, V. G., FAULKNER, R., MIRSKY, A. E.: Acetylation and methylation of histones and their possible role in the regulation of RNA synthesis. Proc. Natl. Acad. Sci. U.S. **51**, 786–794 (1964).
9. AJIRO, K., ZWEIDLER, A., BORUN, T. W.: The identification of the messenger RNA species for individual human histone polypeptide classes. In preparation (1974).
10. ANDERSON, W. B., RUSSEL, T. R., CARCHMAN, R. A., PASTAN, I.: Interrelationship between Adenylate Cyclase Activity, Adenosine 3′:5′ cyclic Monophosphate Phosphodiesterase Activity, Adenosine 3′:5′ cyclic monophosphate levels and growth of Cells in Culture. Proc. Natl. Acad. Sci. U.S. **70**, 3802–3805.
11. ANTOPOVA, E. N., BOGDANOV, YU. F.: Cytophotometry of DNA and Histone in meiosis of *Pyrrhocoris apterus*. Exp. Cell Res. **60**, 40–44 (1970).
12. APPELS, R., WELL, J. R. E.: Synthesis and turnover of DNA bound-histone during maturation of Avian red blood cells. J. Mol. Biol. **20**, 425–434 (1972).
13. ASAO, T.: Regional histone changes in embryos of the newt *Triturus pyrrhogaster* during early development. Exp. Cell Res. **61**, 255–265 (1970).
14. AXEL, R., CEDAR, H., FELSENFELD, G.: Synthesis of globin RNA from duck reticulocyte chromatin *in vitro*. Proc. Natl. Acad. Sci. U.S. **70**, 2029–2032 (1973).
15. BALHORN, R., RIEKE, W. O., CHALKLEY, R.: Rapid electrophoretic analysis for histone phosphorylation. A reinvestigation of phosphorylation of lysine-rich histone during at liver regeneration. Biochem. **10**, 3952–3959 (1971).

16. BALHORN, R., CHALKLEY, R., GRANNER, D.: Lysine-rich histone: A positive correlation with cell replication. Biochemistry **11**, 1095–1098 (1972).
17. BALHORN, R., BORDWELL, J., SELLERS, L., GRANNER, D., CHALKLEY, R.: Histone phosphorylation and DNA synthesis are linked in synchronous cultures of HTC cells. Biochem. Biophys. Res. Commun. **46**, 1326–1333 (1972).
18. BALHORN, R., BALHORN, M., MORRIS, H. P., CHALKLEY, R.: High resolution electrophoresis of tumor histones: Variation in phosphorylation as a function of cell replication rate. Cancer Res. **32**, 1775–1784 (1972).
19. BALHORN, R., TANPHAICHITR, N., CHALKLEY, R., GRANNER, D.: The effect of inhibition of deoxyribonucleic acid synthesis on histone phosphorylation. Biochemistry **12**, 5146–5150 (1973).
20. BASERGA, R., COSTLOW, M., ROVERA, G.: Changes in membrane function and chromatin template activity in diploid and transformed cells in culture. Federation Proc. a32, 2115–2118 (1973).
21. BASES, R., MENDEZ, F.: Dissociation of histone and DNA synthesis in irradiated Hela S-3 cells. Exptl. Cell Res. **69**, 289–294 (1971).
22. BEISSON, J., SONNEBORN, T. M.: Cytoplasmic inheritance of the organization of the cell cortex in *paramecium aurelia*. Proc. Natl. Acad. Sci. U.S. **53**, 275–282 (1965).
23. BILLET, M. A., HINDLEY, J.: A study of the quantitative variation of the histones and their relationship to RNA synthesis during erythropoisis in the adult chicken. European J. Biochem. **28**, 451–462 (1972).
24. BLOCH, D. P., GODMAN, G. C.: A microspectrophotometric study of the synthesis of desoxyribonucleic acid and nuclear histone. J. Biophys. Biochem. Cytol. **1**, 17–28 (1955).
25. BLOCH, D. P., BRACK, S. D.: Evidence for the cytoplasmic synthesis of nuclear histone during spermigonesis in the grasshopper *Chortophaga vividifasciata* (DE GEER). J. Cell Biol. **22**, 327–340 (1964).
26. BLOCH, D. P., MACQUIGG, R. A., BRACK, S. D., WU, J.-R.: The synthesis of Deoxyribonucleic acid and histone in the onion root meristem. J. Cell Biol. **33**, 451–466 (1967).
27. BLOCH, D. P., TENG, C.: The synthesis of deoxyribonucleic acid and nuclear histone of the X-chromosome of the *Rhenia spinosus* spermatocyte. J. Cell Science **5**, 321–332 (1969).
28. BOGANOV, YU. F., LIAPUNOVA, N. A., SHERUDILO, ANTROPOVA, E. N.: Uncoupling of DNA and histone synthesis prior to Prophase I of Meiosis in the cricket *Grillus(Acheta)Domestica* L. Exp. Cell Res. **52**, 59–70 (1968).
29. BONNER, J., CHALKLEY, G. R., DAHMUS, M., FAMBROUGH, D., KUJIMURA, F., HUANG, R. C. C., HUBERMAN, J., JENSEN, R., MARUSHIGE, K., COLENBUSH, H., OLIVER-A, B. M., WIDHOLM, J.: Isolation and characterization of chromosomal nucleoproteins. In: GROSSMAN, L., MOLDAVE, K. (Eds.): Methods in Enzymology, Vol. 12b, p. 3ff. New York: Academic Press 1968.
30. BORUN, T. W., SCHARFF, M. D., ROBBINS, E.: Rapidly-labeled, polyribosome-associated RNA having the properties of histone messenger. Proc. Natl. Acad. Sci. U.S. **58**, 1977–1983 (1967).
31. BORUN, T. W., BASERGA, R.: Unpublished observations.
32. BORUN, T. W., STEIN, G. S.: The synthesis of acidic chromosomal proteins during the cell cycle of Hela S-3 cells. II. The kinetics of residual protein synthesis and transport. J. Cell Biol. **52**, 308–315 (1972).
33. BORUN, T. W., PEARSON, D. B., PAIK, W. K.: Studies of histone methylation during the Hela S-3 cell cycle. J. Biol. Chem. **247**, 4288–4298 (1972).
34. BORUN, T. W., PAIK, W. L., LEE, H. W., PEARSON, D. B., MARKS, D.: Histone methylation and Phosphorylation during the Hela S-3 cell cycle. In: The Control of Proliferation in Animal Cells. First Cold Spring Harbor Symposium 1973.
35. BORUN, T. W., MARKS, D.: Histone F_1 phosphorylation is a necessary but not sufficient cause of DNA replication in the Hela S-3 cell. (In preparation) (1974).
36. BORUN, T. W., PAIK, W. K., MARKS, D.: Histone phorylation and the control of cellular proliferation. In: MEHLMAN, M. A., HANSEN, R. W. (Eds.): Control Processes in Neoplasia. New York: Academic Press 1974.

37. Borun, T. W., Pan, C. J., Ajiro, K.: The isolation of human histone messenger RNA. (In preparation) (1974).
38. Borun, T. W., Gabrielli, Ajiro, K., Zweidler, A., Baglioni, C.: Studies on the translational control of histone synthesis. III. Translation of histone messenger RNA isolated from polyribosomes and postribosomal supernatante of synchronized Hela S-3 cells. (In preparation) (1974).
39. Borun, T. W.: Unpublished observations.
40. Bradbury, E. M., Molgaard, H. V., Stephens, R. M., Boland, L. A., Johns, E. W.: X-ray studies of nucleoproteins depleted of lysine-rich histone. European J. Biochem. **31** 474–482 (1972).
41. Bradbury, E. M., Inglis, R. J., Matthews, H. R., Sarner, N.: Phosphorylation of very-lysine-rich histone in *Physarum polycephalum*. European J. Biochem. **33**,
42. Breindl, M., Gallwitz, D.: Identification of histone messenger RNA from Hela cells. Appearance of histone m-RNA in the cytoplasm and its translation in a rabbit reticulocyte, cell-free system. European J. Biochem. **32**, 381–391 (1973).
43. Bril-Peterson, R., Westerbrink, H. G. K.: A structural basic protein as a counterpart of deoxyribonucleic acid in mammalian spermatazoa. Biochim. Biophys. Acta **76**, 152–154 (1963).
44. Burnett, G. K., Kennedy, E. P.: The enzymatic phosphorylation of proteins. J. Biol. Chem. **211**, 969–980 (1954).
45. Busch, H.: Histones and other Nuclear Proteins. New York: Academic Press 1965.
46. Bustin, M., Cole, R. D.: Bisection of a lysine-rich histone by N-bromosuccinimide. J. Biol. Chem. **244**, 5291–5299 (1969).
47. Bustos-Valdes, S. E., Deisseroth, A., Dounce, A. L.: Metabolic studies of histones and residual proteins of rat liver nuclei. Arch. Biochem. Biophys. **126**, 848–855 (1968).
48. Butler, J. A. V., Davidson, P. F., James, D. W. F., Shooter, K. V.: The histones of calf thymus deoxyribonucleoprotein. I. Preparation and homogeneity. Biochim. Biophys. Acta **13**, 224–232 (1954).
49. Butler, J. A. V., Cohn, P., Simson, P.: The presence of basic proteins in microsomes. Biochim. Biophys. Acta **38**, 386–388 (1960).
50. Butler, J. A. V., Cohn, P.: Studies on histones. 6. Observations on the biosynthesis of histones and other proteins in regenerating rat liver. Biochem. J. **87**, 330–334 (1963).
51. Butler, W. B., Mueller, G. C.: Control of histone synthesis in Hela cells. Biochim. Biophys. Acta **294**, 481–496 (1973).
52. Byrd, Jr., E. W., Kasinsky, H. E.: Histone synthesis during early embryogenesis *Xenopus laevis* (South African clawed toad). Biochemistry **12**, 246–253 (1973).
53. Byvoet (Bijvoet), P.: Metabolic integrity of deoxyribonucleohistones. J. Molec. Biol. **17**, 311–318 (1966).
54. Byvoet (Bijvoet), P.: Metabolic integrity of deoxyribonucleohistones during enzyme induction. Molec. Pharmacol. **3**, 303–305 (1967).
55. Chae, C. B., Williams, A., Krasny, H., Irvin, J. L., Piantadosi, C.: Inhibition of thymidine phsophorylation and DNA and histone synthesis in Ehrlich ascites tumor cells. Cancer Res. **30**, 2652–2660 (1970).
56. Chae, C. B., Smith, M. C., Irvin, J. L.: Effect of *in vitro* histone phosphorylation on template activity of rat liver chromatin. Biochim. Biophys. Acta **287**, 134–153 (1972).
57. Chalkley, G. R., Maurer, H. R.: Turnover of template-bound histone. Proc. Natl. Acad. Sci. (U.S. **54**, 498–505 (1965).
58. Champagne, M., Pouyet, J., Ouellet, L., Garel, A.: Histones d'erythrocytes de poulets. II. Etude physiochemique de l'histone specific. Bull. Soc. Chem. Biol. **52**, 377– (1970).
59. Cognetti, G., Settineri, D., Spinelli, G.: Developmental changes of chromatin nonhistone proteins in sea urchins. Exp. Cell Res. **71**, 465–467 (1972).
60. Cohen, L. H., Gotchel, B. V.: Histones of polytene and non-polytene nuclei of *Drosophila melanogaster*. J. Biol. Chem. **246**, 1841–1848 (1971).
61. Cook, P. R.: Hypothesis on differentiation and the inheritance of gene superstructure. Nature **245**, 23–25 (1973).
62. Cozeolluella, C., Subirana, J. A.: Somatic histones from *Mytilus edulis*. Biochim. Biophys. Acta **154**, 242–244 (1968).

63. CRAIG, N., KELLY, D. E., PERRY, R. P.: Lifetime of messenger RNAs which code for ribosomal proteins in the L'cell. Biochim. Biophys. Acta 246, 493–498 (1971).
64. CRAMPTON, C. F., MOORE, S., STEIN, W. H.: Chromatographic fractionation of calf thymus histone. J. Biol. Chem. 215, 787–801 (1955).
65. CRAMPTON, C. F., STEIN, W. H., MOORE, S.: Comparative studies on chromatographically purified histones. J. Biol. Chem. 225, 363–385 (1957).
66. CROSS, M. E.: Changes in nuclear protein during the cell cycle in cultured mast cells separater by zonal centrifugation. Biochem. J. 128, 1213–1219 (1972).
67. DAS, N. K., ALFERT, M.: Cytochemical studies on the concurrent synthesis of DNA and histone in primary spermatocytes of Urechis caupo. Exp. Cell Res. 49, 51–59 (1968).
68. DAVIDSON, P. F.: Histones from normal and malignant cells. Biochem. J. 66, 703–707 (1957).
69. DAVIDSON, P. F.: Chromatography of histones. Biochem. J. 66, 708–712 (1957).
70. DAVIS, B.: Disc electrophoresis. II. Methods and application to human serum proteins. Ann. N.Y. Acad. Sci. 121, 404–427 (1964).
71. DELANGE, R. J., FAMBROUGH, D. M., SMITH, E., BONNER, J.: Calf and pea histone IV (f2a1). I. Amino acid compositions of the identical COOH terminal 19 residue sequence. J. Biol. Chem. 243, 5906–5913 (1968).
72. DELANGE, R. J., SMITH, E. L., FAMBROUGH, D. M., BONNER, J.: Amino acid sequence of histone IV (f2a1): Presence of -N-acetyllysine. Proc. Natl. Acad. Sci. U.S. 61, 1145–1152 (1968).
73. DELANGE, R. J., FAMBROUGH, D. M., SMITH, E. L., BONNER, J.: Calf and pea thymus histone IV (f2a1). The complete amino acid sequence of call thymus histone IV (f2a1). Presence of -N-acetyllysine. J. Biol. Chem. 244, 319–334 (1969).
74. DELANGE, R. J., SMITH, E. L., BONNER, J.: Calf thymus histone. III. Sequence of the amino and carboxyl terminal regions containing lysyl residues modified by acetylation and methylation. Biochem. Biophys. Res. Comm. 40, 989–993 (1970). See also HOOPER, J. A., SMITH, E. L., SOMMER, K. R., and CHALKLEY, R.: Amino acid sequence of histone III of the testis of the carp, Leuobus bubalus. J. Biol. Chem. 248, 3275–3279 (1973).
75. DICK, C., JOHNS, E. W.: The biosynthesis of the five main histone fractions of rat thymus. Biochim. Biophys. Acta 174, 380–386 (1969).
76. DIGGLE, J. H., PEACOCK, A. R.: The molecular weights and association of histones of chicken erythrocytes. FEBS Letters 18, 138 (1971).
77. ELGIN, S. C. R., HOOD, L. E.: Chromosomal proteins of Drosophila embryos. Biochemistry 12, 4984–4991 (1973).
78. FAMBROUGH, D. M., BONNER, J.: On the similarity of plant and animal histones. Biochemistry 5, 2563–2570 (1966).
79. FAMBROUGH, D. M., FUJIMURA, F., BONNER, J.: Quantitative distribution of histone components in plants. Biochemistry 7, 575–585 (1968).
80. FAMBROUGH, D. M., BONNER, J.: Sequence homology and the role of cysteine in plant and animal arginine-rich histone (f3). J. Biol. Chem. 243, 4434–4439 (1968).
81. FAQUHAR, M, N., MCCARTHY, B. J.: Histone mRNA in eggs and embryos of Stronglo centrotus purpuratus. Biochem. Biophys. Res. Commun. 53, 515–522 (1973).
82. FLAMM, W. G., BIRNSTIEL, M. L.: Inhibition of DNA replication and its effect on histone synthesis. Exptl. Cell Res. 33, 658–660 (1964).
83. GALL, J.: Macronuclear duplication in the ciliated protozoan Euplotes. J. Biophys. Biochem. Cytol. 5, 295–308 (1959).
84. GALLWITZ, D., MUELLER, G. C.: Histone synthesis in vitro by cytoplasmic microsomes from Hela cells. Science 163, 1351–1353 (1969).
85. GALLWITZ, D., MUELLER, G. C.: Histone synthesis in vitro on Hela cell microsomes. The nature of the coupling to deoxyribonucleic acid synthesis. J. Biol. Chem. 244, 5947–5952 (1969).
86. GALLWITZ, D., MUELLER, G. C.: RNA from Hela cell microsomes with properties of histone messenger. FEBS Letters 6, 83–85 (1970).
87. GALLWITZ, D., BREINDL, M.: Synthesis of histones in a rabbit reticulocyte cell-free system directed by a polyribosomal RNA fraction from synchronized Hela cells. Biochem. Biophys. Res. Commun. 47, 1106–1111 (1972).

88. GAUTSCHI, J. R., KERN, R. M.: DNA replication in mammalian cells in the presence of cycloheximide. Exp. Cell Res. **80**, 15–26 (1973).

89. GAVOSTO, T., PEGORARO, L., MASERA, P., ROVERA, G.: Late DNA replication pattern in human haemopoietic cells. A comparitive investigation using high resolution quantitative autoradiography. Exp. Cell **49**, 340–358 (1968).

90. GERNER, E. W., HUMPHREY, R. M.: The cell cycle phase synthesis of non-histone proteins in mammalian cells. Biochim. Biophys. Acta **331**, 117–127 (1973).

91. GERSHEY, E. L., VIDALI, G., ALLFREY, V. G.: Chemical studies of histone acetylation. The occurance of -N-acetyllysine in the f   histone. J. Biol. Chem. **243**, 5018–5027 (1968).

92. GERSHEY, E. L., HASLETT, G. W., VIDALI, G., ALLFREY, V. C.: Chemical studies of histone methylation. Evidence for occurance of 3-methylhistidine in avian erythrocytes histones. J. Biol. Chem. **244**, 4871–4877 (1969).

93. GILMOUR, R. S., PAUL, J.: RNA transcribed from reconstructed nucleoprotein is similar to natural RNA. J. Molec. Biol. **40**, 137–142 (1969).

94. GILMOUR, R. S., PAUL, J.: Tissue specific transcription of the globin gene in isolated chromatin. Proc. Natl. Acad. Sci. U.S. **70**, 3440–3442 (1973).

95. GREENBERG, J. R., PERRY, R. P.: Relative occurance polyadenylic acid sequences in messenger RNA and heterogeneous nuclear RNA of L cells as determined by poly(U) hydroxylapatite chromatography. J. Molec. Biol. **72**, 91–98 (1972).

96. GREENWAY, P. J., MURRAY, K.: Polynucleotides associated with histone preparations. Biochem. J. **131**, 585–591 (1973).

97. GRIFFIN, A. C., WARD, V. C., WADE, J., WARD, D. N.: Synthesis and release of basic proteins or peptides by tumor ribosomes. Biochim. Biophys. Acta **72**, 500–503 (1963).

98. GROSS, K. W., RUDDERMAN, J., JACOBS-LORENA, M., BAGLIONI, C., GROSS, P. R.: Cellfree synthesis of histones by messenger RNA from sea urchin embryos. Nature **241**, 272–274 (1973).

99. GROSS, K. W., JACOBS-LORENA, M., BAGLIONI, C., GROSS, P. R.: Cell-free translation of maternal messenger RNA from sea urchin. Proc. Natl. Acad. Sci. U.S. **70**, 2614–2618 (1973).

100. GRUNSTEIN, M., LEVY, S., SCHEDL, P., KEDES, L.: Messenger RNA for individual histone proteins: fingerprint analysis and *in vitro* translation. In: The organization of genetic material in eucaryotex. 38th Cold Spring Harbor Symposium on Quantitative Biology.

101. GURLEY, L. R., HARDIN, J. M.: The metabolism of histone fractions. I. Synthesis of histone fractions during the life cycle of mammalian cells. Arch. Biochem. Biophys.

102. GURLEY, L. R., HARDIN, J. M.: The metabolism of histone fractions. II. Conservation and turnover of histone fractions in mammalian cells. Arch. Biochem. Biophys.

103. GURLEY, L. R., HARDIN, J. M.: The metabolism of histone fractions. III. Synthesis and turnover of histone F_1. Arch. Biochem. Biophys. **136**, 392–401.

104. GURLEY, L. R., WALTERS, R. A., ENGER, M. D.: Isolation and characterization of f_1 on ribosomes. Biochem. Biophys. Res. Commun. **40**, 428–436 (1970).

105. GURLEY, L. R., WALTERS, R. A.: Response of histone turnover and phosphorylation to X-irradiation. Biochemistry **10**, 1588–1593 (19   ).

106. GURLEY, L. R., WALTERS, R. A., TOBEY, R. A.: The metabolism of histone fractions. IV. Synthesis of histones during the G_1 phase of the mammalian life cycle. Arch. Biochem. Biophys. **148**, 633–641 (1972).

107. GURLEY, L. R., WALTERS, R. A.: The metabolism of histone fractions. V. Relationship between histone and DNA synthesis after X-irradiation. Arch. Biochem. Biophys. **153**, 304–311 (1972).

108. GURLEY, L. R., WALTERS, R. A., TOBEY, R. A.: Histone phosphorylation in late interphase and mitosis. Biochem. Biophys. Res. Commun. **50**, 744–750 (1973).

109. GURLEY, L. R., WALTERS, R. A., TOBEY, R. A.: The metabolism of histone fractions. VI. Differences in the phosphorylation of histone fractions during the cell cycle. Arch. Biochem. Biophys. **154**, 212–218 (1973).

110. GURLEY, L. R., ENGER, M. D., WALTERS, R. A.: The nature of histone F_1 isolated from polysomes. Biochemistry **12**, 237–245 (1973).

111. GURLEY, L.R., WALTERS, R.A., TOBEY, R.A.: Cell cycle-specific changes in histone phosphorylation associated with cell proliferation and chromosomes condensation. J. Cell Biol. (In press) (1974).
112. HANCOCK, R.: Conservation of histones in chromatin during growth and mitosis *in vitro*. J. Molec. Biol. **40**, 457–466 (1969).
113. HARDIN, J.A., EINEM, G.E., LINDSAY, D.T.: Simultaneous synthesis of histone and DNA in synchronously dividing *Tetrahymena pyriformis*. J. Cell Biol. **33** 709–717 (1967).
114. HARRIS, M.: The initiation of deoxyribonucleic acid synthesis in the connective tissue cell with some observations on the function of the nucleolus. Biochemistry J. **72**, 54–60 (1959).
115. HEMPEL, K., LANGE, H.W., BIRKOFER, L.: ε-N-Trimethyllysin, eine neue Aminosäure in Histonen. Naturwissenschaften **55**, 37 (1968).
116. HENRICKS, D.M., MAYER, D.T.: Isolation and characterization of a basic keritinlike protein from mammalian spermatozoa. Exptl. Cell Res. **40**, 402–410 (1965).
117. HEREFORD, L.M., HARTWELL, L.H.: Role of protein synthesis in the replication of yeast DNA. Nature **244**, 129–131 (1973).
118. HIGHFIELD, D.P., DEWEY, W.C.: Inhibition of DNA synthesis in synchronized chinese hamster cells treated in G_1 or early S phase with cycloheximide or puromycin. Exptl. Cell Res. **75**, 314–320 (1972).
119. HNILICA, L.S., JOHNS, E.W., BUTLER, J.A.V.: Observations on the species and tissue specificity of histones. Biochem. J. **82**, 123–129 (1962).
120. HNILICA, L.S.: The specificity of histones in chicken erythrocytes. Experientia **20**, 13–14 (1964).
121. HNILICA, L.S., BESS, C.G.: The heterogeniety of arginine-rich histone. Analyt. Biochem. **12**, 421–436 (1965).
122. HNILICA, L.S.: The structure and biological function of the histones. Cleveland: C.R.C. Press 1972.
123. HOLBROOK, D.J., EVANS, J.H., IRVIN, J.L.: Incorporation of labeled precursors into proteins and nucleic acids in regenerating rat liver. Exptl. Cell Res. **28**, 120 (1962).
124. HOLTZER, H.: Comments on induction during cell differentiation. In: Induktion und Morphogenese. 13th Colloquinum der Gesellschaft für Physiologische Chemie. Berlin-Göttingen-Heidelberg: Springer 1962.
125. HOLTZER, H., MAYNE, R., WEINTRAUB, H., CAMPBELL, G.: Obligatory requirement for DNA synthesis during myogenesis, erythrogenesis and chondrogenesis. In: Biochemistry of Gene Expression in Higher Organisms (POLLACK and LEE, Eds.). Australia and New Zealand Book Co. 287–304 (1973).
126. HODGE, L.D., BORUN, T.W., ROBBINS, E., SCHARFF, M.D.: Studies on the regulation of DNA and proteins in synchronized Hela cells. In: Biochemistry of the Cell Cycle (BASERGA, R., Ed.), Chap. II, p. 15ff. Springfield, Ill.: Thomas 1969.
127. HORI, T-A., LARK, K.G.: Effect of puromycin on DNA replication in Chinese hamster cells. J. Molec. Biol. **77**, 391–404 (1973).
128. HOWARD, A., PELC, S.R.: Synthesis of desoxyribonucleic acid in normal and irradiated cells and its relationship to chromosome breakage. Heredity London (Suppl.) **6**, 261–273 (1953).
129. HUANG, R.C.C., BONNER, J.: Histone, a suppressor of chromosomal RNA synthesis. Proc. Natl. Acad. Sci. U.S. **48**, 1216–1222 (1962).
130. HUBERMAN, J.A., RIGGS, A.D.: On the mechanism of DNA replication in mammalian chromosomes. J. Molec. Biol. **32**, 327–341 (1968).
131. HUST, R.K., JACOBSON, M.: Specification of positional information in retinal ganglion cells of *Xenopus*: assays for analysis of the unspecified state. Proc. Natl. Acad. Sci. U.S. **70**, 507–511 (1973).
132. HYODO, M., FLICKINGER, R.A.: Replicon growth rates during DNA replication in developing frog embryos. Biochim. Biophys. Acta **299**, 29–33 (1973).
133. ISHIKAWA, H., BISHOFF, R., HOLTZER, H.: Mitosis and intermediate sized filiments in developing skeletal muscle. J. Cell Biol. **38**, 538–555 (1968).
134. IWAI, K., ISHIKAWA, K., HAYASHI, H.: Amino acid sequence of slightly lysine rich histone (f2b). Nature **226**, 1056–1058 (1970).

135. JACOB, F., MANOD, J.: Genetic regulatory mechanisms in the synthesis of proteins. J. Molec. Biol. 3, 318–356 (1961).

136. JACOBS-LORENA, M., BAGLIONI, C., BORUN, T.W.: The translation of messenger for histones from Hela cells by a cell-free extract from ouse ascites tumor. Proc. Natl. Acad. Sci. U.S. 69, 2095–2099 (1972).

137. JACOBS-LORENA, M., GABRIELLI, F., BORUN, T.W., BAGLIONI, C.: Studies on the translation of histone messenger RNA by heterologous cell-free systems prepared from cells inactive in DNA synthesis. Biochim. Biophys. Acta 324, 275–281 (1973).

138. JOHNS, E.W., PHILLIPS, D.M.P., SIMSON, P., BUTLER, J.A.V.: Improved fractionation of arginine rich histones from calf thymus. Biochem. J. 77, 631–636 (1960).

139. JOHNS, E.W., BUTLER, J.A.V.: Further fractionation of histones from calf thymus. Biochem. J. 82, 15–18 (1962).

140. JOHNS, E.W.: Studies on histones. 7. Preparative methods for histone fractions from calf thymus. Biochem. J. 92, 55–59 (1964).

141. JOHNS, E.W.: Studies on histones. 8. A degradation product of lysine-rich histone. Biochem. J. 93, 161–163 (1964).

142. JOHNS, E.W.: The electrophoresis of histones in polyacrylamide gels and their quantitation. Biochem. J. 104, 78–82 (1967).

143. JOHNS, E.W., FORRESTER, S.: Studies of nucleoproteins. The binding of extra acidic proteins to DNA during preparation of nuclear proteins. European J. Biochem. 8, 547–551 (1969).

144. JONES, R.B., IRVIN, J.L.: Effect of hydrocortisone on the synthesis of DNA and histones and the acetylation of histones in regenerating liver. Arch. Biochem. Biophys. 152, 828–838 (1972).

145. KEEVERT, J.B., GOROVSKY, M.A.: Studies on histone f_1 in *Tetrahymena* nuclei. 13th Annual Meeting, American Society for Cell Biology, Abstract 233 (1973).

146. KEDES, L.H., GROSS, P.R.: Identification in cleavage embryos of the three RNA species serving as templates for the synthesis of nuclear proteins. Nature 223, 1335–1339 (1969).

147. KEDES, L.H., GROSS, P.R., COGNETTI, G., HUNTER, A.L.: Synthesis of nuclear ad chromosomal proteins on light polyribosomes during cleavage in the sea urchin embryo. J. Molec. Biol. 45, 337–351 (1969).

148. KEDES, L.H., BIRNSTIEL, M.L.: Reiteration and clustering of DNA sequences complementary to histone messenger RNA. Nature 230, 165–169 (1971).

149. KEMPER, B., RICH, A.: Identification of presumptive histone messenger RNA in rapidly labeled polysomal RNA of mouse myelomas. Biochim. Biophys. Acta 319, 364–372 (1973).

150. KINKADE, J.M., COLE, R.D.: The resolution of four lysine-rich histones derived from calf thymus. J. Biol. Chem. 241, 5790–5797 (1966).

151. KINKADE, J.M., COLE, R.D.: A structural comparison of different lysine-rich histones of calf thymus. J. Biol. Chem. 241, 5798–5805 (1966).

152. KINKADE, J.M.: Qualitative species differences and quantitative tissue differences in the distribution of lysine-rich histones. J. Biol. Chem. 244, 3375–3386 (1969).

153. KINKADE, J.M.: Differences in the quantitative distribution of lysine-rich histones in neoplastic and normal tissues. Proc. Soc. Exptl. Biol. Med. 137, 1131–1134 (1971).

154. KLEINSMITH, L.J., ALLFREY, V.G., MIRSKY, A.E.: Phosphoprotein metabolism in isolated lymphocyte nuclei. Proc. Natl. Acad. Sci. U.S. 55, 1182–1189 (1966).

155. KOSSEL, A.L.K.M.L.: Über einen peptonartigen Bestandteil des Zellkerns. Z. Physiol. Chem. 8, 511 (1884).

156. KORINEK, J., SPELSBERG, T.C., MITCHELL, W.M.: mRNA transcription linked to the morphological and plasma membrane changes induced by cyclic AMP in tumor cells. Nature 246, 455–458 (1973).

157. LAKE, R.S., GOIDL, J.A., SALZMAN, N.P.: F_1 histone modification at metaphase in Chinese hamster cells. Exptl. Cell Res. 73, 113–121 (1972).

158. LAKE, R.S., SALZMAN, N.P.: Occurance and properties of a chromatin associated F_1 histone phosphokinase in mitotic Chinese hamster cells. Biochemistry 11, 4817–4826 (1972).

159. LAKE,R.S.: F₁ histone phosphorylation in metaphase chromosomes of cultured Chinese hamster cels. Nature **24**, 145–146 (1973).
160. LAKE,R.S.: Further characterization of histone F₁ phosphokinase of metaphase arrested animal cells. J. Cel. Biol. **58**, 317–331 (1973).
161. LALLIER,R.: Effects of concanavalin A on the development of sea urchin egg. Exptl. Cell Res. **72**, 157–163 (1972).
162. LANGAN,T.: Histone phosphorylation: stimulation by adenosine 3′:5′ monophosphate. Science **162**, 579–580 (1968).
163. LAWRENCE,D.J.R., BUTLER,J.A.V.: The metabolism of histones in malignant tissues and liver of rat and mouse. Biochem. J. **96**, 53–62 (1965).
164. LEE,Y.C., SCHERBAUM,O.H.: Nucleohistone composition in stationary and division synchronized *Tetrahymna* cultures. Biochemistry **5**, 2067–2075 (1966).
165. LEE,H.W., PAIK,W.K., BORUN,T.W.: The periodic synthesis of S-adenosylmethionine: protein methyltransferases during the Hela S-3 cell cycle. J. Biol. Chem. **248**, 4149–4199 (1973).
166. LESTOURGEON,W.M., RUSCH,H.P.: Nuclear acidic protein changes during differentiation in *Physarum polycephalum*. Science **174**, 1233–1236 (1971).
167. LESTOURGEON,W.M., RUSCH,H.P.: Localization of nucleolar ad chromatin residual acid protein changes during differentiatio in *Physarum polycephalum*. Arch. Biochem. Biophys. **155**, 144–158 (1973).
168. LINDSAY,D.T.: Histnes from developing tissues in the chicken: heterogeniety. Science **144**, 420–422 (1964).
169. LINDSAY,D.T.: Electrophoretically identical histones from the ribosomes and chromosomes of chicken liver. Arch. Biochem. Biophys. **113**, 68–694 (1966).
170. LOUIE,A.J., SUNG,M.T., DIXON,G.A.: Modification of histones during spermiogeesis in trout. III. Levels of phosphorylated species and kinetics of phosphorylation of histone. IIb. J. Biol. Chem. **248**, 3335–3340 (1973).
171. LUCK,J.M., COOK,H.A., ELDRIDGE,N.T., HALEY,M.I., KUPKE,D.W., RASSMUSSEN,P.S.: on the fractionation of calf thymus histones. Arch. Biochem. Biophys. **65**, 449–467 (1956).
172. LUCK,J.M., RASSMUSSEN,P.S., SATAKE,K., TSVETIKOV,A.N.: Further studies on the fractionation of calf thymus histone. J. Biol. Chem. **233** 1407–1414 (1958).
173. LYNCH,W.E., BROWN,R.F., UMEDA,T., LANGRETH,S., LIEBERMAN,I.: Synthesis of deoxyribonucleic acid by isolated nuclei. J. Biol. Chem. **245**, 3911–3916 (1970). .
174. MAIZEL,J.V. Acrylamide gel electrophorograms by mechanical fractionation: radioactive adenovirus proteins. Science **151**, 990–998 (1967).
175. MALLETTI,L.A., NEBLETT,M., EXTON,J.A., LANGAN,T.: Phosphorylation of lysine-rich histone in isoated perfused liver. J. Biol. Chem. **248**, 6289–6291 (1973).
176. MARKS,D., PAIK,W.K., BORUN,T.W.: The relationship of histone phosphorylation to deoxyribonucleic acid replication and mitosis during the Hela S-3 cell cycle. J. Biol. Chem. **248**, 5660–5667 (1973).
177. MARZLUFF,W.F.; McCARTY,K.S., TURKINGTON,R.W.: Insulin dependent synthesis of histone in relation to the mammalian epithelial cell cycle. Biochim. iophys. Acta **190**, 517–526 (1969).
178. MAURITZEN,C.wM., STARBUCK,W.C., SAROJAL,S., TAYLOR,C.W., BUSCH,H.: The fractionation of arginine-rich histones from fetal calf thymus by exclusion chromatography. J. Biol. Chem. **242**, 2240–2245 (1967).
179. McALLISTER,H.C.,JR., WAN,Y.C., IRVIN,J.L.: Electrophoresis of histones and histone fractions on polyacrylamide gels. Analyt. Biochem. **5**, 321–329 (1963).
180. McLEISH,J.: Comparative microphotometric studies of DNA and arginine in plant nuclei. Chromosoma **10**, 686γ10 (1959).
181. MACGILLIVARY,A.J.: The histones of some human tissues. Biochem. J. **110**, 181–185 (1968).
182. MIRSKY,A.E., POLLISTER,A.H.: Chromosin, a desoxyribonucleoprotein comples of the cell nucleus. J. Gen. Physiol. **30**, 117–197 (p. 129) (1947).
183. MIRSKY,A.E.: The structure of chromatin. Proc. Natl. Acad. Sci. U.S. **68**, 2945–2948 (1971).

184. MIRSKY, A. E., SILVERMA, N. B., PANDA, N.: Blocking histones of the accesibility to DNA in chromatin: addition of histones. Proc. Natl. Acad. Sci. U.S. **69**, 3243–3246 (1972).
185. MIRSKY, A. E., SILVERMAN, B.: The effect of selective extraction of histones on template activities of chromatin by use of exogenious DNA and RNA polymerases. Proc. Natl. Acad. Sci. U.S. **70**, 1973–1975 (1973).
186. MIRSKY, A. E.: Personal communication.
187. MOAV, B., NEMBER, M.: Histone synthesis. Assignment to a special class of polyribosomes in sea urchin embryos. Biochemistry **10**, 881–888 (1971).
188. MOHBERG, J., RUSCH, H. P.: Nuclear histones in *Physarum polycephalum* during growth and differentiation. Arch. Biochem. Biophys. **138**, 418–432 (1970).
189. MUELLER, G. C., KAJAWARA, K., STUBBLEFIELD, E., RUECKERT, R. P.: Molecular events in the reproduction of animal cells. I. The effect of puromycin on the duplication of DNA. Cancer Res. **22**, 1084–1090 (1962).
190. MURRAY, K.: The occurance of epsilon N-methyl lysine in histones. Biochemistry **3**, 10–15 (1964).
191. MURRAY, A. W., FROSCIO, M., KEMP, B. E.: Histone phosphotases and cyclic nucleotide stimulated protein kinase from human lymphocytes. Biochem. J. **129**, 995–1002 (1972).
192. NEELIN, J. M., NEELIN, E. M.: Zone electrophoresis of calf thymus histones in starch gel. Canad. J. Biochem. Physiol. **38**, 355–363 (1960).
193. NEELIN, J. M., BUTLER, G. C.: A composition of histones from chicken tissues by zone electrophoresis in starch gel. Canad. J. Biochem. Physiol. **39**, 485–491 (1961).
194. CALLAHAN, P. X., LAMB, D. C., MURRAY, K.: The histones of chicken erythrocyte nuclei. Canad. J. Biochem. Physiol. **42**, 1743–1752.
195. NEIDLE, A., WAELSCH, H.: Histones: species and tissue specificity. Science **145**, 420–422 (1964).
196. NEMER, M., LINDSAY, D. T.: Evidence that S phase polysomes of early sea urchin embryos may be responsible for the synthesis of chromosomal proteins. **35**, 156–160 (1969).
197. OLIVER, D. R., CHALKLEY, R.: An electrophoretic analysis of *Drosophila* histones. Exptl. Cell Res. **73**, 303–310 (1972).
198. OLIVER, D. R., BALHORN, R., GRANNER, D., CHALKLEY, R.: Molecular nature of F_1 histone phosphorylation in cultured hepatoma cells. Biochemistry **11**, 3921–3915 (1972).
199. ONTKO, J. A., MOOREHEAD, W. R.: Histone synthesis after inhibition of deoxyribonucleic acid replication. Biochim. Biophys. Acta **91**, 658–660 (1964).
200. OPPENHEIMER, B. S., OPPENHEIMER, E. T., DANISHEFSKY, I., STOUT, E. P., EIRICH, F. R.: Further studies of polymers as carcinogenic agents in animals. Cancer Res. **15**, 333–340 (1955).
201. ORD, M. G., STOCKEN, L. A.: Metabolic properties of histones from rat liver and thymus gland. Biochem. J. **98**, 888–897 (1966).
202. ORD, M. G., STOCKEN, L. A.: Variations in the phosphate content and the thiol/disulfide ratio of histones during the cell cycle. Studies with regenerating rat liver and sea urchins. Biochem. J. **107**, 403–410 (1968).
203. ORD, M. G., STOCKEN, L. A.: Further studies on the phosphorylation and the thiol/disulfide ratio of histones in growth and development. Biochem. J. **112**, 81–89 (1969).
204. PAIK, W. K., KIM, S.: Epsilon-N-dimethyllysine in histones. Biochem. Biophys. Res. Comm. **27**, 479–483 (1967).
205. PAIK, W. K., KIM, S.: Omega -n-methylarginine in histones. Biochem. Biophys. Res. Comm. **40**, 224–229 (1970).
206. PAIK, W. K., KIM, S.: Protein methylation. Science **174**, 114–119 (1971).
207. PALAU, J., BUTLER, J. A. V.: Trout liver histones. Biochem. J. **100**, 779–783 (1966).
208. PAINTER, R. B., SCHAEFER, A. W.: The rate of synthesis along replicons of different kinds of mammalian cells. J. Molec. Biol. **45**, 467–479 (1969).
209. PANYIM, S., CHALKLEY, R.: High resolution acrylamide gel electrophoresis of histones. Arch. Biochem. Biophys. **130**, 337–346 (1969).
210. PANYIM, S., CHALKLEY, R., SPIKER, S., OLIVER, D.: Constant electrophoretic mobility of the cysteine containing histone in plants and animals. Biochim. Biophys. Acta **214**, 216–221 (1970).

211. PANYIM, S., BILEK, D., CHALKLEY, R.: An electrophoretic comparison of vertebrate histones. J. Biol. Chem. **246**, 4206–4215 (1971).

212. PAUL, J., GILMOUR, R.S.: Organ-specific restriction of transcription in mammalian chromatin. J. Molec. Biol. **34**, 305–316 (1968).

213. PAWSE, A.R., ORD, M.G., STOCKEN, L.A.: Histone kinase and cell division. Biochem. J. **122**, 713–719 (1971).

214. PEDERSON, T., ROBBINS, E.: Absence of translational control of histone synthesis during the Hela cell cycle. J. Cell Biol. **45**, 509–513 (1970).

215. PERRY, R.P., KELLY, D.: Messenger RNA turnover in mouse L cells. J. Molec. Biol. **79**, 681–696 (1973).

216. PHILLIPS, D.M.P.: The presence of acetyl groups in histones. Biochim. J. **87**, 258–263 (1963).

217. PHILLIPS, D.M.P., JOHNS, E.W.: A fractionation of histones of group F2a from calf thymus. Biochem. J. **94**, 127–130 (1965).

218. PIHA, R.S., CUENOD, M., WAELSCH, H.: Metabolism of histones in liver and brain. J. Biol. Chem. **241**, 2379–2404 (1966).

219. PIPKIN, J.L., LARSON, D.A.: Characterization of the very-lysine-rich histones of active and quiescent anther tissues of *Hippeastrum belladonna*. Exptl. Cell Res. **71**, 249–260 (1972).

220. PIPKIN, J.L., LARSON, D.A.: Changing patterns of nucleic acid, basic and acidic proteins in generative and replicative nuclei during pollen germination and pollen tube formation in *Hippeastrum belladonna*. Exptl. Cell Res. **79**, 28–42 (1973).

221. PLATZ, R.D., HNILICA, L.S.: Phosphorylation of non-histone chromatin proteins during sea urchin development. Biochem. Biophys. Res. Commun. **54**, 222–227 (1973).

222. PRESCOTT, D.M., KIMBALL, R.F.: Relation between RNA, DNA and protein synthesis in the replicating nucleus of Euplotes. Proc. Natl. Acad. Sci. U.S. **47**, 686–693 (1961).

223. PRESCOTT, D.M.: Regulation of cell reproduction. Cancer Res. **28**, 1815–1820 (1968).

224. PRESCOTT, D.M.: Structure and reproduction of eucaryotic chromosomes. Advan. Cell Biol. **1**, 57–117 (1970).

225. PRESCOTT, D.M.: The synthesis of total macronuclear protein, histone and DNA during the cell cycle in *Euplotes erystomus*. J. Cell Biol. **31**, 1–9 (1966).

226. QUAGLIAROTTI, G., OGAWA, Y., TAYLOR, C.W., SAUTIERE, P., JORDAN, J.J., STARBUCH, W.C., BUSCH, H.: Structural analysis of the glycine arginine rich histone. II. Sequence of the half of the molecule containing the aromatic amino acids. J. Biol. Chem. **244**, 1796–1802 (1969).

227. RAFF, R.A.: Polar lobe formation by embryos of *Ilyanassa obsoleta*. Exptl. Cell Res. **71**, 455–459 (1972).

228. RANIERI, R., SIMEIAMAN, G., BOUTWELL, R.K.: Stimulation of the phosphorylation of mouse epidermal histones by tumor promoter agents. Cancer Res. **33**, 134–139 (1973).

229. RASMUSSEN, P.S., MURRAY, K., LUCK, J.M.: On the complexity of calf thymus histones. Biochemistry **1**, 79–89 (1962).

230. RAYMOND, S., WEINTRAUB, L.: Acrylamide gel as a supporting medium for zone electrophoresis. Science **130**, 711 (1959).

231. REICHMAN, M., PENMAN, S.: Stimulation of polypeptide initiation *in vitro* in Hela cells. Proc. Natl. Acad. Sci. U.S. **70**, 2678–2682 (1973).

232. ROBERTS, J.H., STARK, P., SMULSON, M.: Stimulation of DNA synthesis by adenosine diphosphoribosylation of Hela nuclear proteins during the cell cycle. Biochem. Biophys. Res. Commun. a**52**, 43–50 (1973).

233. ROBBINS, E., BORUN, T.W.: The cytoplasmic synthesis of histones in Hela cells and its temporal relationship to DNA replication. Proc. Natl. Acad. Sci. U.S. **57**, 409–416 (1967).

234. RUDDERMAN, J.V., BAGLIONI, C., CROSO, T.R.: Histone mRNA and histone synthesis during embryogenesis. Nature **247**, 36–37 (1974).

235. SADGOPAL, A., BONNER, J.: The relationship between histone and DNA synthesis in Hela cells. Biochim. Biophys. Acta **186**, 349–357 (1969).

236. SANGER, R., LANE, D.: Molecular basis of maternal inheritance. Proc. Natl. Acad. Sci. U.S. **69**, 2410–2413 (1972).

237. SAUTIERE, P., STARBUCK, W. C., ROTH, C., BUSCH, H.: Structural analysis of the glycine-arginine rich histone (f2a1). I. Sequence of the carboxyl terminal portion. J. Biol. Chem. **243**, 319–334 (1968).
238. SAUTIERE, M. P., TYROU, D., LAINE, B., MIZON, J., LAMBETIN-BREYNAERT, M. D., RUGGIN, P., BISERTE, D.: Structure primaire de l'histone riche en arginine et en lysine du thymus de veau. Compt. Rend. **274**, 1422 (1972).
239. Schochetman, G., Perry, R. P.: Early appearance of histone messenger RNA in polyribosomes of cultured L cells. J. Molec. Biol. **63** 591–596 (1972).
240. SEALE, R. L., ARONSON, A. I.: Chromatin associated proteins of the developing sea urchin embryo. I. The kinetics of synthesis and characterization of nonhistone proteins. J. Molec. Biol. **75**, 633–645 (1973).
241. SEALE, R. L., ARONSON, A. I.: Chromatin associated proteins of the developing sea urchin embryo. II. Acid-soluble proteins. J. Molec. Biol. **5**, 647–658 (1973).
242. SEED, J.: The synthesis of DNA, RNA and nuclear protein in normal and tumor strain cells. I. Fresh embryo cells from humans. J. Cell Biol. **28**, 233–248 (1966).
243. SEED, J.: The synthesis of DNA, RNA and nuclear protein in normal and tumor strain cells. II. Fresh embryo mouse cells. J. Cell Biol. **28**, 249–256 (1966).
244. SHAPIRO, A. L., VINUELA, E., MAIZEL, J. V., JR.: Molecular weight estimations of polypeptide chains by electrophoresis in SDS polyacrylamide gels. Biophys. Biochem. Res. Commun. **28**, 815–820 (1967).
245. SHAPIRO, I. M., LEVINA, L. Y.: Autoradiographic study on the time of nuclear protein synthesis in human leukocyte blood culture. Exptl. Cell Res. **47**, 75–85 (1967).
246. SHAPIRO, I. M., LEVINA, L. Y.: Incorporation of proteins labeled with ^{3}H-lysine into chromosomes of human leukocytes cultures *in vitro*. Chromosoma **25**, 30 (1968).
247. SHEROD, D., JOHNSON, G., CHALKLEY, R.: Phosphorylation of mouse ascites tumor cell lysine-rich histone. Biochemistry **9**, 4611–4615 (1970).
248. SHEPHERD, G. R., HARDIN, J. M., NOLAND, B. J.: Methylation of lysine residues in histone fractions in synchronized mammalian cells. Arch. Biochem. Biophys. **143**, 1–5 (1971).
249. SIEBERT, G., ORD, M. G., STOCKEN, L. A.: Histone phosphokinase activity in nuclear and cytoplasmic cell fractions from normal and regenerating rat liver. Biochem. J. **122**, 721–725 (1971).
250. SILVERMAN, B., MIRSKY, A. E.: Addition of histones to histone depleted nuclei: effect on template activity towards DNA and RNA polymerases. Proc. Natl. Acad. Sci. U.S. **70**, 2637–2641 (1973).
251. SIMPSON, R. T., REECK, G. R.: A comparison of the proteins of condensed and extended chromatin fractions of rabbit liver and calf thymus. Biochemistry **12**, 3853–3858 (1973).
252. SKOULTCHI, A., GROSS, P. R.: Maternal histone messenger RNA: detection by molecular hybridization. Proc. Natl. Acad. Sci. U.S. **70**, 2840–2844 (1973).
253. SMITH, J. A., STOCKEN, L. A.: Characterization of a non-histone protein isolated from histone F$_1$ preparations. Biochem. J. **131**, 859–861 (1973).
254. SMITH, J. A., STOCKEN, L. A.: Identification of poly (ADPR) convalently bound to histone F$_1$ *in vivo*. Biochem. Biophys. Res. Commun. **54**, 297–300 (1973).
255. SOLOMOS, T., LATIES, G. G.: Cellular organization and fruit ripening. Nature **245**, 390–392 (1973).
256. SONNENBERG, B. P., ZUBAY, G.: Nucleohistone as primer for RNA synthesis. Proc. Natl. Acad. Sci. U.S. **54**, 415–420 (1965).
257. SPAULDING, J., KAJIWARA, K., MUELLER, G. C.: The metabolism of basic proteins in Hela cell nuclei. Proc. Natl. Acad. Sci. U.S. **56**, 1535–1542 (1966).
258. STEDMAN, E., STEDMAN, E.: Cell specificity of histones. Nature **166**, 780–781 (1950).
259. STEDMAN, E., STEDMAN, E.: The basic proteins of cell nuclear. Phil. Trans. Roy. Soc. B **235**, 565 (1957).
260. STEIN, G. S., BORUN, T. W.: The synthesis of acidic chromosomal proteins during the cell cycle of Hela S-3 cells. I. The accelerated accumulation of acidic residual nuclear protein before the initiation of DNA replication. J. Cell Biol. **52**, 292–307 (1972).
261. STELLWAGEN, R. H., COLE, R. D.: Danger of contamination in chromatographically prepared arginine-rich histones. J. Biol. Chem. **243**, 4452–4455 (1968).

262. STELLWAGEN, R. H., COLE, R. D.: Comparison of histones obtained from mammary gland at different stages of development and lactation. J. Biol. Chem. a243, 4456–4462 (1968).
263. STELLWAGEN, R. H., COLE, R. D.: Chromosomal proteins. Ann. Rev. Biochem. 38, 951–990 (1969).
264. STELLWAGEN, R. H., COLE, R. D.: Histone biosynthesis in the mammary gland during development and lactation. J. Biol. Chem. 244, 4878–4887 (1969).
265. STEVELY, W. S., STOCKEN, L. A.: Variations in phosphate content in histone F_1 in normal and irradiated tissues. Biochemistry J. 110, 187–191 (1968).
266. STOUT, J. T., PHILLIPS, R. L.: Two independently inhibited electrophorietic varients of the lysine-rich histones of Maize (Zea mays). Proc. Natl. Acad. Sci. U.S. 70, 3043–3047 (1973).
267. SUBIRANA, J. A., PALAU, J., COZEOLLUELLA, C., RUIZ-CANILO, A.: Very lysine rich histone of echinoderms and molluscs. Nature 228, 992–993 (1970).
268. SUMMERBELL, D., LEIVIS, J. H., WOLPERT, L.: Postitional information in chick limb morphogenesis. Nature 244, 492–496 (1973).
269. SWIFT, H.: The constancy of desoxyribosenucleic acid in plant nuclei. Proc. Natl. Acad. Sci. U.S. 36, 643–653 (1950).
270. SZYAN, I., BURDMAN, J. A.: The relationship between DNA synthesis and the synthesis of nuclear proteins in rat brain. Effect of hydrocortisone acetate. Biochim. Biophys. Acta 299, 294–353 (1973).
271. TAKAI, S., BORUN, T. W., MUCHMORE, J., LIEBERMAN, I.: Concurrent synthesis of histone and desoxyribonucleic acid in liver after partial hepatectomy. Nature 219, 860–861 (1968).
272. TELLER, D. C., KINKADE, J. M., COLE, R. D.: The molecular weight of lysine-rich histone. Biophys. Biochem. Res. Commun. 20 739–744 (1965).
273. TIDWELL, T., ALLFREY, V. G., MIRSKY, A. E.: The methylation of histones during regeneration of the liver. J. Biol. Chem. 243, 707–715 (1968).
274. TOMKINS, G. M., GELEHRTER, T. P., GRANNER, D., MARTIN, D., SAMUELS, H. A., THOMPSON, E. B.: Control of specific gene expressions in higher organisms. Science 166, 1474–1480 (1969).
275. UMAÑA, R., UPDIKE, S., RANDALL, J., DOUNCE, A. L.: Histone metabolism. In: The Nucleoproteins (BONNER, T'so, Eds.). San Francisco: Holden-Day 1964.
276. VIDALI, G., GERSHEY, E. L., ALLFREY, V. G.: Chemical studies of histone acetyllation. The occurance of epsilon-N-acetyllysine in calf thymus histone. J. Biol. Chem. 243, 6361–6366 (1968).
277. VIDALI, G., BOFFA, L. C., ALLFREY, V. G.: Properties of an acidic histone-binding protein fraction from cell nuclei. Selective precipitation and deacetylation of histones F2a1 and F3. J. Biol. Chem. 247, 7365–7373 (1972).
278. WASSERMAN, G. B.: Molecular genetics and developmental biology. Nature 245, 163–165 (1973).
279. WEBER, K., OSBORN, M.: The reliability of molecular weight determinations by dodecyl sulfate-polyacrylamide gel electrophoresis. J. Biol. Chem. 244, 4406–4412 (1969).
280. WEINBERG, E. S., BIRNSTIEL, PURDOM, I. F., WILLIAMSON, R.: Genes coding for polyribosomal 9S RNA of sea urchins: conservation divergence. Nature 240, 449–453 (1972).
281. WEINTRAUB, H.: A possible role for histone in the synthesis of DNA. Nature 240, 449–453 (1972).
282. WESSELS, N. K., SPONNER, B. S., ASH, J. F., BRADLEY, M. O., LUDUENA, M. A., TAYLOR, E. L., WRENN, J. T., YAMADA, K. M.: Microfiliments in developmental processes. Science 171, 135–143 (1971).
283. WHITTAKER, J. R.: Segregation during ascidian embryogenesis of egg cytoplasmic information for tissue-specific enzyme development. Proc. Natl. Acad. Sci. U.S. 2096–2100 (1973).
284. WILLIAMSON, D. H.: Replication of the nuclear genome in yeast does not require concurrent protein synthesis. Biochem. Biophys. Res. Commun. 52, 731–740 (1973).
285. WOLPERT, L., CLARKE, M. R. B., HORNBRUCH, A.: Positional signalling along Hydra. Nature 239, 101–105 (1972).

286. Woodward, J., Rasch, E., Swift, H.: Nucleic acid and protein metabolism during the mitotic cycle in *Viva faba*. J. Biophys. Biochem. Cytol. **9**, 445–462 (1960).
287. Yamada, M., Sugimura, T.: Effects of deoxyribonucleic acid and histone on number and length of chains of poly ADPR. Biochemistry **12**, 3303–3308 (1973).
288. Yarbro, J. W., Niehaus, W. G., Barnum, C. P.: Effect of hydroxyurea on regenerating rat liver. Biochem. Biophys. Res. Commun. **19**, 592–597 (1965).
289. Zauderer, M., Liberti, P., Baglioni, C.: Distribution of histone messenger RNA among free and membrane-bound polyribosomes of a mouse myeloma line. J. Molec. Biol. **79**, 577–586 (1973).
290. Zampetti-Bosseler, F., Malpoix, P., Fievez, M.: Differential inhibition of protein synthesis in the nucleus and cytoplasm of Hela cells in the presence of actinomycin, puromycin and hydroxyurea. European J. Biochem. **9**, 21–26 (1969).
291. Zubay, G. (Ed.): Papers in Biochemical Genetics. First Ed., p.466. New York: Holt, Rienhart and Winston 1968.
292. Zweidler, F.: Strukturen und Replication der Chromosomen. Ph. D. thesis. Universität Zürich: Truninger Druckanstalt 1965.
293. Zwiedler, A., Borun, T. W.: In preparation (1974).

Cell Changes in Neurospora

R. E. NELSON, C. P. SELITRENNIKOFF, and R. W. SIEGEL

Department of Biology, University of California, Los Angeles, CA, USA

I. Introduction

Replicating cells may be predictably altered by exogenous signals so that their progeny come to express a novel phenotype. Differentiation, a process so familiar and fundamental to biology, is exemplified by the vegetative life cycle of *Neurospora crassa*. There are three characteristic stages, and each consists of but a single basic cell type. Upon proper stimulation, the proliferation of recumbent vegetative hyphal cells is discontinued, and instead aerial hyphae are produced. Aerial hyphae grow and then eventually produce resting vegetative spores called macroconidia. Figure 1 summarizes the relationships among the three cell types. These cell changes are at least superficially reminiscent of development in higher multicellular organisms; replicating stem cells are induced to produce a population of intermediary cells which pass through a critical number of cell cycles before they go on to form morphologically and functionally distinct, terminally differentiated cells in which mitotic activity has ceased. In an attempt to exploit what appears to be a particularly clear and simple model for eukaryotic cell differention, we have elected to examine aspects of the processes of aerialogenesis and macroconidiogenesis.

 N. crassa and certain other lower plants (SESSOMS and HUSKEY, 1973) offer a combination of advantages for studies of differentiation in a eukaryote not available in more complex organisms. Large populations of cells can be cultured on

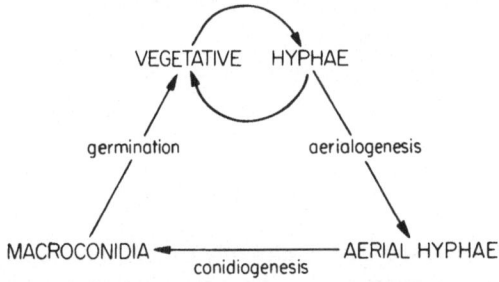

Fig. 1. Vegetative life cycle of *Neurospora crassa*. Methods that permit the temporal sequencing of these stages and their characterization are given in Section II-B

simple synthetic media and easily quantified. The suitability of *N. crassa* for conventional genetic analyses and biochemistry is well known. For developmental genetics, it is particularly important to have on hand genes whose expressions are limited to a single cell type. The power of such "phase-specific" loci in determining whether biochemical changes associated with development have a cause-and-effect relationship has been pointed out by NASRALLAH and SRB (1973); these authors reported the first phase-specific gene in *Neurospora* and others will be described here. *Neurospora* offers special advantages for the isolation of phase-specific mutants because strains blocked in the development of aerial hyphae and of macroconidia are otherwise viable and suitable for crosses. And phenotypic descriptions are simplified, in part at least, by the fact that the development of the aerial hyphae and macroconida does not depend upon cell migration, cell unions, or other forms of cell traffic. Finally, the aerial hyphae and the macroconidia evidently carry complete genetic endowments because both cell types can directly produce vegetative hyphae from which the complete organism can be formed. For this reason, mechanisms of differentiation that call for somatic mutation and permanent or long-lasting aneuploidy can be discounted and more subtle models for differential genic expression must be considered.

We have been attempting to formulate answers to several specific and interlocked questions:

1. Is the differentiation of aerial hyphae and of macroconidia dependent upon the synthesis of nucleic acids and proteins or, instead, can the rapid appearance of these new structures be ascribed to a "preformed system" of organelles and macromolecules present in the vegetative hyphae prior to the time of induction?

2. Can certain of the molecular and morphological cell changes which uniquely accompany the formation of a particular cell type be precisely identified so that phase-specific developmentally programmed functions are brought to light?

3. Are the structural genes for such phase-specific functions accessible, and can the products of genes that control a phase-specific event be identified?

4. Given a "kit" of structural genes whose expressions are limited to but a single phase-specific cell change, will clues as to their coordinate regulation be obtained by the isolation of other mutants with pleiotropic effects?

We present here results relevant to the first three questions and, in a preliminary way, to the fourth. That material is introduced by a description of pertinent details of the vegetative life cycle in *Neurospora*. Several excellent reviews cover conidiation (TURIAN and BIANCHI, 1972) and genetic regulation in *Neurospora* (METZENBERG, 1972) and other fungi (CALVO and FINK, 1971; GROSS, 1969)—matters of interest to our own work.

II. Life Cycle

A. Cell Types

The anatomy of *Neurospora* is remarkably simple. The vegetative mycelium is the product of numerous microscopic filaments which have elongated, branched, fused, and intertwined with each other. Daughter cells are simply apical or lateral extensions of mother cells. Each vegetative hypha is divided by cross walls into a linear sequence of multinucleate membrane-bound protoplasmic bodies. A central perforation in each cross wall permits limited intercellular communication. The aerial hyphae are constructed in a similar way. Several considerations suggest that the hyphal units demarcated by incomplete septa can act as independent or self-sufficient entities and hence are the functional equivalents of prototypic cells: (a) When a very short hyphal bit, consisting of one to three units, is isolated on nutritive medium it may be capable of independent growth. (b) If the hyphal wall is mechanically disrupted, cytoplasm flows out of the injured cell, the internal pressure is reduced, and as a sure and rapid response to the lowered turgor, the pores in the cross walls immediately adjacent to the site of injury are plugged and the two neighboring cells survive (WILSON, 1961). (c) Although aerial hyphae are protoplasmically continuous with the vegetative hyphae from which they arise, the two structures have been found to be biologically and biochemically distinct.

Vegetative hyphae are the predominant elements formed by *Neurospora*. Hyphal elongation is virtually the only mechanism for growth of the organism. Under suitable conditions, vegetative hyphae will differentiate aerials for production of asexual spores or specific structures for sexual reproduction and subsequent sexual spore formation. In stationary cultures that have begun to exhaust the available food supply, small alterations occur in the activities of certain enzymes, thus distinguishing actively growing hyphae from the "mature" hyphae (ZALOKAR, 1959a). These and other changes (ZALOKAR, 1959b) that take place during hyphal growth suggest alternative and readily interconvertible metabolic states rather than cellular differentiation. As we shall see later, it is probable that all vegetative hyphae are equipotential with respect to the formation of aerials and so in the present context (at least) they consist of a single basic cell type.

It is not surprising that the status of aerial hyphae as differentiated structures has remained problematic, for they have not been the subject of intensive research. The initial question as to whether the aerials are merely reoriented vegetative hyphae that are enabled to form conidia because of their superior locale is answered by a group of recent observations. Actively growing vegetative and aerial hyphae were prepared from a single culture of vegetative hyphae. To do so, a mycelial mass was cut into two parts; one part was induced to form aerials (using methods described in Section II-B) and the other was submerged in fresh liquid growth medium to permit the development of new vegetative hyphae. When hyphae prepared in these ways had grown to 400–500 μ in length, they were harvested and then tested for their capacities to (a) form thriving cultures on nutritive agar media, (b) form heterocaryons, and (c) produce restricted or "colonial" growth on medium supplemented with sorbose. Reports from a number of laboratories led us to anticipate that vegetative hyphal isolates would regularly yield viable cultures upon transfer, participate with high frequency in heterocaryon formation, and be sorbose-sensitive. Those results were found. However, the aerials behaved differently. They inaugurated relatively few viable cultures and participated less frequently in the formation of heterocaryons. Finally, all of the aerial hyphae that proved to be viable on sorbose-supplemented media differed from the vegetative hyphal isolates in that they were resistant to sorbose for at least the initial six hours of growth. It should be noted that the sorbose result excludes the possibility that the aerial crop is a mixed population consisting of differentiated aerial hyphae as well as vegetative hyphae newly arising from the original mycelial mass. The conclusion that vegetative hyphae and aerial hyphae are composed of fundamentally distinct cell types is fully supported by studies on the distribution of the enzyme NADase described in Section III.

A second question concerning the basic biology of aerial hyphae is their "commitment" to eventually develop conidiophores and macroconidia. The distinction between aerials and conidiophores is made clear below. In bacteria and lower plants the operational definition of "commitment to sporulation" is simply the point in development when that course cannot be reversed by transferring the organism to full growth medium. Accordingly, the time of commitment is determined by periodically transferring cells from incomplete media (which induce sporulation) to a medium suitable for vegetative growth. While an entirely comparable test for isolated aerial hyphae can hardly be envisioned, we have found that when aerial hyphae were isolated and then transferred to growth medium, conidiogenesis was always aborted; some hyphae died while others grew and soon acquired all the properties of normal vegetative hyphal isolates. Even conidiophores, harvested after the appearance of presumptive macroconidia, developed as vegetative hyphae or died; all failed to liberate mature conidia. Control aerials, isolated from the vegetative mycelium on glass coverslips and never exposed to growth medium, formed conidiophores and conidia within a few hours. Thus there is at present no evidence for the irreversible commitment of aerial hyphae to conidiogenesis; and it has been shown that under certain conditions aerials may undergo redifferentiation. The observation that the capacity of aerials to form conidia is not dependent upon their protoplasmic continuity with the vegetative hyphae demonstrates that their uniqueness, whatever its physical basis, is self-maintained in the course of many cell and nuclear doublings.

Only a brief sketch of the anatomy and composition of the macroconidia is necessary for present purposes. We can confine our remarks to evidence supporting the view that conidiogenesis involves the differential expression of the genome. It is unfortunate that most of what is known about the conidia contributes little to a fundamental understanding of conidial development. Macroconidia are small dormant cells phenotypically adapted to serve as agents for the survival and dispersal of *Neurospora* by wind or other supervenient agents. In proper environments conidia germinate to produce vegetative hyphae. It may be supposed that in nature they occasionally function as male gametes. The vesicles are typically subspherical, or shaped like footballs truncated at both ends, about 7 μ in length, each containing an average of three nuclei. Ultrastructural details of the conidia distinguish them from hyphae; and cell wall fractionation reveals subtle differences (see Turian and Bianchi's review, 1972). In accord with dormancy, oxydative phosphorylation is absent and respiration is much reduced (Greenswalt et al., 1972). Conidia may lack polysomes (Henney and Storck, 1964), certain proteins have a different distribution of amino-terminal amino acids (Rho and de Busk, 1971) and the t-RNA methylases are unique (Wong et al., 1971). Finally, galactosamine, a component of the vegetative hyphae, is absent from the conidia (Gratzner, 1972; Schmit et al., 1973).

B. Aerialogenesis and Conidiogenesis

The conditions under which *Neurospora* are ordinarily cultivated are such that vegetative hyphae, aerial hyphae, and macroconidia propagate simultaneously. The resulting composite of different cell types and processes is not convenient for developmental studies. Turian (TURIAN and MATIKIAN, 1966) has exploited the use of media which much retard the formation of one life-cycle stage and favor another, in order to study metabolic shifts important for development (COMBÉPINE and TURIAN, 1970); since these media do not restrict the organism to a particular process, they are not entirely satisfactory for our purposes. But by following the important clues provided by other workers, chiefly STINE (STINE and CLARK, 1967), experimental situations were reported (SIEGEL et al., 1968) which limit Neurospora to the production of vegetative hyphae and permit the formation of aerial hyphae and then conidia only when cultures are transferred to new conditions. In this way it has been possible to temporally separate, and hence to isolate, a particular process and particular kind of cell for intensive study.

Vegetative hyphae are grown in darkness on liquid nutritive media supplemented with Tween 80 in stationary flasks; these conditions normally prohibit the production of aerial hyphae and conidia. Whenever the vegetative hyphal cells are washed so as to exclude abruptly an external source of nutrients, then transferred to sealed finger bowls and exposed to light and proper aeration, the reproduction of vegetative hyphae is promptly replaced by the generation of aerial material. However, in order to realize optimal development the vegetative hyphal mass should be about 39 mg dry weight. Aerialogenesis, preceded by a two-hour lag period, is featured by cell and nuclear multiplication resulting in the vertical elongation of the hyphae. Soon the aerials branch and, optimally, about 10^6 of these branches develop further as conidiophores, each yielding about 100 macroconidia. The remaining branches never sporulate and their function in the economy of the organism remains uncertain. The period during which aerials yield conidiophores (see below) appears to be quite narrow; we have not observed these structures among aerials less than 2 hrs old; nor do new conidiophores arise from aerials several hours older. When the induction of aerialogenesis is carried out under conditions of low relative humidity, the propagation of aerial hyphae virtually ceases before the onset of conidiogenesis (UREY, 1971). Moreover, the two processes can be separated even more clearly; if vegetative hyphae are induced in the presence of nitrogen, aerialogenesis proceeds in a normal way but conidiogenesis is blocked. Figure 2 shows the events which occur in finger bowls subsequent to induction under conditions of high relative humidity. It can be seen that aerialogenesis and conidiogenesis are fairly well separated in time. The figure also reveals that there is no substantial increase in the dry weight of the aerial hyphae between the seventh and tenth hours and that the aerial mass decreases thereafter. Since this is the period of most active conidiation, the direct conversion of aerial hyphal cells into macroconidia without compensatory replacement of materials from the vegetative hyphae is indicated; as noted before, the aerials may be functionally independent of the vegetative hyphae.

It appears that all vegetative hyphae are capable of forming aerials. Hyphal outgrowths from newly germinated conidia, as well as hyphae at least 20 genera-

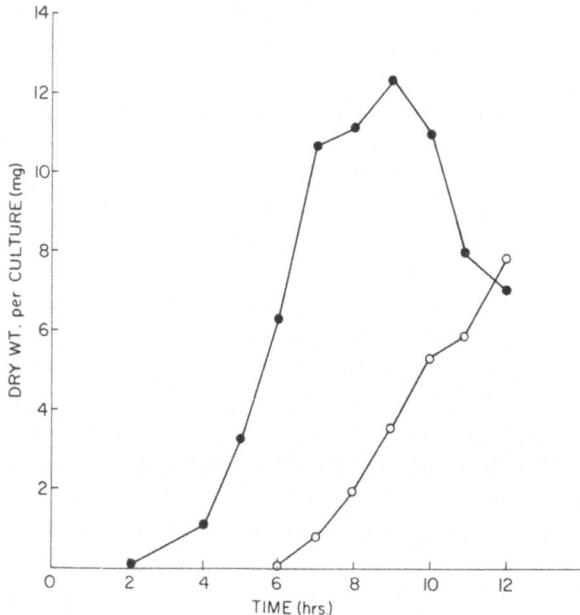

Fig. 2. Development of aerial hyphae and conidia following induction (0 hrs) of vegetative hyphae. ○—○ dry weight of conidia (1 mg = 3.13×10^7 conidia); ●—● dry weight of the aerial hyphae

tions old, can be readily induced. Log and stationary phase vegetative hyphae provide similar conidial yields. For these reasons, competence is seemingly independent of both a critical number of prior cell divisions and of "physiological state." However, we have recovered several single-gene mutants which produce ostensibly normal vegetative hyphae but do not initiate aerial hyphae upon induction. This situation is in marked contrast with what is known for *Aspergillus*; in that fungus competence (which permits the development of conidiophores upon induction) occurs at a particular stage in the life cycle, and this stage is under genic control (Axelrod et al., 1973).

Optimal aerial development is brought about by the removal of all exogenous nutrients and, as expected, increase in the dry weight of the aerial hyphal mass is accompanied by a decrease in the mass of the vegetative hyphae; eventually, 30% of the starting material is converted into aerial hyphae and conidia.

Conidiation illustrates two quite general aspects of cell differentiation, morphogenesis and the shutting down of metabolism and cell activity.

In these ways the process of conidiation is very different from the process of aerial hypha growth; and hence it is surprising that both processes may occur simultaneously among the branches of a single aerial hypha. This complexity can be avoided by environmental conditions, as already noted, or by the use of mutants (e.g. "crisp-snowflake") which fail to develop nonconidiogenic aerial branches. Conidiophores originate as branches of the aerial hyphae that are more narrow in girth than others (Bianchi and Turian, 1967). They appear during the fourth to sixth hour after induction, and during the next two to three hours undergo a dramatic sequence of morphogenetic changes. The initial step in the differentiation of the conidia is the migration of one (or a few) nuclei to within 3 μ of the terminal tip of the conidiophore. Next the tip swells, becomes almost spherical in shape, and soon forms an apical "bud"; this in turn grows and buds so that eventually a chain of presumptive macroconidia is produced. These contain division products of the original nuclei as well as nuclei that have subsequently migrated into the elongating chain. The final step in conidiation is the construction of complete cross walls or septa between adjacent conidia and the cessation of the nuclear divisions.

We have isolated a number of phase-specific mutant strains that fail to complete these steps and are thus aconidial. All produce vegetative hyphae indistinguishable from wild type with respect to growth and morphology. They also form aerial hyphae and conidiophores comparable to wild type, but there the similarity ends. One mutant is blocked early in conidiogenesis so that nuclear migration is rare or absent; another, blocked somewhat later, regularly fails to produce the initial macroconidial bud (MATSUYAMA, unpubl.); finally, two loci controlling the process of septation have been discovered and will be discussed in Section IV. Phase-specific aconidial mutants are rare and exceptional; the overwhelming majority of mutants selected initially for their aconidial characteristic also produce defective vegetative and/or aerial hyphae. Looking at this point in reverse, mutants selected for gross hyphal morphologic abnormalities are often aconidial. Evidently the products of many genes contribute to the normal functioning of diverse cell types. While this phenomenon is neither novel nor surprising, it prompts two closely related questions: (a) Are the gene products necessary for normal hyphal growth and morphology sufficient for the differentiation of the conidia? (b) Is there an absolute requirement for the synthesis of macromolecules after induction in order that aerials and conidia differentiate? The first question has already been answered in part and is considered again in Section IV. The second question is answered by the experiments described next.

C. Macromolecular Synthesis

The possibility that aerial hyphae develop from a preformed system of molecules and organelles merely released, rearranged, or activated for expression by induction is raised by the facts that (a) aerials grow at the expense of vegetative hyphal cells and (b) that mutations upsetting aerialogenesis are generally found to have detectable effects on the properties of vegetative hyphae. Precedence for such a view comes from the work of SOLL and SONNEBORN (1971), who find that starved cells of *Blastocladiella* pass through a sequence of morphologic changes in the absence of all macromolecular synthesis and turnover. The observation that competence is independent of a minimum number of cell divisions and of physiological state poses an initial difficulty for the hypothesis. Other kinds of experiments show decisively that marcromolecular syntheses are required for the differention of aerials and conidia.

Addition of amino-acid analogs or cycloheximide to vegetative hyphae soon after induction blocks protein synthesis, as judged by incorporation studies

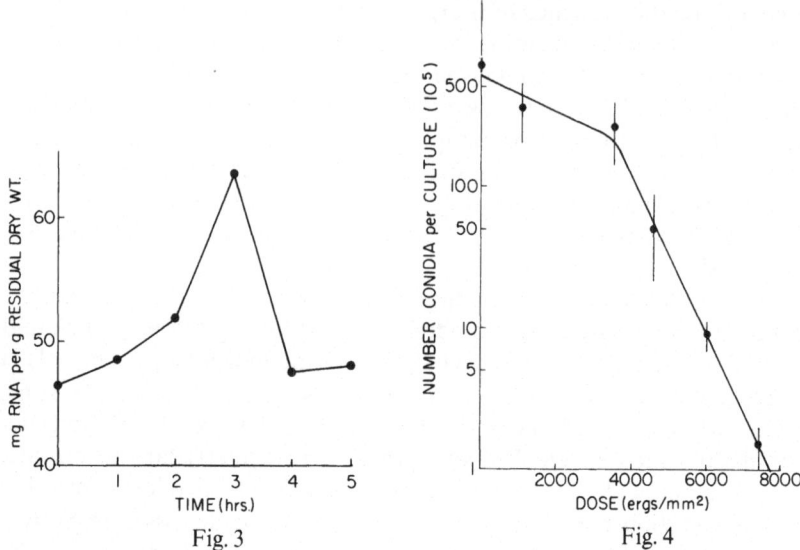

Fig. 3   Fig. 4

Fig. 3. Fluctuations in total RNA content per unit weight of culture following induction (average of two experiments)

Fig. 4. Inactivation of conidiogenesis with ultraviolet light. Vegetative hyphae of the UV-sensitive strain *uvs-3* were irradiated at induction and the numbers of macroconidia produced were determined after 12 hrs. *uvs-3* is altered in photoreactivation repair (Schroeder, 1970)

Table 1. The numbers of aerial hyphae and macroconidia per culture following treatment with FUdR at induction

Treatment[a]	Aerials[b]	Macroconidia[c]
FUdR, 10^{-2} M	3.0×10^3	$< 10^4$
FUdR, 10^{-3} M	3.6×10^3	6.0×10^4
FUdR, 10^{-4} M	4.4×10^3	2.0×10^5
Control	6.8×10^3	7.7×10^7

[a] Drug present throughout development of aerial hyphae and macroconidia.
[b] Aerials were asseyed 4 hrs after the start of induction.
[c] Macroconidia were assayed 8 hrs after the start of induction.

(Urey, 1971), and abnormally few aerial hyphae and conidia are produced. The inhibition of transcription by treatment with actinomycin (Totten and Howe, 1972) dramatically reduces or abolishes the yield of aerial hyphae. In full accord with this result, the onset of aerialogenesis is accompanied by a significant increase in the cellular concentration of RNA (Fig. 3) and by increases in the specific activities of several enzymes (Urey, 1971). We find that vegetative hyphae exposed to 5-fluorodeoxyuridine 6 hrs prior to the time of induction subsequently produce virtually no aerials at high doses (10^{-2} M) or very few at lower concentrations (10^{-4} M). Since the drug is a mitotic inhibitor, binding to thymidilate synthetase and subsequently blocking DNA synthesis, it is reasonable to suppose that the induction of aerial hyphae is necessarily associated with the generation of new DNA and nuclei. Table 1 shows that treatment started after the time of induction reduces both the numbers of aerials and the macroconidia formed. High intensity ultraviolet light preferentially blocks the syntheses of nucleic acids. UV irradiation, delivered during a one-minute period at the onset of induction, interferes with both aerialogenesis and conidiation. Figure 4 shows that at high doses the number of conidia formed may be reduced by a factor of at least 500. The probability that conidia will germinate following their production from cultures exposed to high doses of UV is reduced only about fourfold. Thus the UV effect is most readily attributable to interference with the rate of those processes peculiar to conidiogenesis. These results, together with the fact that conidiophores first appear 4 hrs after induction, suggest that aerial hyphae must pass through a critical number of mitotic cycles before they are competent to differentiate conidia.

There is a direct correlation between the numbers of aerial hyphae and the numbers of conidia formed by a culture. For this reason it is difficult to determine whether an inhibitor interfering with aerialogenesis has a direct effect on conidiogenesis. The problem is solved in part by treatments started late in development, after the conidiophores have appeared and the first mature conidia have been produced. Under those conditions, UREY (1971) reports, the cycloheximide and DL-ethionine-mediated inhibition of protein synthesis is correlated with a 90–99% reduction in the numbers of conidia eventually assayed. Turning to transcription, we find that if actinomycin is applied directly to developing conidiophores, previously grown on the surfaces of glass cover slips, there is a radical reduction in the numbers of conidia formed. Controls treated with distilled water are not affected. However, if the drug is added via the vegetative hyphae to cultures ready to conidiate, then the conidia are produced in normal numbers; this result confirms those reported by earlier workers (TURIAN and BIANCHI, 1972). They treated the vegetative hyphal mass with RNA inhibitors at a series of time points during the development of aerials and conidia and found that early treatments interfered with development but cultures subjected to drugs during later stages were unaffected. Their provisional inference, that conidiogenesis is dependent upon m-RNA's synthesized early in aerialogenesis and then stored or masked, seems questionable in light of the effects of direct drug treatment of conidiophores.

The foregoing results reveal that *Neurospora* is a biological conformist. Development requires the transcription of ribonucleic acids and in agreement with the most recent results on the water mold *Achlya* (KANE et al., 1973) there is no evidence for stable m-RNA. That synthesis of proteins, some of which may be novel, is required cannot be interpreted to mean that the macromolecules described above as phase-specific and characteristic of aerials and conidia are essential for the differentiation of these cell types. The dependence upon DNA synthesis and the continuation of mitotic cycles relates development in *Neurospora* to a number of model systems (WESSELLS, 1968) and distinguishes it from others (e.g. MOSCONA, 1971; BRADLEY, 1973).

It is not unreasonable to suppose that *Neurospora* might dismantle the machinery for synthesis of amino acids as a specific response to sudden starvation or to other conditions under which that machinery would be of no immediate use. There is some evidence leading to the suspicion that the syntheses of certain amino acids are in fact suspended during aerialogenesis and the production of conidia. Auxotrophs, which have an absolute requirement for an amino acid during growth of vegetative hyphae, form normal crops of aerials and conidia in absence of amino acid supplementation after the time of induction. For example, vegetative hyphae of leucine-1 will grow at normal rates in the presence of 90–100 µg/ml leucine; resulting mycelial mats were washed and deprived of all nutrients and then, upon induction, were found to give rise to normal crops of aerials and conidia. But when mycelia grown and washed in the same way were transferred to minimal medium (lacking leucine) there was no outgrowth of vegetative hyphae. Thus, under these conditions leucine utilized for the development of the aerial mass is unavailable for the production of vegetative hyphae. However, when vegetative hyphae from the auxotroph were transferred to conditions suitable for induction of aerials for 2 hrs—the normal lag period—then called on to form fresh vegetative hyphae in liquid medium lacking leucine, they were found to do so. (Care was taken to exclude the formation of vegetative hyphae from nascent aerials). These observations suggest that during the lag period either an increase in the activities of proteases occurs (MAHADEVAN and MAHADKAR, 1970; MITCHELL and SABAR, 1966; DRUCKER, 1972) to make leucine available by hydrolysis of vegetative hyphal proteins, or (far less likely) the establishment of some undiscovered pathway for leucine synthesis is established. In any case, aerial development in this strain is not dependent upon the usual course of leucine biosynthesis; the most interesting hypothesis is that the process is normally turned off. The source of amino acids following induction could be fruitfully approached by enzyme assays and by the use of appropriate mutants.

III. Genetic Alteration of a Luxury Molecule

The special functions that characterize terminally differentiated cells are often dependent upon the presence of certain molecular species found in no other cell type. Thus luxury molecules are distinguished from other molecules whose ubiquitous presence strongly suggests that they perform housekeeping functions. The discrimination between these two classes of molecules might profitably be based upon genetic analyses as well as upon studies of distribution. It can be predicted that mutational loss of a luxury molecule will not bring about lethality as the absence of a housekeeping molecule does; although survival under conditions of stress might be threatened. Such a mutational definition of a luxury molecule is feasible in *Neurospora*. Indeed, it is of interest to discover how a cell lacking a certain luxury molecule might behave.

NADase is a luxury molecule. Its discovery in *Neurospora* by KAPLAN et al. (1951) led to the association of the enzyme with events culminating in the development of macroconidia (ZALOKAR and COCHRANE, 1956; STINE, 1968; UREY, 1971). These results suggested an important function for the enzyme in macroconidiation or in conidial germination. We have confirmed UREY's (1971) report that

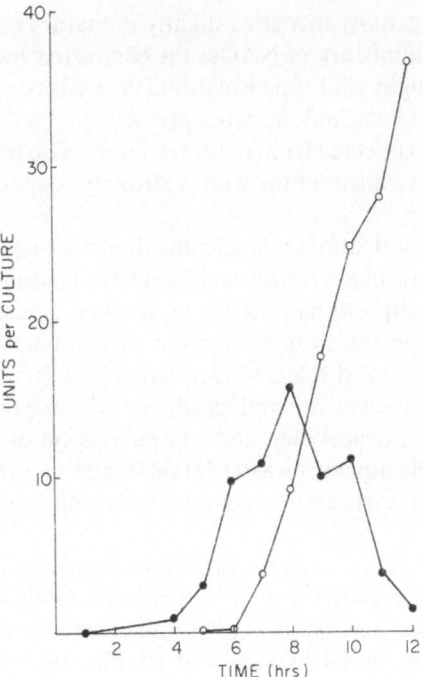

Fig. 5. Distribution of NADase activity in the aerial mass following induction (0 hrs). ●—● enzyme activity recovered from aerial hyphae; ○—○ enzyme activity recovered from mature conidia. One unit equals one micromole NAD hydrolyzed per min at 25° C

enzyme activity is readily recovered from aerial hyphae and macroconidia, and we find no activity associated with the growth of vegetative hyphae. The activity which accompanies conidia rapidly disappears when the spores are transferred to growth media. We could not detect NADase in sexual spores. Figure 5 describes changes in total NADase activity during development of aerial hyphae and conidia. Initially all of the activity is associated with the aerials; then, as the numbers of conidia increase, most of the activity becomes located in these cells, with little remaining in the aerial hyphae. Other observations demonstrate concordance of aerial hyphae and enzyme, strongly suggesting that the formation of conidiogenic hyphae is a necessary and sufficient condition for the appearance of NADase. (a) Three mutants lacking aerials produce no enzyme, and a temperature-sensitive mutant forms aerials and enzyme only under permissive conditions (HOCHBERG and SARGENT, 1973). (b) Wild-type vegetative hyphae that are abruptly starved, exposed to light, and aerated but allowed to remain submerged in liquid produce neither aerials nor NADase (UREY, 1971). When aerial development is blocked by inhibitors after induction, enzyme cannot be detected. (c) The aconidial mutant "fluffy" has NADase activity (STINE, 1968), as do the phase-specific aconidials described in Section II. The double mutant "crisp-snowflake" produces very short aerial hyphae and lacks nonconidiogenic aerial branches—yet the yield of NADase is high. It may be supposed that the bulk of the activity characteristic of mature conidia represents enzyme already present in aerials at the time of the origin of the conidia, although activity increases slightly in mature conidia that are stored (Stine, 1967). A natural history of NADase is completed by noting that we have found activity for all eight wild-type isolates of *N. crassa* so far examined, for the related species *N. sitophila* and *N. tetrasperma*, and finally for ten additional isolates of *Neurospora* collected from natural sources. The importance of NADase to the survival of the organism in the wild is strongly suggested by its widespread occurrence.

NADase hydrolyses nicotinamide adenine dinucleotide (NAD) to nicotinamide and adenosine diphosphate ribose, and NADP to nicotinamide and adenosine 2′-monophosphate, 5′-diphosphate ribose with equal efficiency. Activity can be conveniently assayed by the cyanide method of Colwick et al. (1951), by the colorometric titration of acid released (Menegus and Nelson, unpubl.), and by the chromatography of substrate and products. All enzyme activity associated with mature conidia is extracellular, and a large fraction of the total activity can be solubilized by simple aqueous wash (Zalokar and Cochrane, 1956). Enzyme associated with hyphal cells can be measured following dispersion in an aqueous solution.

It is likely that the sharp increase in NADase activity associated with post-induction development represents *de novo* synthesis. Consistant with this idea is the finding that enzyme activity is neither enhanced nor eroded when extracts from vegetative hyphae, aerial hyphae, and conidia are combined. Critical evidence will depend upon the identification of an enzyme protein that permits the strict association of changes in activity correlated with changes in the amounts of protein; in that connection, enzyme from the aerial mass is being purified (Menegus, pers. comm.).

We have developed a bioassay for the recognition of NADaseless mutants, using *Haemophilus* as the test organism. *Haemophilus* requires exogenous NAD for growth. At high concentrations of exogenous NAD *Neurospora* and *Haemophilus* can be cultured together; at low concentrations of NAD the bacteria will not grow in the immediate vicinity of wild-type fungal cells that are liberating NADase. On the other hand, NADaseless mutants would not be expected to prevent the growth of neighboring *Haemophilus* cells. When we used "colonial" strains of *Neurospora*, a large number of mutagenized isolates were rapidly scored for the NADase phenotype. In this way five independent mutants whose singular difference from wild type appeared to be the absence of NADase activity were recovered from 500000 surviving mutagenized isolates. All proved to be single-gene defects. High resolution recombination studies and *in vivo* and *in vitro* complementation indicated that all five occur within a single gene. This new gene, *nada*, was localized on the right arm of linkage group IV. We were also able to isolate a number of single-gene mutants differing from the *nada* strains in that they produce low—but characteristic—levels of NADase and are morphologically abnormal; the mutations were subsequently localized elsewhere in the genome. We appreciate the possibility that these loci include a number of genes which specifically regulate developmental events.

nada appears to be the structural gene for NADase. It is most significant that one of the five *nada⁻* strains has been found to produce temperature-sensitive activity. If carefully manipulated, the mutant shows about 2% of wild-type activity when measured at $25°$ C; 50% of this activity is inactivated in 45 sec by pretreatment at $45°$ C. The half-life of the wild-type enzyme at $45°$ C is 29 min. (The four remaining mutants produce less activity than that resolved by present techniques, 0.01% of wild-type). Second, there is no evidence that the $nada^-$ defect alters either the regulation or the synthesis of NADase, since extracts from $nada^-$ strains have no influence on the activity extracted from wild-type and the mutations are recessive to wild-type in heterocaryons; the amount of enzyme measured for a hererocaryon is directly proportional to the ratio of $nada^+ : nada^-$ nuclei. Finally, $nada^+$ and $nada^-$ strains grow equally well on a number of disaccharides, indicating that $nada$ mutations do not significantly modify other extracellular glycohydrolases. Identification of immunologically cross-reacting material from $nada^-$ cultures would complete the evidence that $nada$ is the structural gene for NADase, and CRM studies are in progress.

What is the function of NADase in the natural history of *Neurospora?* When it is absent there is no noticeable alteration in the development of aerial hyphae or macroconidia, or in the process of conidial germination. Nor are growth and fertility of the $nada^-$ strains upset. The intracellular pools of oxidized and reduced pyridine dinucleotides present in $nada^+$ and $nada^-$ conidia are similar (BRODY, per. comm.); hence there is no evidence for direct communication between these endogenous substrates and the enzyme. Aqueous washes of $nada^+$ and $nada^-$ conidia contain very low but similar levels of NAD. However, $nada^-$ conidia liberate an NADase-sensitive substance(s) which supports the growth of *Haemophilus* in the absence of exogenous NAD. The foregoing results lead to the question of whether NAD and NADP are the natural substrates for NADase and again emphasize our ignorance of the function of the enzyme. By way of apology, we refer to one condition that obliges *Neurospora* to make use of the $nada^+$ product. Conidia from auxotrophs which require nicotinamide for growth (e.g. $nic^- nada^+$) will germinate on media supplemented with NAD while the double mutant will not. Otherwise, NADase is a luxury molecule with a vengence!

IV. Genes for Cell Morphogenesis

Cell differentiation is regularly accompanied by changes in cell morphology. The view that cell shape and form are under genic control is supported by numerous studies, and in certain cases (e.g., sickling of red blood cells; determination of cell shape in *E. coli*) something is known of the molecules which specify cell morphology. As one approach to the mechanism of cell morphogenesis, it would be of interest to uncover a sample of those genes which are directly and solely concerned with the process. With this as a foundation, investigations of the expressions of these genes and their control could be pursued. Mutations that block differentiation of the macroconidia can be brought to light. To reinforce a point made earlier, the biology of *Neurospora* offers the opportunity to discriminate between mutations with phase-specific effects and others that primarily affect mother cells or delay their expressions for many cell cycles.

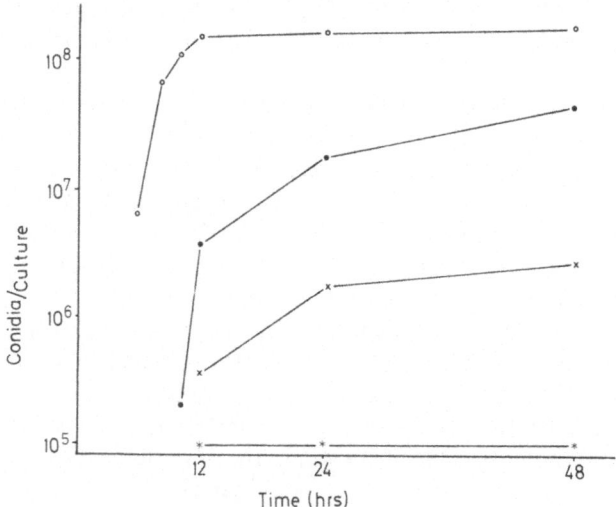

Fig. 6. Production of macroconidia from wild type (O—O), *csp-1* (●—●), *csp-2* (×—×), and the double mutant *csp-1*, *csp-2* (*—*) following induction

Attention has centered on three strains carrying single-gene mutations which interfere with the final steps in conidiogenesis, namely the production of free spores from conidiophores that have already produced chains of "proconidia". It was reasoned that since this function was unique to conidiation, mutants in which the process was altered stood a good chance of being phase-specific. The practical advantage of such "late" mutants is that they can be rapidly and macroscopically recognized whereas "early" mutants cannot. The first mutant was recovered from the progeny of wild-type conidia previously exposed to ethyl-methane-sulfonate. Among 3500 colonies characterized by normal growth on minimal medium and the production of macroscopically visible clumps of conidia, five failed to liberate individual conidia when their culture vessels were inverted and tapped (SELITREN-NIKOFF and NELSON, 1973). The inference that these would be defective for some aspect of conidial septation was confirmed by microscopic examination. Crosses showed that each strain carried a different single-gene mutation. Upon careful examination, four of the strains were found to produce hyphae that differed from wild type in very subtle ways (SELITRENNIKOFF and ABE, 1972) and were set aside. The remaining mutant appears to be defective only with respect to conidial separation and hence has been designated *csp-1* ("conidial-separation defective"). The second mutant was generously provided by Prof. G. DEBUSK who discovered it among a group of UV-treated isolates. *csp-2*, like *csp-1*, forms chains of macroconidia but few individual spores. The third mutant was isolated from the standard crisp-snowflake stock (NELSON et al., 1973), but its origin is otherwise uncertain; since the phenotype is identical to that of *csp-2*, with which it is allelic, the properties of the strain need not be described further. The mutants complement each other and assort independently. *csp-1* is located on the left arm of linkage group I, close to the centromere, and *csp-2* is on the proximal portion of the right

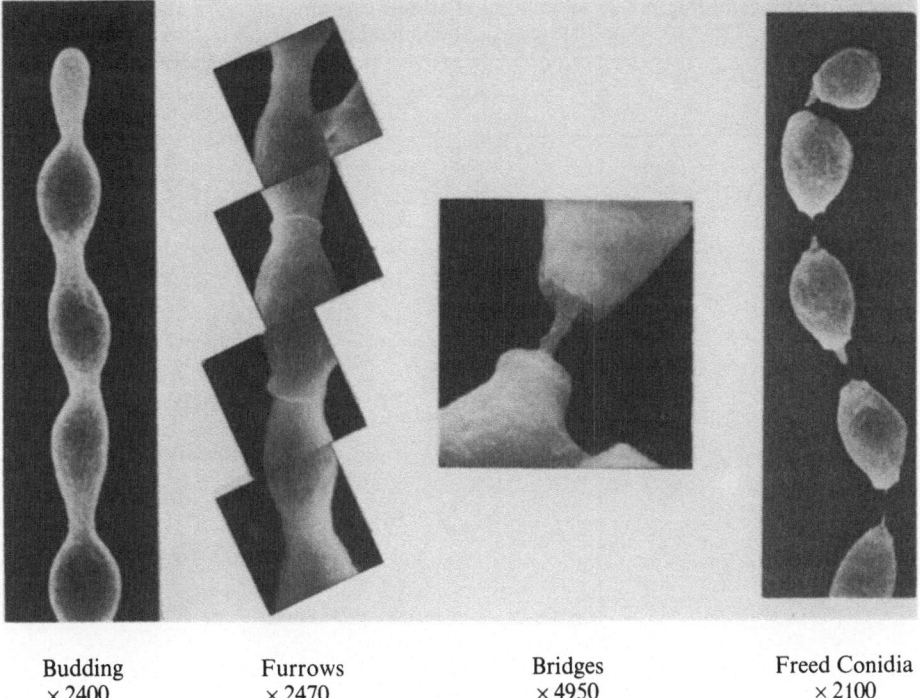

Budding	Furrows	Bridges	Freed Conidia
× 2400	× 2470	× 4950	× 2100

Fig. 7. Scanning electron micrographs showing the final events of conidiogenesis in wild type

arm of linkage group VII. In heterocaryons, all are recessive to their wild-type alleles.

The most obvious phenotypic defect of the mutants is the failure to liberate normal numbers of free conidia. Those conidia that are produced appear to be entirely normal, and therefore the mutations are considered to be examples of leaky genes. Figure 6 compares the mutants with wild type and shows that the double mutant yields fewer conidia than either parental strain. Scanning electron microscopy affords a rather precise description of external changes during the late stages of conidiation. Unlike wild type, which is shown in Fig. 7, conidiophores of *csp-1* rarely produce furrows and bridges (even 24 hrs after induction). Examination of *csp-2* shows that this strain is blocked somewhat earlier; chain formation in *csp-2* and the double mutant *csp-1 csp-2* is normal, but later stages occur so infrequently that they cannot be found. To summarize, all three strains are defective for the production of the proper interconidial septa which allow chains of presumptive conidia to fall apart into individual spores.

The idea that the *csp* functions are restricted to the process of conidiogenesis is based on evidence indicating that the vegetative and sexual life cycles are otherwise indistinguishable from those of wild-type strains. Table 2 demonstrates that the growth rates of vegetative hyphae conform with that of wild type. Neither light nor scanning electron microscopy of the hyphae discloses differences. When the mutants are induced for form aerial hyphae, these structures are produced on

Table 2. Dry weight (mg per 25 ml culture) of strains carrying the *csp* mutations and wild type after growth on minimal medium

Strain	Time (hrs)		
	48	72	96
csp-1	33	68	85
csp-2	29	65	79
csp-1, csp-2	29	70	92
wild type	30	70	92

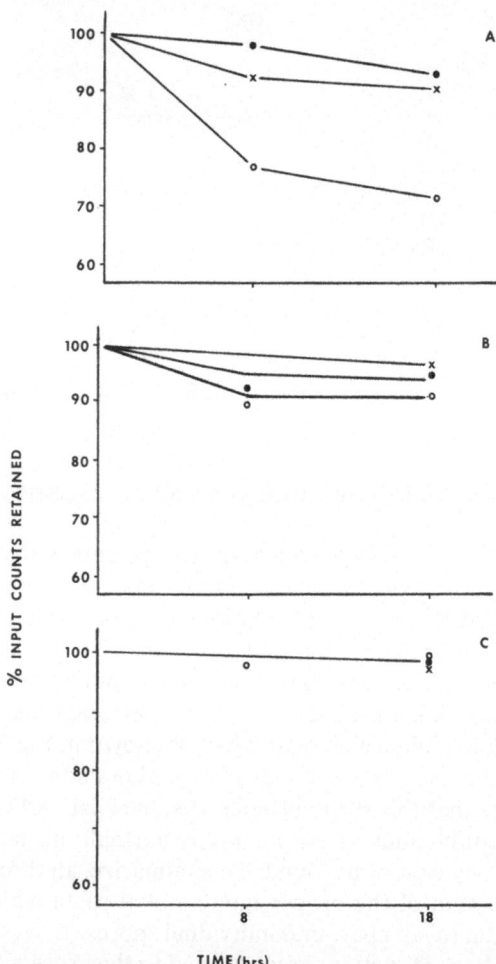

Fig. 8 A–C. Autolytic activity of ^{14}C-glucose labeled cell walls and cell-wall protein purified from induced cultures. (A) At 9.5 hrs following induction, a cell wall fraction was prepared from the aerial mass and allowed to self-digest for 18 hrs. (B) Same as (A) except that the cell wall fraction was prepared from vegetative hyphae at induction (0 hrs). (C) Same as (A) except that the cell wall fraction was prepared from the aerial mass at 5 hrs following induction
○—○ wild type; ●—● *csp-1*; ×—× *csp-2*

the time schedule found for wild type, and in comparable numbers, lengths, and morphology. Conidiophores and the structures produced early in the process of conidiation are normal. If the chains of conidia produced by the mutants are subjected to mechanical disruption, the spores are freed and will germinate in a normal manner. Finally, the sexual cycles culminating in the formation of ascospores are normal in all respects. The conclusion that the three mutations identify phase-specific genes is strongly indicated.

Initial inquiries directed toward eventual characterization of the *csp* gene products give new support to this conclusion and demonstrate a difference between the cell walls of the *csp* mutants and wild type. Labeled cell walls together with proteins physically associated with cell wall material were isolated from cultures grown in ^{14}C-glucose. The method was modified from that described by MAHADEVAN and MAHADKAR (1970) and is to be described in detail in a later publication. Preparations were incubated at 25° C for 18 hrs and the residual radioactivity in the cell wall was measured during that period. The relative amounts of endogenous autolytic activities in the mutants and wild type were determined at various points in the developmental cycle. The results shown in Fig.8A demonstrate significant autolytic activity in the aerial mass of wild type but little, if any, in the mutants. There was virtually no evidence for endogenous autolytic activity in vegetative hyphae at the time of induction (Fig.8B) or in aerial hyphae prior to the time of conidiogenesis (Fig.8C). Since the onset of autolytic activity and conidiogenesis are correlated in wild type and since mutants that fail to liberate conidia lack such activity, it is tempting to suppose that the *csp* genes specify a conidial "septase(s)". This predicts that temperature-sensitive revertants of the conidial-separation phenotype will display temperature-sensitive endogenous autolytic activity, a result which would exclude more complex explanations of the present data. Such mutants are now being sought. If recovered, the release of cell wall counts could be used as an assay for the isolation and characterization of the products of the *csp* genes.

V. Concluding Remarks

The molecular bases for adaptive changes in the metabolic apparatus have been intensively studied in procaryotes and extended somewhat less successfully to the more "difficult" cells provided by higher organisms. The philosophy or the faith that an understanding of those regulatory events would give insight into the basic nature of cellular differentiation is commonplace and is certainly justified to the extent that metabolic shifts usually go hand in hand with cell morphogenesis. It is also appreciated that models based on adaptive processes do not immediately account for deterministic phenomena or long-lasting heritable cell changes. Perhaps the most important lesson taught by the penetrating investigations of adaption is that penetrating explanations for cell differentiation, in the strict sense, must start with the discovery of genes that specify particular functions, with the aim of learning of their control.

In accord with this view, we characterized the most general features of cell changes in *Neurospora* and then looked specifically for molecular and morphological changes that uniquely accompany the formation of a particular cell type. As the next step we sought phase-specific mutations in order to identify genes which seem to interfere with these particular functions. It seems obvious that the role of a phase-specific function in differentiation is clarified by the detection of its structural gene, preferably through the use of conditional mutants, and that such analyses are favored over others which simply show that cell development and a number of functions are simultaneously blocked by drug treatments or mutation. Future work needs to be directed toward assembling a collection of differentiation-related functions together with the structural genes for these. Having garnered structural genes whose expressions are limited to a single phase-specific event, one can then profitably seek to identify the loci that regulate these events and hence control processes critical for cell changes.

Continued study of cellular differentiation in *Neurospora* should open up fruitful avenues of approach toward solutions of general problems, such as the coordinate versus step-wise activation of genes and the designation of particular genic functions as necessary for development rather than simply as correlates of development. We will refer to this again shortly. First it is appropriate here to call attention to the concept that development as exemplified by *Neurospora* may be fundamentally distinct from embryogenesis in one important respect; as is the general case for events which lead to sporogenesis in microorganisms, development of aerials and conidia procedes from the breakdown of preexisting protoplasm rather than from specially stored or exogenously supplied nutrients. *Neurospora* typifies catabolic rather than anabolic development, a point previously stressed by BRODY (1973). Since development and cellular differentiations are one in the same in this organism, its utility as an all-purpose model requires circumspection. However, it is clear that the vegetative cycle in *Neurospora* has something in common with certain processes known for higher forms; for example, the breakdown of calcified cartilage and associated cartilage cells could provide materials utilized during the subsequent production of bone-forming cells and bone; and it may be supposed that bone turnover or renewal in growing mammals and even in the adult could involve catabolic development. Of course catabolic development does not lessen the usefulness of *Neurospora* as a model for the switch of one well-defined specialized cell type to another within clonal cultures; EGUCHI and OKADA (1973) recently demonstrated that proliferating cells derived from pigmented retinae of chick embryos form lens tissue.

There has been much general interest in those macromolecules whose concentrations or localizations change in concert with cellular differentiation. Once the observations that designate such molecules as phase-specific are completed, it is important to clarify the roles of these molecules. Four possible interpretations of the specific presence (or absence) of a given molecule need to be considered. First, the molecule may play a direct and necessary role in the differentiation and functioning of the cell in which it appears. Second, the molecule may be nonessential to the cell in which it appears but it could be critical for the development of future cell types. Third, the molecule may not be necessary for differentiation but its appearance is an integral part of the developmental program; it is a luxury molecule. Fourth, the appearance of the molecule may be simply fortuitous, that is, part of a set of functions distinctly separate from those related to the cell changes but which merely happen to be set into motion by the conditions of induction. Of course if the molecule is extrinsic to cell change its appearance will have nothing to do with the developmental program under investigation. (For example, carotenoid synthesis is photo-induced in *Neurospora* and thus accompanies the formation of aerial hyphae and conidia among cultures illuminated in finger bowls; aerials and conidia can be induced in darkness and the cultures then remain colorless.) The opportunity to distinguish among these interpretations is afforded by *Neurospora* and is demonstrated by our analyses.

Acknowledgements

The authors are grateful to Professor Stuart Brody for many helpful suggestions offered during the preparation of this review. The work was supported by National Science Foundation grants to R.W.S.

References

AXELROD, D. E., PASTUSHOK, M., BATY, B.: A gene controlling inducibility of development. Genetics **74** (Suppl.), p. 13 (1973).

BIANCHI, D. E., TURIAN, G.: Nuclear division in *Neurospora crassa* during conidiation and germination. Experientia **23**, 192–194 (1967).

BRADLEY, M. O.: Arginase activity patterns and their regulations during embryonic development in liver and kidney. Develop. Biol. **33**, 1–17 (1973).

BRODY, S.: Metabolism, cell walls, and morphogenesis. In: Cell Differentiation. New York: Academic Press (In press) 1974.

CALVO, J. M., FINK, G. R.: Regulation of biosynthetic pathways in bacteria and fungi. Ann. Rev. Biochem. **40**, 943–968 (1971).

COLWICK, S. P., KAPLAN, N. O., CIOTTI, M. M.: The reaction of pyridine nucleotide with cyanide and its analytical use. J. Biol. Chem. **191**, 447–459 (1951).

COMBÉPINE, G., TURIAN, G.: Activités de quelques enzymes associés à la conidiogenèse du *Neurospora crassa*. Arch. Mikrobiol. **72**, 36–47 (1970).

DRUCKER, H.: Regulation of exocellular proteases in *Neurospora crassa*: induction and repression of enzyme synthesis. J. Bacteriol. **110**, 1041–1049 (1972).

EGUCHI, G., OKADA, T. S.: Differentiation of lens tissue from the progeny of chick retinal pigment cells cultured *in vitro*: A demonstration of a switch of cell types in clonal culture. Proc. Natl. Acad. Sci. U.S. **70**, 1495–1499 (1973).

GRATZNER, H. G.: Cell wall alterations associated with hyperproduction of extracellular enzymes in *Neurospora crassa*. J. Bacteriol. **111**, 443–446 (1972).

GREENAWALT, J. W., BECK, D. P., HAWLEY, E. S.: Chemical and biochemical changes in mitochondria during morphogenetic development of *Neurospora crassa*. In: AZZONE, G. F., et al. (Eds.): Biochemistry and Biophysics of Mitochondrial Membranes, pp. 541–558. New York: Academic Press 1972.

GROSS, S. R.: Genetic regulatory mechanisms in the fungi. Ann. Rev. Genet. **3**, 395–424 (1969).

HENNEY, H. R., JR., STORCK, R.: Polyribosomes and morphology in *Neurospora crassa*. Proc. Natl. Acad. Sci. U.S. **51**, 1050–1055 (1964).

HOCHBERG, M. L., SARGENT, M. L.: NADase levels in various strains of *Neurospora crassa*. Neurosp. Newsl. **20**, 21 (1973).

KANE, B. E., REISKIND, J. B., MULLINS, J. T.: Hormonal control of sexual morphogenesis in *Achlya*: dependence on protein and ribonucleic acid syntheses. Science **180**, 1192–1193 (1973).

Kaplan, N. O., Colowick, S. P., Nason, A.: Neurospora diphosphopyridine nucleotidase. J. Biol. Chem. **191**, 473–483 (1951).

Mahadevan, P. R., Mahadkar, U. R.: Role of enzymes in growth and morphology of *Neurospora crassa:* cell-wall bound enzymes and their possible role in branching. J. Bacteriol. **101**, 941–947 (1970).

Metzenberg, R. L.: Genetic regulatory systems in *Neurospora.* Ann. Rev. Gen. **6**, 111–132 (1972).

Mitchell, R., Sabar, N.: Autolytic enzymes in fungal cell walls. J. Gen. Microbiol. **42**, 39–42 (1966).

Moscona, A. A.: Control mechanisms in hormonal induction of glutamine synthetase in the embryonic neural retina. In: Hamburgh, M., Barrington, E. J. W. (Eds.): Hormones in Development, pp. 169–189. New York: Appleton-Century-Crofts 1971.

Nasrallah, J. B., Srb, A. M.: Genetically related protein variants specifically associated with fruiting body maturation in *Neurospora.* Proc. Natl. Acad. Sci. U.S. **70**, 1891–1893 (1973).

Nelson, R. E., Chandler, T., Selitrennikoff, C. P.: *cr sn:* the significance of macroconidiation for mutant hunts. Neurosp. Newsl. **20**, 33–34 (1973).

Rho, H. M., DeBusk, A. G.: NH$_2$-Terminal residues of *Neurospora crassa* proteins. J. Bacteriol. **107**. 840–845 (1971).

Schmit, J. S., Martins, S., Brody, S.: Molecular prepackaging during conidia formation in *Neurospora crassa.* (Abstr.) Ann. Meeting of Am. Soc. Microbiol. 1973.

Schroeder, A. L.: Ultraviolet-sensitive mutants of *Neurospora.* II. Radiation Studies. Molec. Gen. Genetics **107**, 305–320 (1970).

Selitrennikoff, C. P., Abe, V.: New *os-2* and *os-5* alleles. Neurospora Newsl. **19**, 23 (1972).

Selitrennikoff, C. P., Nelson, R. E.: A screening technique for the isolation of macroconidiation mutants. Neurosp. Newsl. **20**, 34 (1973).

Sessoms, A. H., Huskey, R. J.: Genetic control of development in *Volvox:* Isolation and characterization of morphogenetic mutants. Proc. Natl. Acad. Sci. U.S. **70**, 1335–1338 (1973).

Siegel, R. W., Matsuyama, S. S., Urey, J. C.: Induced macroconidia formation in *Neurospora crassa.* Experientia **24**, 1179–1181 (1968).

Soll, D. R., Sonneborn, D. R.: Zoospore germination in *Blastocladiella emersonii:* cell differentiation without protein synthesis. Proc. Natl. Acad. Sci. U.S. **68**, 459–463 (1971).

Stine, G. J.: Germination and enzyme activities affected by the aging of *Neurospora* conidia in water. Neurosp. Newsl. **11**, 7–8 (1967).

Stine, G. J.: Enzyme activities during the asexual cycle of *Neurospora crassa.* II. NAD- and NADP-dependent glutamic dehydrogenase and nicotinamide adenine dinucleotidase. J. Cell Biol. **37**, 81–88 (1968).

Stine, G. J., Clark, A. M.: Synchronous production of conidiophores and conidia of *Neurospora crassa.* Canad. J. Microbiol. **13**, 447–453 (1967).

Totten, R. E., Howe, H. B., Jr.: Temperature-dependent actinomycin D effect on RNA synthesis during synchronous development in *Neurospora crassa.* Biochem. Genet. **5**, 521–532 (1971).

Turian, G., Bianchi, D. E.: Conidiation in *Neurospora.* Botan. Rev. **38**, 119–154 (1972).

Turian, G., Matikian, N.: Conidiation of *Neurospora crassa.* Nature **212**, 1067–1068 (1966).

Urey, J. C.: Enzyme patterns and protein synthesis during synchronous conidiation in *Neurospora crassa.* Develop. Biol. **26**, 17–27 (1971).

Wessells, N. K.: Problems in the analysis of determination, mitosis, and differentiation. In: Fleischmajer, R., Billingham, R. E. (Eds.): Epithelial-Mesenchymal Interactions, pp. 132–151. Baltimore, Md.: Williams and Wilkins 1968.

Wilson, J. F.: Microsurgical techniques for Neurospora. Am. J. Botany **48**, 46–51 (1961).

Wong, R. S. L., Scarborough, G. A., Borek, E.: Transfer-RNA methylases during germination of *Neurospora crassa.* (Abstr.) Bacteriol. Proc. **61** (1971).

Zalokar, M.: Enzyme activity and cell differentiation in *Neurospora.* Am. J. Botany **46**, 555–559 (1959a).

Zalokar, M., Cochrane, V. W.: Diphosphopyridine nucleotidase in the life cycle of *Neuro-* (1959b).

Zalokar, M., Cochrane, V. W.: Diphosphopyridine nucleotidase in the life cycle of *Neurospora crassa.* Am. J. Botany **43**, 107–110 (1956).

Subject Index

Page numbers in bold face refer to systematical treatment of the subject

Acetylaminofluoren
  given concomitantly with partial
    hepatectomy 234
Acetylcholinesterase 69
Achlya (water mold)
  no evidence for stable mRNA 299
Actin 3, 5, 18
Actin gene 18
Actinomycin D 210
  cell attachement 75
  inhibition of transcription in *Neurospora*
    and cell differentiation 298
  partial hepatectomy 222
Activity
  merestimatic, and differentiation in
    plants 152
Adenosine monophosphate, cyclic (cAMP)
  8-bromo derivative, stable biologically
    active form of
      competitive inhibitor of cAMP for
        binding sites on the phosphodi-
        esterases 185, 192
      effect on histone F_1 phosphorylation
        185, 274
  dibutyrylcyclic AMP, effect on
    metabolically blocked cultures of
    C. crescentus 144, 146f.,
  myogenic cells, in the cell cycle of 13
  phosphodiesterases 184, 192
  regulation of cell division 185
  tyrosine-aminotransferase, induction of
    202
Adenylyl-cytokinin, 6-substituted 183, 192
Adrenal cortex
  start of function und development of a
    group of enzymes in liver 203
Aerialogenesis
  Neurospora crassa 291, **295 ff.**
Alanine transaminase
  in regenerating liver 222
Albumin
  in liver cells 212
  production, influenced by hormones
    214

Albumin
  synthesizing cells, distribution in the liver
    lobule 216
Aldolase
  in embryonic and regenerating liver 222
Allelic forms
  tropomyosin and myosin 19
Alternative steady states
  systems of chemical reactions 170
Amblystoma
  neurulae 50
Amino acid analogs
  inducing blocks of protein synthesis and
    development of abnormally few aerial
    hyphae and conidia *(Neurospora)*
    297
Amino acids, aminoterminal
  different distribution in the proteins of
    macroconidia *(Neurospora)* 294f.
Amoeba proteus
  critical mass for cytokinesis 102
Amphibian retina
  clonal cell culture 54
Amphibians
  ploidy of hepatocytes 209
Amputating
  pieces of cytoplasma of *Amoeba proteus*
    102
Analysis
  acetylated histone, triton gel system
    266f.
  activity of NAD-ase 302
  bioassay for recognition of NADaseless
    mutants *(Neurospora)* 302
  DNA-content of cells with the
    Feulgen method 65
  lipopolysaccharide from cell wall of
    C. crescentus 141
  spectrophotometric of the DNA-
    content of neural tube cells 65
Aneuploidy 167
Antherea
  correlation between cell division and
    juvenile hormon sensitivity 116

Results and Problems in Cell Differentiation

A Series of Topical Volumes
in Developmental Biology

Editors: W. Beermann, Tübingen
J. Reinert, Berlin
H. Ursprung, Zürich

Springer-Verlag
Berlin
Heidelberg
New York

Volume 1
The Stability of the Differentiated State
Edited by H. Ursprung
56 figures. XI, 154 pages. 1968
ISBN 3-540-04315-2 Cloth DM 59,—
ISBN 0-387-04315-2 (North America)
Cloth $21.00

This volume brings together authoritative reviews on the subject from the point of view of tissue culture (both in vitro and in vivo), regeneration, dedifferentiation, metaplasia, clonal analysis, and somatic hybridization. The ten contributing authors are actively working in the respective areas of interest and have written short, concise accounts of their own and related work. The book is intended for advanced students who want to keep up with recent developments in these areas of basic biology which are so important for the understanding of the development of higher organisms, both normal and malignant. The book as a whole uncovers similarities in experimental systems as widely different as plants and insects and provides new insights into fundamental processes.

Volume 2
Origin and Continuity of Cell Organelles
Edited by J. Reinert and H. Ursprung
135 figures. XIII, 342 pages. 1971
ISBN 3-540-05239-9 Cloth DM 72,—
ISBN 0-387-05239-9 (North America)
Cloth $24.90

A great deal of precise information is available on structure and function of cell organelles. But much controversy exists on the questions of their origin, and the mode of their replication. This book contains authoritative reviews relevant to this question on the following components of cells: mitochondria, vacuoles, desmosomes, centrioles, polar granules, Golgi apparatus, endosymbionts, plastids, microtubules, lysosomes, and membranes in general.

Volume 3
Nucleic Acid Hybridization in the Study of Cell Differentiation
Edited by H. Ursprung
29 figures. XI, 76 pages. 1972
ISBN 3-540-05742-0 Cloth DM 36,—
ISBN 0-387-05742-0 (North America)
Cloth $13.40

The method of molecular hybridization of nucleic acids has made it possible to obtain more precise information on quantitative and—in some cases at least—qualitative composition of the genome. Both beginners and advanced students will find the book helpful for understanding the principles of the method, the various techniques that are used, and the biological implications of the results already obtained.

Volume 4
Developmental Studies on Giant Chromosomes
Edited by W. Beermann
110 figures. XV, 227 pages. 1972
ISBN 3-540-05748-X Cloth DM 59,—
ISBN 0-387-05748-X (North America)
Cloth $20.00

This volume is devoted to the question of structural and functional organization of the eucaryotic genome in relation to cell differentiation as revealed by current work on the genetic fine structure and synthetic activities of giant polytene chromosomes and their subunits, the chromomeres.

Volume 5
The Biology of Imaginal Disks
Edited by H. Ursprung and R. Nöthiger
56 figures. XVII, 172 pages. 1972
ISBN 3-540-05785-4 Cloth DM 46,—
ISBN 0-387-05785-4 (North America)
Cloth $16.00

The book summarizes the current state of knowledge concerning the biochemistry and ultrastructure of the imaginal disk. It reviews experiments of the mechanisms of cell determination as studied in imaginal disks.

Volume 6
W. J. Dickinson and D. T. Sullivan
Gene-Enzyme Systems in Drosophila
32 figures. XI, 163 pages. 1975
ISBN 3-540-06977-1 Cloth DM 58,—
ISBN 0-387-06977-1 (North America)
Cloth $23.80

This book brings together for the first time biochemical, genetic, and developmental work, on enzymes in Drosophila. It documents the usefulness of gene-enzyme systems as tools in the investigation of numerous problems, particularly those relative to the organization and regulation of the eukaryotic genome. Research methods are described as well as the impact of the findings upon other fields such as population genetics, evolution, biochemistry, and physiology.

Prices are subject to change without notice

Progress in Molecular and Subcellular Biology

Editorial Board:
F. E. Hahn, Washington
T. T. Puck, Denver
G. F. Springer, Evanston
W. Szybalski, Madison
K. Wallenfels, Freiburg

Managing Editor:
Fred A. Hahn, Washington

Springer-Verlag
Berlin
Heidelberg
New York

Volume 1

By B. W. Agranoff, J. Davies,
F. E. Hahn, H. G. Mandel, N. S. Scott,
R. M. Smillie, C. R. Woese
32 figures. VII, 237 pages. 1969
ISBN 3-540-04674-7   Cloth DM 68,—
ISBN 0-387-04674-7 (North America)
Cloth $19.90

Volume 1 contains contributions by
Woese, Davies and Mandel who are
concerned either theoretically or
experimentally with the nature of the
genetic code and its correct translation. Smillie and Scott write about the
biosynthesis of chloroplasts and the
photoregulation of the underlying
basic processes. Agranoff reviews cri-
tically current ideas concerning the
molecular basis of higher nervous
activities. The managing editor, Hahn,
makes a critical analysis of the origin,
conceptual content and probable future developments in molecular
biology.

Volume 2

**Proceedings of the Research
Symposium on Complexes of
Biologically Active Substances with
Nucleic Acids and Their Modes of
Action**
Held at the Walter Reed Army
Institute of Research, Washington,
16-19 March, 1970
158 figures. IX, 400 pages. 1971
ISBN 3-540-05321-2   Cloth DM 78,—
ISBN 0-387-05321-2 (North America)
Cloth $24.90

These 28 contributions are a definitive
account of the most recent research
results on selected aspects of the title
topic. The major topical subdivisions
are: Antibiotics and Nucleic Acids —
Antimalarials and Nucleic Acids —
Alkaloids, Natural Polyamines and
DNA — Intercalation into Supercoiled
DNA — Synthetic Drugs and Dyes
Binding to Nucleic Acids — Anti-
mutagens — Carcinogens and Nucleic
Acids — The Natural State of DNA.
Among the contributors are
D. M. Crothers, G. F. Gause, W. and
H. Kersten, L. S. Lerman,
H. R. Mahler, P. O. P. Ts'o, J. Vino-
grad, and M. Waring.
The editor, F. E. Hahn, introduces
Volume 2 with a review of the history,
current state and perspectives of the
field of nucleic acid complexes.
The book will interest research scien-
tists active in the study of nucleic acid
complexes and molecular biologists,
molecular pharmacologists, bio-
chemists and biophysicists seeking
comprehensive and up-to-date information in this specialized field.

Volume 3

By A. S. Braverman, D. J. Brenner,
B. P. Doctor, A. B. Edmundson,
K. R. Ely, M. J. Fournier, F. E. Hahn,
A. Kaji, C. A. Paoletti, G. Riou,
M. Schiffer, M. K. Wood
58 figures. VII, 251 pages. 1973
ISBN 3-540-06227-0   Cloth DM 72,—
ISBN 0-387-06227-0 (North America)
Cloth $29.90

The volume is concerned with the
transccription and translation of
genetic information and with products
of both processes in prokaryotic and
eukaryotic organisms. Reverse transcription is reviewed with respect to
its theoretical importance to molecular
biology and its practical importance
to cancer research. Translation, i.e.
protein biosynthesis, is analyzed in the
light of knowledge of numerous
inhibitors of individual reaction steps
in the overall sequence. The transcription of transfer RNA is reviewed in
detail as one aspect of transcription
of genetic information. Thalassemia
and Bence-Jones proteins have been
treated as selected examples of molecular pathology in eukaryotes. The
nature of mitochondrial DNA in neoplastic cells is reviewed as one con-
tribution to the molecular biology of
cancer. The book signifies the advance-
ment of molecular biology from the
study of bacteria and their viruses to
the consideration of events and
entities in eukaryotic organisms.

Prices are subject to change
without notice